PRINCIPLES OF PHYTOPATHOLOGY

Second Edition

Clare B. Kenaga
Purdue University

Prospect Heights, Illinois

For information about this book, write or call:
Waveland Press, Inc.
P.O. Box 400
Prospect Heights, Illinois 60070
(708) 634-0081

First published by Waveland Press, Inc., 1986

ISBN 0-88133-268-2

Printed in the United States of America

7 6 5 4 3

To My Wife

Doris I. Kenaga

PREFACE

The purpose of this book is several-fold. It is intended: (1) to present in a logical manner salient and up-to-date concepts of the broad field of plant pathology; (2) for use in undergraduate, introductory plant pathology courses (wherein the students may be, but are usually not, majoring in plant pathology); (3) to be essentially self-explanatory, especially when accompanied by discussion sessions concerning the subject matter and laboratory periods devoted to the study of various plant diseases, etc.; (4) to present principles and important concepts in a manner that focuses the reader's attention upon them; (5) to present sufficient material for courses of 3 to 5 credit-hours, allowing instructors considerable latitude as to the material they wish to use.

Logical And Up-to-date

Undergraduate Use

Self-explanatory

Focus Attention On Principles And Important Concepts

The first eight chapters in this book deal with a short historical background, disease concepts, terminology, and the classes of causal agents. Air pollution is discussed within the context of noninfectious diseases. A chapter, devoted to market pathology, is included lest it be forgotten that disease losses also occur after harvest. A somewhat traditional discussion of the influence of environment is followed by chapters on inoculum and pathogen variation. An orthodox discussion of host resistant mechanisms is followed by unorthodox discussions of pathogens present in the host, and host responses to infection. The latter is treated in a physiological vein. A non-mathematical discussion of plant disease epidemics leads to disease control chapters on breeding for resistance, inoculum reduction (including chemical controls), and quarantines.

In this, the second edition, areas which are considerably revised and expanded include the Mollicutes and related organisms, Ch. 3); air pollution toxicity, (Ch. 8); and fungicides, including their use and mammalian toxicity, (Ch. 19). Other sections with important revisions and/or additions include taxonomic charts of major groups of fungal plant pathogens, (Ch. 4); multicomponent viruses, and viroids, (Ch, 5); disease complexes involving nematodes, (Ch. 6); discussion of monocultures and crop uniformity, (Ch. 17); strategies of breeding, including specific and general resistance, use of tolerance, varietal blends and multiline varieties, (Ch. 18); discussions regarding changing cultural practices (such as no-till planting), variety deployment, and biological controls including mycorrhizae, (Ch. 19).

In order to achieve the objectives set forth, definitions, important concepts, and principles (rules) are underlined in the body of the text. The reasons are readily apparent. As the book is read, the reader is made immediately aware of the importance of these words or phrases and pauses in his or her reading to concentrate upon them. The underlining effectively outlines the most important points which aid in subsequent study of the subject matter. Many side-notes are placed in wide outer margins to assist the learner during reading of the text and during recall of subject matter. Many of these side-notes are purposely redundant, making use of the honored axiom in education that redundancy is an aid to learning.

Side Notes Assist In Recall And Are Often Redundant

All students are encouraged to underline additional concepts and make further side-notes. Ample space has been provided for this purpose. In an earlier (1967) limited-edition text written specifically for the introductory course in plant pathology at Purdue University, I first utilized the concepts of underlining and the use of side-notes. This met with the overwhelming approval of the majority of the students. To these willing students whose thoughts and comments have aided me in the preparation and revision of this book, I extend my sincere thanks.

Students Are Encouraged To Make Further Side-Notes And Underlining

To those colleagues who have read various parts of this manuscript and have given valuable suggestions, criticisms and counsel, I express my gratitude: G. B. Bergeson, J. M. Ferris, J. F. Hennen, R. M. Lister, E. B. Williams, and A. J. Ullstrup. I am deeply indebted to R. W. Samson for his valuable counsel in numerous discussions during the preparation of this manuscript, and for the many photographs that he has generously allowed me to use.

A special acknowledgment of debt and gratitude is expressed to R. N. Campbell for his many highly constructive comments, criticisms and suggestions. To D. H. Scott go my sincere thanks for permission to place in the glossary his list of plants sensitive and tolerant to air pollutants. Colleagues who have generously contributed illustrations, and to whom I am also indebted are: J. B. Bancroft, E. E. Butler, C. E. Bracker, G. B. Cummins, R. J. Green, A. F. El-Hadidi, R. O. Hampton, A. C. Hill, A. Kelman, R. M. Lister, E. G. Sharvelle, J. F. Spears, A. J. Ullstrup, and J. F. Worley. A sincere debt of gratitude is also expressed to G. B. Cummins, who as Chairman of the Botany and Plant Pathology Department, Purdue University, encouraged and supported the writing of the interim text and the first edition of the present text. My thanks are expressed also to the present departmental Chairman, M. L. Tomes, who has also supported the writing of both editions of this text.

Clare B. Kenaga

CONTENTS

INTRODUCTION TO PHYTOPATHOLOGY

ECONOMIC IMPORTANCE OF PLANT DISEASES

Phytopathology, perhaps better known as plant pathology, is the science of plant disease. The importance of this branch of biology cannot be overestimated. The health of plants is vital to society and human progress. Due to the need to feed an ever larger world population, plant health becomes more important with each passing year. This text treats some of the major principles and problems involved in protecting the health of domestic plants.

Phytopathology Is The Science Of Plant Disease

In studying the diseases of man and his domesticated animals, the physician and the veterinarian deal with individuals. For the most part, however, single diseased plants are usually relatively unimportant; and, therefore, phytopathology is essentially a community science. Stated another way, phytopathology deals mainly with large numbers of plants.

Phytopathology Is A Community Science

The estimated average annual loss to plant diseases in the United States from 1951 to 1960 was about $4.5 billion which includes $0.25 billion for disease control measures. Approximately 100,000 plant diseases are present on the North American continent. Of course, the vast majority of these diseases occur on an occasional basis and are relatively unimportant. When presented with vast acreages of susceptible hosts, however, some of these hitherto unimportant diseases may become epidemics of major importance. For example, in 1970 the United States had a $1 billion loss because of an epidemic of southern corn leaf blight. Fully fifteen percent of the corn crop was lost by this previously minor disease.

Under natural conditions, plants usually occur in mixed stands in which several plant types compete for space, light, moisture and nutrients. This arrangement tends to minimize the attack of infectious agents on one type of plant. Upon examination, however, evidence of disease is readily found within natural plant communities, yet they rarely become truly epidemic. There is clear evidence that these diseases vary in incidence and severity from year to year due to environmental influences. Over a period of many years, naturally occurring infectious disease agents and their natural host plants establish a balanced relationship of coexistance. The host plants periodically or annually suffer a limited amount

Rule -- Host And Pathogen Coexistence

of damage, but not sufficient to prohibit their reproduction and subsequent regrowth. This is a matter of natural selection, or survival of the fittest. It is an evolutionary process requiring genetic adjustments by both pathogen and host plant. This balanced relationship takes nature many, many years or even centuries to accomplish.

Modern Agronomic Practices Offer Opportunity For High Incidence Of Disease

Monoculture

The greatest need for plant disease control is in connection with those crops that are cultivated. When several hundred people are housed within a single square mile, the area may be said to be crowded, and extra care must be taken to prevent the development of serious public health problems. In comparison, one acre of wheat may contain approximately one million individual plants, all of which may be nearly identical from a genetical standpoint. When individual plants that are all of nearly identical genetic origin are crowded together, it offers optimum conditions for a plant disease. This crowding is a deliberate modern agronomic technique designed to promote maximum agricultural production through the use of improved varieties and soil management, but it provides highly favorable conditions for devastating attacks by some plant pathogens.

SIGNIFICANCE OF PLANT HEALTH

Man Dependent Upon Plants

Man is completely dependent upon plants. Plants are his source of food, either directly or indirectly. In the latter case, they are the food source for the animals that in turn serve as a food source for man. Plants also supply lumber, drugs, beverages, spices, fibers and chemical extracts. Fossil plants are the basis of the world's fuel reserves and thus are its principal current source of energy. Indeed, man occupies this planet as a guest of the plant kingdom.

Only Plants Synthesize Food

Plants are of greatest significance as synthesizers of foods vital to the health and well-being of man and animals. They alone among living forms utilize solar energy and by photosynthesis convert elementary substances into organic compounds essential for animal metabolism and growth. Man and animals are transformers, rather than synthesizers, of the basic food materials. Therefore, the ultimate population of the world must in the foreseeable future be governed by the availability of appropriate plant species.

Although vast land areas are planted to nonfood crops, more than 80 percent of the world's agricultural effort is directed toward food and animal feed production. Even though 300 plant species are widely cultivated for food and feed around the world, 95 percent of the world's annual food production is derived from fewer than a dozen crops--wheat, rice, corn, potatoes, sweet potatoes, sugar cane, cassava, beans, coconut and bananas.

When the magnitude of the world's dependence on ample and regular supplies of plants and plant products for every aspect of human existence is realized, the importance of "plant health" is readily apparent. Plants are prone to attack by a large variety of plant pathogens, approximately 8,000 species of fungi, 175 species of bacteria, 300 plant viruses, and 500 species of nematodes. Plant pathologists have long known that: each and every crop plant species is potentially subject to its particular diseases. Unless the mode of action of plant pathogens is understood and methods are devised for their control, the detriment to society in health, comfort, and economic well-being will be enormous. A sound understanding of the principles of phytopathology and the application of these principles to the problems of plant disease control is necessary to maintain the level of productivity of plants, which is essential to the welfare of mankind.

Rule -- Partiular Diseases

Knowledge Of Principles Vital

Selected References

Le Clerg, E.L. 1964. Crop losses due to plant diseases in the United States. Phytopathology. 54:1309-1313.

Paddock, W.C. 1967. Phytopathology in a hungry world. Annual Review of Phytopathology. 5:375-390.

Stakman, E.C. 1964. Opportunity and obligation in plant pathology. Annual Review of Phytopathology. 2:1-12.

CHAPTER 1

BRIEF HISTORICAL BACKGROUND

1-1 EARLY BELIEFS

Plant diseases were undoubtedly a handicap in man's attempts to feed himself even before he started raising his own plants. The Old Testament tells us of plant diseases wrought on man because of his sins. Theophrastus, the Father of Botany, was familiar with the plant diseases of his time. Some three hundred years before Christ, he wrote of plant troubles we can recognize as scorch, rot, scab, and rust. Indeed, so devastating were the cereal rusts that the early Romans worshipped a god of rust called "Robigus" whom they honored annually as a means of rust prevention.

Theophrastus: The Father Of Botany

Ancient times gave way to the intellectual "black-out" of the Middle Ages. The early sparks of understanding were all but extinguished by superstitions, avoidance of reason, and religious dogmas that marked this era. The theory of spontaneous generation of lower animals and plants, passed down from century to century, had become firmly established in the minds of men as a hard and fast law of nature.

Spontaneous Generation

Thus the scholars of the Middle Ages, though impressed by the appearance of plant diseases, were quite confused as to the factors that brought them about. The idea of spontaneous generation was firmly held until the middle of the nineteenth century, and it was a dominant factor in delaying the development of our present concepts of parasitism and disease.

In the sixteenth, seventeenth and eighteenth centuries, some prominent botanists became skeptical of spontaneous generation. However, they offered little if any evidence to disprove this belief, and their views were simply not accepted. Early botanists began to categorize plants, and even some of the fungi, into rough taxonomic catalogues. By the turn of the eighteenth century, the dogma of the constancy of species had become ingrained in the minds of men, and this concept was fostered by the religious authorities of the day who found it to coincide well with the prevalent religious teachings.

Dogma Of Constancy Of Species

Linnaeus: Classification Of Plants

Linnaeus in 1753 published his system of classifying plants in the book entitled "Species Plantarum". It was he who established the presently used system of naming living things with the Latin binomial system. Linnaeus firmly believed the dogma of constancy of species and the theory of spontaneous generation.

1-2 BEGINNINGS IN SCIENTIFIC STUDIES

1-2.1 FUNGI

We do not know the exact toll that plant diseases took from the European peasant and landowner. We can only infer what happened by reading accounts of great destructive diseases that took place from time to time, attended by disaster, famine, and tragic suffering. Death from the "holy fire", which we now attribute to the eating of ergot infected grain, was a constant menace to people of Europe during the Middle Ages. When the black ergot sclerotia of cereal crops were milled into flour, baked into bread, and eaten, the result was human poisoning so terrible as to be almost beyond description. The gangrenous type of ergotism caused the fingers and toes, and sometimes the ears and nose to become necrotic, and death followed if the victim continued to eat the poisonous bread. This disease occurred several times in epidemic form in the sixteenth century through the nineteenth century in both Germany and France. The epidemics in Germany were of the convulsive type that attacks the nervous system, causing convulsions, delirium and death. Ergot poisoning did not become as destructive in North America because the bread crop here was mainly wheat, whereas in Europe the bread crop was mainly rye, and ergot is much more severe on rye than on wheat.

Fig. 1.1 Ergot of cereal crops.

Prevost: Fungi The Cause Of Bunt

In 1807, Prevost, a Swiss professor, observed the germination of spores of the wheat-bunt organism. He theorized that this organism penetrated the young wheat plant and was the actual cause of bunt. This revolutionary idea was soundly refuted by the French Academy and did not gain favorable attention until 40 years later. Prevost was the first person to clearly demonstrate that microorganisms could cause disease. He was the first to promote what is now called the germ theory.

Germ Theory

Tulasne Brothers

In Paris in 1847, two brothers, Louis and Charles Tulasne, published studies on the wheat-bunt fungus that confirmed Prevost's observations of spore germination and the relationship of the fungus to the wheat plant. Their observations also concerned several other smut and rust organisms.

De Bary

The fine studies of the Tulasne brothers were followed in 1853 by an impressive paper in three parts devoted to the fungi, including the smuts and rusts. This fine work was by Anton De Bary who, trained in medicine, had turned to botany as a career at an early age. In Part III of his paper, De Bary placed the weight of his judgment and his evidence squarely behind the claims of Prevost and openly challenged the theory of spontaneous generation.

Irish Famine

During the 1800's, the importance of plant diseases increased in Europe along with the increase in population and intensity of plant culture. An extremely tragic event led to the beginning of real knowledge of plant diseases and to the development of the science of plant pathology. This was the Irish famine of 1845 and 1846. Two circumstances were responsible for the depth of this tragedy. First, the peasant population had become almost wholly dependent on their potato gardens for food. Second, the potato crop on this island of eight million people was almost completely destroyed by late blight in these two years. Starvation

Fig. 1.2A A potato leaflet showing symptoms of late blight, incited by the fungus *Phytophthora infestans*. In the presence of continued cool moist weather, this disease will completely destroy all leaf and stem tissue. *(Courtesy of R. W. Samson)*

stalked through every peasant home in Ireland. The physical misery and spiritual anguish suffered because of the devastation caused by this one disease is hard to imagine. Ireland lost almost a third of its population between 1845 and 1860 as a direct result of the outbreak of late blight.

Fig. 1.2B A potato tuber infected with the late blight fungus. The tuber is cut in half to show the rotting of the tissue. *(Courtesy of R. W. Samson)*

A million people died from starvation or from disease following malnutrition. The scourge started an emigration from Ireland, and by 1851, 1,640,000 people had emigrated.

The epidemic in Ireland was part of a pandemic--that is, the disease suddenly became widespread and of epidemic proportions almost simultaneously in several European countries and in the United States as well. The disease had not appeared in these regions more than 2 or 3 years previously. In the meantime, the pathogen had increased so immensely and had become so widely distributed that when the weather became extremely favorable, as happened during the years of the pandemic, it attacked rapidly and in force over a wide area.

Why did this sudden outbreak overwhelm Ireland and affect other countries much less? The answer is both political and agricultural. Primarily, the Irish peasants were politically subdued and held in a pitiful and miserable state of poverty, which was lower than in other countries affected by the pandemic. Secondly, the Irish peasants relied upon the filling, productive and easily grown potato for their main food.

The food resources of the peasants of other countries were more varied, so the destruction of a single crop, the potato, did not have such a profound importance.

Plant Pathology Born Of Catastrophe

Germ Theory Accepted

After this grimmest of epidemics abated, its consequences remained. The disease had become a fixture in potato culture and was more or less evident year after year. Everyone from politician to preacher, physician to layman, landowner to scientist, turned his attention to discover the cause of this plant disease. However, it remained for De Bary to describe the nature and the life history of the causal pathogen of this disease. Modern plant pathology was born out of this world catastrophe. The old theory of spontaneous generation gave way reluctantly to the germ theory.

1-2.2 BACTERIA

Bacteria Cause Plant Disease

In 1877, Burrill, at the University of Illinois, reported that bacteria could cause a plant disease. He showed that fire blight of pear was incited by a bacterial organism.

Smith-Fischer Controversy

In the early 1890's, E. F. Smith proved that bacteria were the cause of several plant diseases including a wilt of cucurbits. The validity of Smith's work on bacterial diseases was openly challenged in a book written by Alfred Fischer, a German, who had studied with De Bary. Fischer claimed Smith's data were unreliable. The controversy that arose was classical. In publications, Smith answered Fischer, Fischer replied back, and Smith finalized the argument with two rebuttal papers. This incident illustrates how difficult it was for results based on sound data to be accepted, even by the scientific community, when they were in opposition to the deep-seated beliefs of the times.

1-2.3 VIRUSES

Mayer Transmits Tobacco Mosaic In Plant Sap

Virus diseases of plants were studied for many years before their nature was known. A Dutchman, Adolf Mayer, in 1886 published a report on a disease of tobacco that he named "mosaic". Mayer discovered that when he macerated the tissue of a diseased leaf and injected the juice into the midrib of a healthy plant, the latter, after about 10 days, began to show typical symptoms. These appeared, not on the inoculated leaf, but on the youngest leaves of the growing plant; and as new leaves expanded, they were also diseased. This is the first known record of the experimental transmission of a virus disease. He found that continuous heat at 141° F. did not alter infectivity of the juice, but several hours at 177° F. destroyed it.

In 1891, E. F. Smith published his studies on the peach-yellows disease. He was unable to transmit the inciting entity by transfer of plant juices as Mayer had done; however, he was able to transmit the disease by grafting a bud from a diseased tree to a healthy one.

Smith Transmits Peach-yellows By Bud Graft

In 1892, a Russian, Dmitrii Ivanowski, reported that he had passed the infectious juice of tobacco mosaic through a bacteria-proof filter and found that the filtrate was still infectious! This was the first positive proof that the causal agent was not a bacterium.

Martinus Willem Beijerinck, a Dutch investigator, in 1898 published his findings that a minute amount of infectious sap of mosaic tobacco plant was sufficient to inoculate several plants. He demonstrated that the contagious entity was increased in some way in the infected plant. He called the infectious entity a virus. He also found the virus would remain infectious in infected leaves that were dried down and kept in the herbarium for two years.

Tobacco Mosaic: Multiplies In Plants; Can Be Dried; Called A Virus

The question then arose, "was the virus living or non-living?" The debate over this issue was extreme and heated. In both the peach-yellows disease and in tobacco mosaic, the causal entities were transmissible and both increased in the host. The tobacco mosaic was filterable through bacteria-proof filters, and was capable of existing for a time outside of a living host in a dried state; the causal entity of peach-yellows, however, was not.

In 1894-95, Hashimoto in Japan, proved that a leafhopper was important in causing dwarf disease of rice. However, it was thought this disease was caused by the leafhopper itself. Twelve to fourteen years later, it was found the leafhopper did not cause the disease directly, but transmitted it to healthy plants after feeding on diseased plants for five days. This was the first demonstration that insects could serve as vectors for plant viruses.

Insects Are Vectors

It remained for Stanley, at the Rockefeller Institute, in 1935, to crystalize tobacco mosaic virus. His crystals were not changed by ten successive recrystalizations. The molecules were quite large and 100 times more infectious than a suspension made from ground leaves of diseased tobacco plants. Stanley believed his crystals were pure protein in nature. At the present time, we know that virus particles are not pure protein, but the protein is a sheath containing an inner core of ribonucleic acid (RNA) in plant and some animal viruses, and deoxyribonucleic acid (DNA) in bacteriophages, and few plant viruses.

Virus Is A Protein Plus DNA Or RNA

In 1958, a plant virus was found to be transmitted by a plant parasitic nematode. This virus was previously described

Nematodes Are Vectors

as "soil-borne" because the method of transmission was by some unknown means in the soil. Now a number of plant viruses are known to have nematode vectors.

1-2.4 NEMATODES

Diseases of plants provoked by tiny, soil-borne parasitic eelworms, or nematodes, have been known for over two centuries. However, only a few species of plant parasitic nematodes were described prior to 1850, and their vast importance as agents of plant disease had yet to be revealed.

Cobb: Father Of Nematology

Nathan A. Cobb is frequently referred to as the "Father of Nematology" in the United States. At the turn of the twentieth century, he took advanced training in Europe, then accepted a position for eight years in Australia. He then joined the United States Department of Agriculture in Hawaii. In both positions he distinguished himself as an outstanding plant pathologist and pioneering scientist. During these years he also studied many aspects of nematology. Cobb then transferred to Washington, D. C., where he devoted most of the rest of his career to nematology. His leadership was largely responsible for the creation of the Division of Nematology, in 1929, and he was also installed as its first director. Cobb quickly surrounded himself with colleagues and students of exceptional caliber, most of whom were destined to become leaders and pioneers in their own right. Cobb's publications included many detailed works on taxonomy and morphology, and he devised many new techniques, a number of which are still being used today. During this period, all of the studies in nematology in the United States were centered in the Division of Nematology.

Soil Fumigants Provide Nematode Control

Economical and practical control of nematodes was afforded for the first time in the early 1940's by the discovery and availability of low cost soil fumigant chemicals. When injected into the soil, these chemicals volatilized and diffused through the air spaces of the soil, killing many of the nematodes. This allowed growers to control diseases caused by nematodes and provided the opportunity to demonstrate and dramatize the damage they caused. A stimulus was thereby provided for increased study into all aspects of nematology.

Christie & Perry Find New Group Of Parasitic Nematodes

Prior to 1951, the only nematodes proven to cause plant disease were found either wholly or partially inside plant roots. However, in 1951, Christie and Perry found a group of nematodes which damaged plants by feeding on roots while never entering them or becoming attached to them. It is now known that many nematodes damage plants in this manner.

Since the middle of the 1950's, disease complexes have been shown to exist whereby the presence of a parasitic nematode in combination with another pathogen may cause a destructive plant disease. A number of disease complexes are now known, and involve the interaction of nematodes with bacteria, fungi, viruses, and other nematodes

Disease Complexes

Selected References

Keitt, G. W. 1959. History of plant pathology. pp. 61-97, In J. G. Horsfall, and A. E. Dimond, (ed.), Plant pathology. Vol. I. Academic Press, New York.

Large, E. C. 1940. The advance of the fungi. Holt, New York. (paperbook edition by Dover, New York)

Orlob, G. B. 1964. The concepts of etiology in the history of plant pathology. Pflanzenschutz-nachrichten. "Bayer" 17:185-268.

Parris, G. K. 1968. A chronology of plant pathology. Johnson and Sons, Starkville, Miss.

Raski, D. J. 1959. Historical highlights of nematology. pp. 384-394, In C. S. Holton, et al., (ed.), Plant pathology problems and progress, 1908-1958. The University of Wisconsin Press. Madison, Wisconsin.

Stakman, E. C., and J. G. Harrar. 1957. Principles of plant pathology. The Ronald Press Co., New York. pp. 6-34.

Stevenson, J. A. 1959. The beginnings of plant pathology in North America. pp. 14-23, In C. S. Holton, et al., Plant pathology problems and progress, 1908-1958. The University of Wisconsin Press, Madison, Wisconsin.

Thorne, G. 1961. Principles of nematology. McGraw-Hill Book Co., New York. pp. 2-21.

Walker, J. C. 1969. Plant pathology. Ed. 3. McGraw-Hill Book Co., New York. pp. 14-46, 552-565.

Whetzell, H. H. 1918. An outline of the history of phytopathology. Saunders, Philadelphia.

CHAPTER 2

PLANT DISEASE CONCEPT, DEFINITIONS, SYMPTOMS AND CLASSIFICATION

2-1 PLANT DISEASE CONCEPT

Plant Disease Definition

To the grower or layman a plant disease is any abnormality of a plant, its parts or products, that reduces its economic value or esthetic quality. However, this statement expresses man's concern strictly from his own economic viewpoint. A proper definition of plant disease is: "a harmful alteration of the normal physiological and biochemical development of a plant," (National Academy of Science, 1968). (Also see Ch. 16, heading 16-4)

Fig. 2.1 Apple scab, caused by *Venturia inaequalis*. Scab disfigures the fruit and reduces its ability to be stored due to loss of moisture through the superficial lesions. *(Courtesy of R. W. Samson)*

It is best to think of plant disease as a series of harmful processes operating over a period of time. A plant disorder that occurs suddenly as a result of an external force is usually thought of as an injury. An injury is a disorder that occurs swiftly, over a short time period and is a one-time phenomenon.

Injury--A Swiftly Occurring Disorder

Diseases vary greatly in the amount and type of damage that occurs. A number of plant diseases of minor importance from the plant's viewpoint are of major importance from man's viewpoint. Some diseases simply disfigure part or all of their hosts, while other diseases totally destroy their hosts. Most diseases are between these two extremes.

Plant Disease Variable

When we think of plant disease, we should think of it as variable, changing, dynamic as opposed to fixed, rigid or definite. It varies with the duration and amount of change from the normal, and with the extent and final overall effect. Also, plant disease should be thought of as being composed of physiological changes and morphological (structural) changes. The physiological changes are abnormal, invisible processes that may be initiated as a pathogen attempts to enter and continues until the terminal stage of disease is reached. Some physiological changes precede the morphological changes. The morphological changes increase until they become visible, at which time they are called symptoms. There is a progression of both physiological and morphological changes from onset (start) of disease until its termination. This leads to the obvious principle: plant diseases are dynamic biological processes.

Rule -- Dynamic Nature Of Plant Disease

Disease may be temporary or permanent. An example of the former is partial wilting in dry soil on a hot day, with recovery when water is added or the temperature decreases. However, the permanent disorders are more clearly discernible as being "diseases," and cause the damage that all too often limits crop production. Therefore, these are the diseases that are of concern to plant pathologists.

Plant Disease Is Temporary Or Permanent

2-2 DEFINITIONS AND TERMINOLOGY

Each branch of science has its own terminology that is generally peculiar to it. A thorough knowledge of the definitions and terminology in a given scientific field is basic and absolutely essential for understanding that field. Plant pathology is no different in this respect. Before we can communicate intelligently concerning plant pathology, attention must be focused on the definitions and terminology peculiar to this field. The terms here will be used repeatedly throughout this text and in most texts dealing with this branch of biology.

2-2.1 CAUSES OF DISEASE

Noninfectious Disease Agents

A noninfectious plant disease cannot be transmitted from a diseased plant to a healthy plant and cause disease in the latter. There are a number of agents which act singly or in combination to cause noninfectious plant diseases. Various conditions of weather, nutrient deficiencies and excesses are some of these agents. Although damage by wind, hail and lightning is usually thought of as plant injury and not disease, these agents are often discussed under noninfectious disease agents because they cause plant disorders of a noninfectious nature.

Infectious diseases are caused by agents that are capable of being transmitted from a diseased plant to a healthy plant and of causing disease in the latter when the environment is favorable. Homeowners, growers and students new to the study of plant pathology, often think of infectious agents as preying upon the plants they attack and devouring plant tissue much as an insect might feed upon a leaf. What is seldom understood is that living green plants react in various ways to the presence of infectious agents. In fact, when certain of these reactions are of sufficient strength and suddenness, they may confer resistance to the plant.

Infectious Disease Agents

Infectious disease agents include fungi, bacteria, viruses, nematodes and parasitic seed plants. When one such agent enters the plant disease "picture", there is a mistaken tendency to regard it as the sole cause of the disease. We know that a specialized environment is often necessary to "condition" the plant, causing it to become susceptible to the agent. Infectious disease agents, generally require a rather narrow range of favorable environment in order to infect the plant. The ensuing interaction between the plant and the agent, which produces plant disease, is also conditioned by environmental influences. The environment may determine whether disease develops at all and, if so, at what rate and with what degree of severity. Therefore, it is obvious that the infectious disease agent is not the sole cause of disease. Because of this, we refer to the infectious disease agent involved as the incitant, causal organism, or causal agent, implying that it incited disease under the influence of other factors.

A causal agent and a favorable environment constitute two of the three requisite factors for the occurrence of infectious plant disease, the third factor is, of course, a susceptible host. The classical presentation of the interacting factors necessary for the occurrence of plant disease, is the plant disease triangle, as shown in the following illustration.

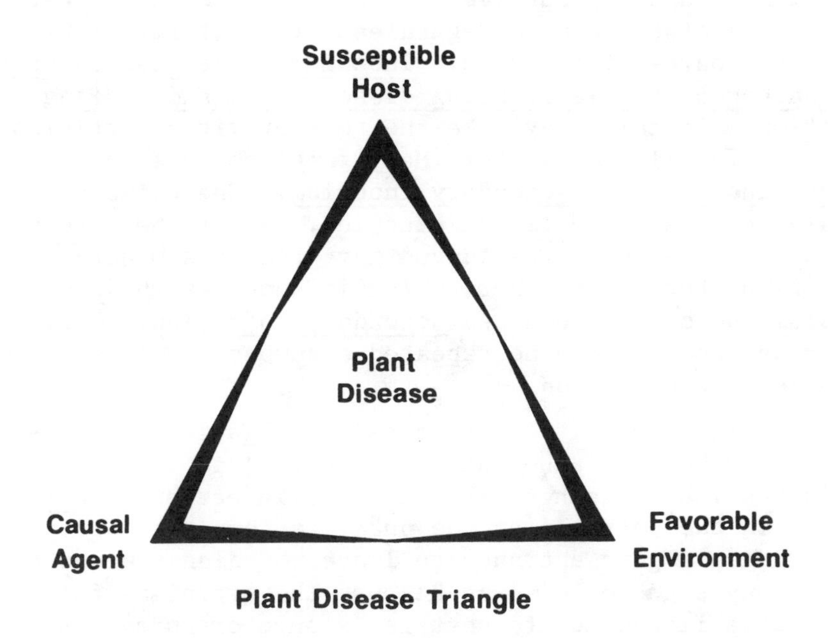

Plant Disease Triangle

Many parasites are causal agents of plant diseases. The term parasite, refers to an organism which is partially or wholly dependent upon living tissue for its existence. A saprophyte is an organism which subsists upon inorganic and dead organic matter. A pathogen is an agent that incites infectious disease. Usually, this is either a bacterium, fungus, nematode, parasitic seed plant, or virus. The terms parasite and pathogen are not synonymous, since an agent may be a parasite and not incite disease. Pathogenicity is the capacity of a pathogen to incite disease, and pathogenesis is the period in disease development from the time of infection to the final reaction in the host.

Terms

Parasite

Saprophyte

Pathogen

Pathogenicity

Pathogenesis

At least one common name has been given to each important plant disease. The common name is usually descriptive of some phase of the disease, e.g., late blight of potato, stem rust of wheat, clubroot of cabbage, etc. The generic names of the causal organisms are attached to the common names of some diseases, e.g., fusarium wilt of tomato. Except for viruses, most plant disease incitants have been given a specific name within a given genus according to the universally accepted system of plant nomenclature: this is the Latin binomial. Since the generic name and specific name are Latin words, they are printed in italics, or underlined when not in italics.

Latin Binomial

When a parasite is a part of the causal complex, it must overwinter (in the temperate zone) in some manner. The overwintering stage may be in or on seed of the host, in overwintering plants, in insect vectors, in perennial wild hosts,

Terms

Source Of Primary Inoculum

in debris of infected plants of the previous season, or in the soil. Fungi may survive as mycelia, dormant spores, and various other forms. Regardless of its form, it becomes the source of inoculum. Inoculum is defined as portions of a pathogen capable of being disseminated and causing infection. Inoculum may take the form of virus particles, bacteria, fungal spores etc. Many fungi porduce both primary inoculum and secondary inoculum. The primary inoculum causes the primary infection. After the disease develops in the host, the fungus may produce another spore crop of one form or another, which is known as the secondary inoculum and serves to cause secondary infection. This latter spore cycle may be repeated a number of times during a single growing season.

Primary Infection

Secondary Inoculum And Infection

Penetration

The initial activity of a parasitic organism in relation to the host is penetration. However, it is important to distinguish between penetration and infection. Penetration is the entrance of the parasite into the host. Many organisms gain entrance to leaves of plants without causing any signs of disease because the organisms fail to become established due to host resistance or unfavorable environment. These organisms die without proceeding further. Thus, we may have penetration without infection. Infection implies the establishment of a pathogen inside the host. There may be an interval of time between infection and the appearance of symptoms, this is known as the incubation period. The disease cycle is the sequence of events that occur between the time of infection and the final expression of disease. This may be a lengthy process involving the production of secondary inoculum and a progression of symptoms in the host. While the disease cycle is intimately associated with the causal organism, it is distinct from the life cycle of the organism. The causal organism may be able to complete a part or all of its life cycle as a saprophyte. Also, it may require an alternate host in order to complete its life cycle, or be transmitted by an insect vector.

Infection

Incubation Period

Disease Cycle

Life Cycle

2-2.2 LEVELS OF PARASITISM

It has long been recognized that there are many categories of parasitism, ranging from a symbiotic relationship to a highly pathogenic relationship. Remember, a parasite subsists in whole or in part upon living tissue, this does not mean that a parasite must be a pathogen. For instance, two organisms may live in close association with each other and be parasitic upon each other, yet they may not be pathogenic, but necessary for the development of each other. Such a relationship is called symbiosis. The classic relationship to illustrate this is in the lichens where fungi and algae live with one another. The photosynthetic capacity of the alga furnishes carboydrates for itself and

Symbiosis

the fungus. The fungus furnishes materials from the substrate by means of its hydrolytic enzymes. These minerals and organic nutrients are essential for the development of the fungus and the alga.

Parasitic phenomena are exceedingly complex. However, several levels of parasitism have been delineated. The highest level of parasitism is that of the obligate parasite, which refers to an organism that can grow only in living tissue and not on dead organic or inorganic matter. We commonly consider rust fungi, powdery mildew fungi, and downy mildew fungi as being among this group even though some members (especially the rusts) may be grown in a limited manner on specialized media. A facultative saprophyte is an organism that is ordinarily parasitic but may live saprophytically under certain conditions. Many fungi within this category are easily grown on artificial media but multiply in the field only as parasites. A facultative parasite is an organism that is ordinarily saprophytic but may live parasitically under certain conditions. Many organisms within this category are considered as "weak parasites". An obligate saprophyte is an organism that can develop only on dead organic and inorganic matter. These four categories are useful in that they channel our thinking of microorganisms as to various levels of parasitism with the acknowledged drawback that some of them straddle the line between categories.

Obligate Parasite

Facultative Saprophyte

Facultative Parasite

Obligate Saprophyte

2-2.3 HOST PLANT CONDITION

For plant diseases in which parasites are a part of the causal complex, the diseased plant is termed the host. Extremes of weather, various fertilizer and soil conditions, and other external factors may have an adverse effect on plants making them more vulnerable to attack by one or more pathogens than they otherwise would be. We speak of this phenomenon as predisposition, and the plants thus affected are called predisposed.

Host

Predisposition

Two more important terms to consider are resistance and susceptibility. Resistance refers to the ability of a host plant to suppress or retard the activity of a pathogen or other injurious factor. Resistance may be an inherited character, and it may also be influenced by predisposition or age of the plant. For example, young succulent tissue is generally less resistant than is mature tissue. Also, some conditions of environment will not promote the proper maturation of plant tissues. They therefore remain in a

Resistance

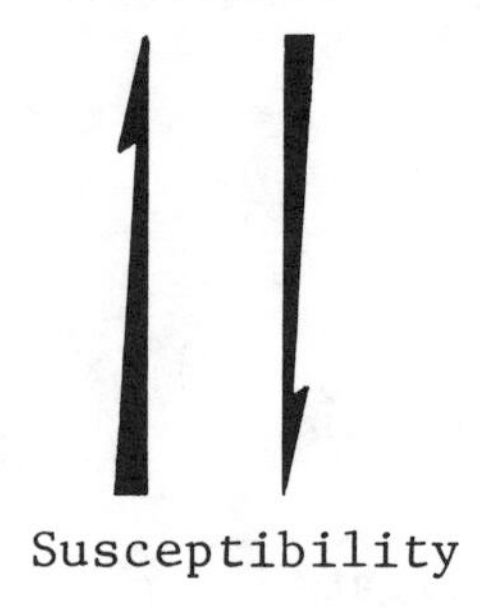

Susceptibility

succulent condition and are generally less resistant than if they had been grown under conditions that promote the normal hardening of tissues. Susceptibility may be regarded as the converse of resistance, and refers to a condition of a host plant in which it is normally subject to attack by one or more pathogens. There is an infinite number of gradations between resistance and susceptibility. The higher the resistance, the lower the susceptibility, and vice-versa. However, regardless of whether a plant is resistant or susceptible to one pathogen, this does not imply there is a similar relationship to other pathogens.

2-3 PLANT DISEASE SYMPTOMS

Most plant diseases involve profound physiological and morphological changes within the host. These changes vary with the host, environment, and causal agent. This section concerns the morphological changes of plant disease.

A Common Problem

Perhaps the most common problem of students in their initial study of plant pathology is the failure to recognize and distinguish between plant disease, its cause, and symptoms. A plant disease is a process that develops over a period of time; it is caused by a noninfectious agent or by an infectious agent that is profoundly influenced by environment; and it is usually detected by the visible alterations of the plants which are called symptoms.

Rule -- Changing Symptoms

Since plant disease is not static, but is dynamic, active and constantly changing, plant disease symptoms change as the disease progresses. The complete series of symptoms that occurs as a disease progresses is known as the symptom complex. This includes all of the visible changes that occur in the infected plant from the onset of visible change until the recovery or death of the plant or plant part. The severity of the disease often influences the symptom complex. For many plant diseases, the pathogen involved may not be determined by symptom expression alone. For example, a bacterial pathogen may cause the same symptoms as a fungal pathogen; a virus may cause the same symptoms as accidental herbicide injury; and one fungal pathogen may cause the same symptoms as a different fungal pathogen. Some particular infectious diseases may produce symptoms which are unique and distinct, and the pathogen is easily identified. The symptoms of other infectious diseases may not be distinct, and therefore it is often necessary to isolate and identify the specific pathogen involved before the cause of the plant disease can be established.

Symptom Complex

Most symptoms of plant disease may be conveniently placed in these major categories:

Symptoms

- Abnormal coloration of host tissue
- Wilting of host
- Death of host tissue
- Defoliation and fruit drop
- Abnormal growth increase of host
- Stunting of host
- Replacement of host tissue

2-3.1 ABNORMAL COLORATION OF HOST TISSUE

Plant tissues that are normally green may take on a deeper green, a lighter green color, or various shades of red. When normally green tissue is yellowish, it may be caused by insufficient light, and is termed etiolation; if it is caused by a factor other than light, the condition is called chlorosis. If the tissue is white, the symptom is called albinism.

Chlorosis

Leaves often show a variation in color. If the leaf veins become whitish or translucent while the rest of the leaf remains normally green or nearly so, the condition is said to be veinclearing. When leaf veins remain normally green or nearly so, but tissue between the veins becomes chlorotic, the symptom is called interveinal chlorosis. Leaves also may take on a definite varigated pattern of all shades of green to yellow, but with sharply defined borders; this is called mosaic. When the varigation is less pronounced, and the light and dark areas have diffuse boundaries, the condition is more apt to be called mottling. The terminology is not precise.

Veinclearing

Interveinal Chlorosis

Mosaic

Stems, flowers and fruit may also take on abnormal coloration. Normally brown stems may take on a lighter or darker appearance, flowers and fruit may become varigated in color, and normally green stems may appear reddish or purplish.

Leaf spots are symptomatic of many plant diseases. Leaf spots are areas of diseased tissue and are called lesions. Some pathogens cause leaf spots which are small and rather uniformly restricted in size. This occurs when the pathogens are capable of growing through the host leaf tissue a short and limited distance. These lesions do not increase substantially in size, but instead are localized, and are called local lesions. Other pathogens produce leaf spots that are not restricted in size. The pathogens continue to grow as long as the environment is favorable. Therefore, the resulting lesions are variable in size and often grow together to cover a large surface area of the leaf.

Fig. 2.2 Local lesions, some of which have coalesced to form large lesions.

Leaf Spots

Leaf spots may be few in number or numerous. They may be any shade of green, yellow or red. They may be concentrically zoned with various shades of color. Each spot may be separate and distinct, or two or more may grow together. In the latter instance, the spots are said to coalesce. Spots which appear on the upper side of a leaf may or may not appear the same on the lower side. On broadleaf (dicotyledonous) plants, spots tend to be rounded or angular. Veins of the leaf tend to restrict the development of some spots and cause the formation of angular spots. On narrowleaf (monocotyledonous) plants, spots tend to be elliptical or linear.

2-3.2 WILTING OF HOST

Plant Death Follows Permanent Wilting

Wilting may be caused by a shortage of soil water or an infectious agent. In either case, the condition may be temporary or permanent. The first appearance of wilting is usually temporary. Wilting may occur during the heat of the day, with recovery during the cool of the night. Death of the plant always follows permanent wilting. Plants in any stage of growth from the seedling stage to the mature plant may be subject to wilting.

2-3.3 DEATH OF HOST TISSUE

Dead Conditions

Death of host tissue may be subdivided according to the manner in which the disease is manifested. The resulting conditions are necrosis, rotting, canker formation, and mummification.

Necrosis: is the death of plant tissue, and does not imply its disintegration by decay or rotting although this may occur. The coloration of necrotic tissue is tan, brown, gray, or black. Typically, it is dark brown. This is frequently the final phase of a symptom complex. The entire plant, a branch or branches, or only the centers of leaf spots may die. When an entire plant or a large branch or branches die, the condition is called extensive necrosis. Sometimes necrotic tissue of local lesions drop out of leaves leaving a ragged or perforated appearance. This condition is called shot hole.

Fig. 2.3A Alternaria leafspot of tomato, caused by *Alternaria solani*. Note the zonate "target spots". *(Courtesy of R. W. Samson)*

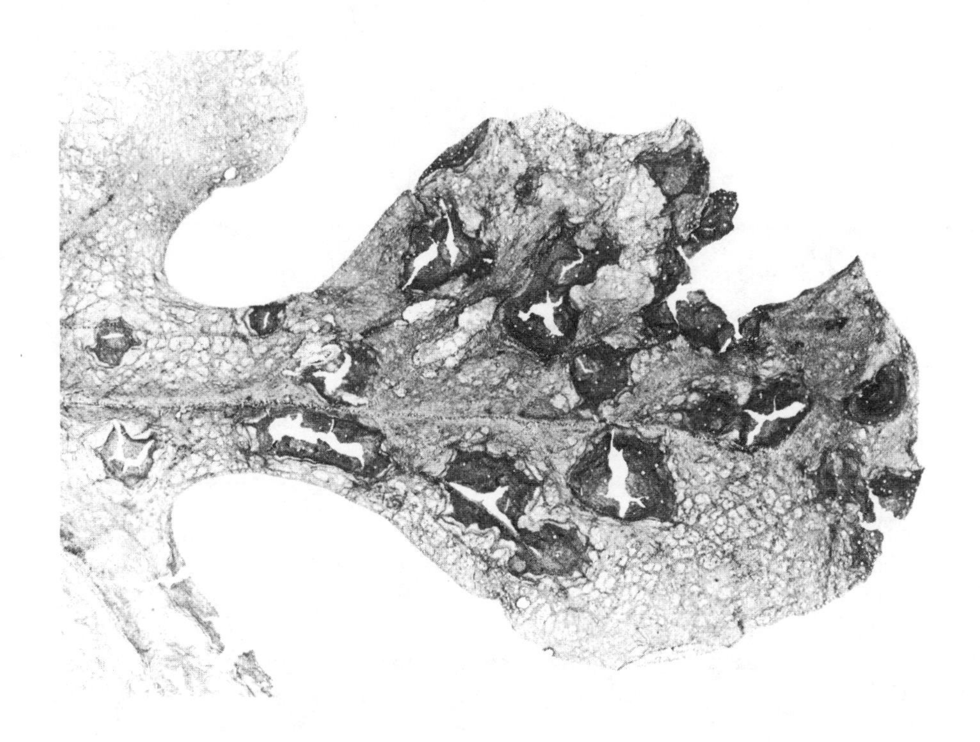

Fig. 2.3B Anthracnose of watermelon, caused by *Colletotrichum lagenarium*. The ragged appearance is because of the disintegration of the necrotic tissue, and it is characteristic of this disease. *(Courtesy of R. W. Samson)*

Dry Rot

Soft Rot Wet

Rotting: is the disintegration and decomposition of plant tissue, and may be of two general types, a dry rot or soft rot. These are fairly descriptive terms. A dry rot is a firm dry decay; and a soft rot is soft, usually watery, and often odoriferous. The type of rot often depends upon the plant structure involved. Fleshy plant parts with a high water content are more subject to soft rot, while less fleshy parts are more subject to dry rot. All plant parts including stems, roots, fleshy leaves, floral buds and fruit may be subject to rotting.

Seedlings may become infected and die prior to emergence from the soil. This is called pre-emergence damping-off. Also, seedlings that have emerged from the soil may be infected and die from a rotting of the stem near the soil line. When the stem is rotted through, the seedling falls over. This is called post-emergence damping-off.

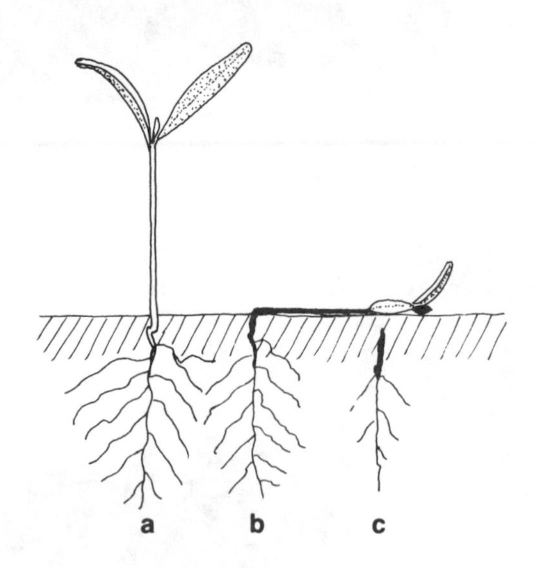

Fig. 2.4 Dark red table beet seedlings.

a. healthy
b. post-emergence damping-off
c. pre-emergence damping-off

Root Rot
Stem etc

There are many infectious agents that cause root rots. These may attack plants at almost any stage of growth. The rotting of all kinds of fruits is common, and so is the decay of modified stems such as tubers, rhizomes, bulbs and corms. The stems of corn plants may become rotted at or near the soil line and greatly weaken the stem. This is called stalk rot. When ears of corn become rotted, the affliction is termed an ear rot. The common names of rots are rather descriptive, and generally include the plant part or organ affected.

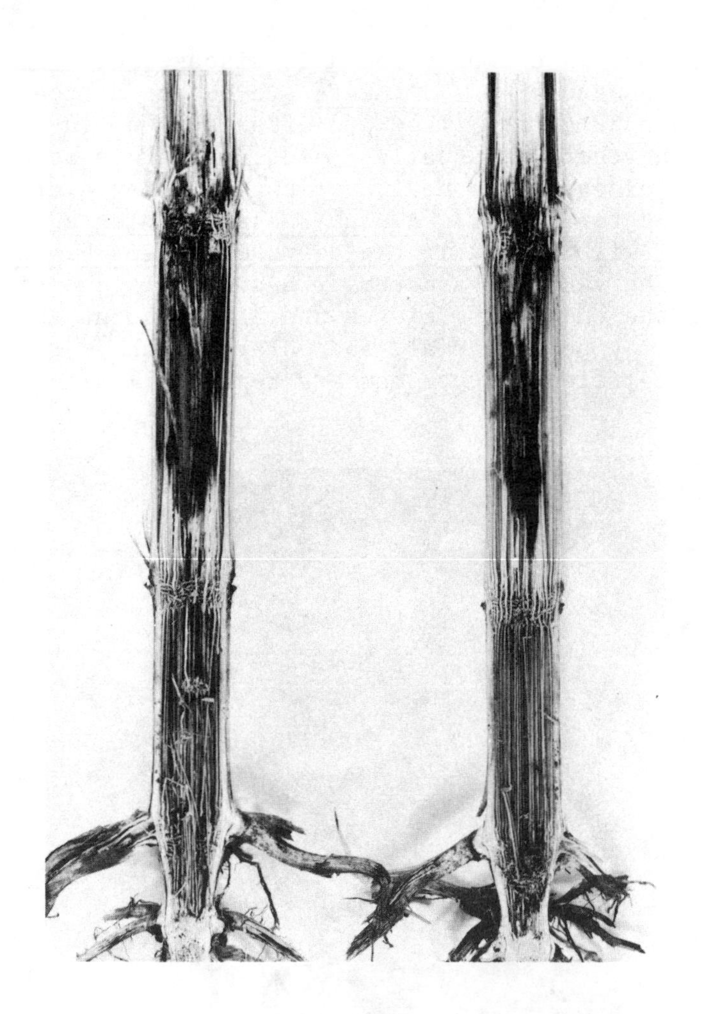

Fig. 2.5A Stalk rot of corn, caused by the fungus *Macrophomina phaseoli*. Notice the disintegration of the pith. Numerous black sclerotia are present on the vascular strands. *(Courtesy of A. J. Ullstrup)*

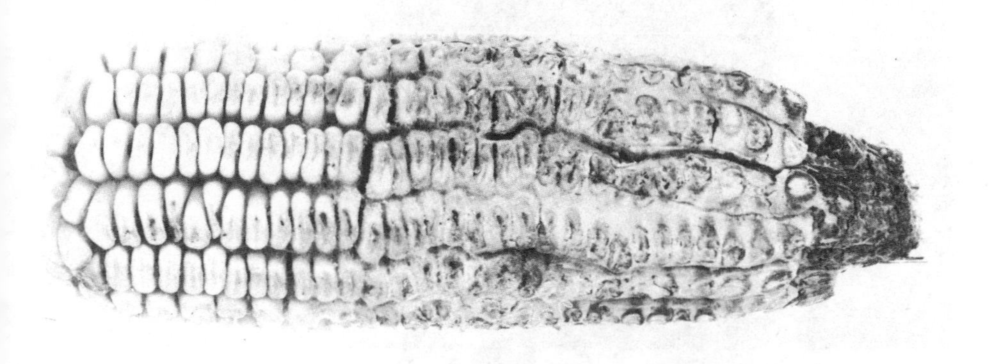

Fig. 2.5B Gibberella ear rot, caused by the fungus *Gibberella zeae*. The kernels are cemented together by a mass of pink mycelium. Corn infected with this fungus causes emesis in pigs. *(Courtesy of A. J. Ullstrup)*

Canker formation: is commonly found in diseases of perennial shrubs and trees. Cankers are necrotic areas in twigs, branches or stems. They are often sunken in appearance, and there is usually a distinct margin at the border of the diseased and healthy tissue. Often a crack will develop at the border. A canker that girdles a branch or stem will result in the death of tissue beyond the canker. The sunken character of a canker is due to the death of the underlying tissue and a subsequent shrinking. Cankers may be perennial, and enlarge each year until the branch is girdled. Other cankers reach full size in one year.

Fig. 2.6 Dothiorella canker on pin oak, caused by *Dothiorella quercina*. *(Courtesy of R. W. Samson)*

Mummification: is the initial rotting of fruit by a fungal pathogen which ramifies the fruit transforming it into a dry shriveled "mummy". The mummy is highly resistant to decay by other organisms, and generally serves as a means of overwintering for the fungal pathogen. Such mummies generally bear fruiting bodies of the pathogen in the spring. Mummified fruit may drop to the ground or remain attached to the host throughout the winter.

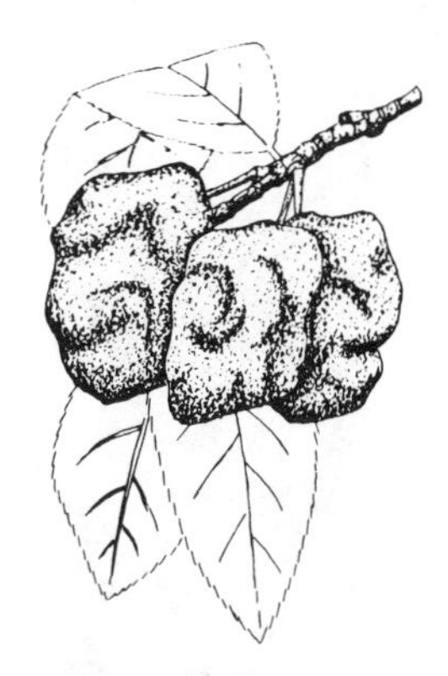

Fig. 2.7
Mummified plums

2-3.4 DEFOLIATION AND FRUIT DROP

Diseased leaves and fruit of broadleaf plants often drop prematurely. Some plant species and varieties are more subject to leaf drop than others. Also, some infecttious agents are more prone to cause leaf drop than others. Leaf drop is the terminal symptom for a number of diseases of foliage. However, leaf drop does not usually occur unless the leaves are severely infected. Both leaf drop and fruit drop may be caused by non-parasitic factors as well as by parasites.

2-3.5 ABNORMAL GROWTH INCREASE OF HOST

Almost any plant part may become subject to abnormal growth of tissue by the interaction of a host and parasite. Overgrowth, and sometimes malformed tissue is the result of hyperplasia (increase in number of plant cells), or hypertrophy (increase in size of plant cells), or both. The net result may be a twisting, curling and puckering of leaves, or the formation of galls on roots, stems, leaves or floral parts. Sometimes excessive branching occurs that is referred to as witches'-brooms. An excessive branching of the root system may also occur.

Hyperplasia

Hypertrophy

2-3.6 STUNTING OF HOST

Stunting is synonymous with dwarfing and either term may be used. The entire plant may be stunted or only a part or parts of it. This condition results from hypoplasia (reduction in cell number), or hypotrophy (reduction in cell size), or both.

Hypoplasia

Hypotrophy

Replacement

2-3.7 REPLACEMENT OF HOST TISSUE

The replacement of host organs by a parasite is especially common in diseases directly affecting the blossom or fruit. The pathogen may completely or incompletely replace the normal host tissue. The structures which replace the host tissue often are overwintering and/or reproductive forms of the parasite. Replacement may also involve a change in color size and shape. Affected tissue is often enlarged and malformed. In some diseases, floral organs are converted into leafy structures. By definition, replacement is caused only by parasitic agents and not by non-parasitic agents.

Fig. 2.8 Corn smut *(Ustilago maydis)* on tassel. This is an example of replacement of host tissues with the pathogen and involves malformed, overgrowths of the host. The black galls contain masses of smut spores. *(Courtesy of A. J. Ullstrup)*

2-4 SIGNS OF INFECTIOUS AGENTS

In many infectious diseases caused by certain fungi and parasitic seed plants, the pathogen itself is only partially embedded in, or attached to the host, and the exposed portion is clearly visible. This is called a "sign", and should not be confused with disease symptoms. Signs of fungal pathogens include fruiting bodies, spores and mycelium. Nematodes may be seen attached to the root system of host plants. The aerial stems, leaves, flowers and fruit of parasitic seed plants are attached to their hosts and are clearly visible.

The Sign

NOT A SYMPTOM

Fig. 2.9A An apple twig infected with powdery mildew fungus, *Podosphaera leucotricha*. The whitish, cottony covering over the leaves is the fungus itself. This is the "sign". The disease "symptoms" include the twisted and distorted leaves, and stunted growth. *(Courtesy of R. W. Samson)*

Fig. 2.9B Late blight of tomato, incited by *Phytophthora infestans*. This is the same fungus that causes late blight of potato. In this picture the stem lesion and dead leaves are "symptoms". No "sign" is present because the pathogen is not visible. *(Courtesy of R. W. Samson)*

Fig. 2.9C Late blight of potato. The white zone around the dark lesion is a cottony growth of the fungus which is bearing spores. The dark center of the lesion is a "symptom", and the whitish zone of the fungus is a "sign".
(Courtesy of R. W. Samson)

2-5 KOCH'S POSTULATES

Robert Koch

In 1882, a German, Robert Koch, published a monumental work on tuberculosis. In this paper he published experimental methods of inquiry which were based in part on his earlier study of anthrax. Although Koch's papers on anthrax and tuberculosis were epoch-making revelations, we are exceptionally interested in his experimental methods of inquiry which have become rules for all who study pathogenic organisms. The rules are usually called Koch's postulates. In summary form they are:

Koch's Postulates: Proof Of The Cause Of Disease

1. The organism believed to have caused the disease must always be present in the host when the disease occurs.
2. The organism believed to cause the disease must be isolated from the host and grown in pure culture.
3. The organisms obtained from pure culture, when inoculated into healthy hosts, must produce the characteristics of the disease.
4. The organism believed to cause the disease must be re-isolated, grown in pure culture, and compared with the organism first injected. (This addition to the postulates was added at a later time by Erwin F. Smith).

Koch's postulates are utilized to develop proof of pathogenicity whenever the pathogen can be isolated and cultured. Modifications must be made whenever nematodes, viruses, obligate parasitic fungi, and noninfectious causal agents are involved in the disease. If Koch's postulates cannot be followed exactly, the experimental conditions are modified so as to parallel the postulates as closely as possible. This requires strict tests involving adequate uninoculated or untreated control plants being subjected to the same rigors of the experiment as the inoculated or treated plants. Furthermore, it is usually necessary to duplicate the experiment in order to establish validity.

2-6 CLASSIFICATION OF PLANT DISEASES

Classification By Causal Agents

There are several systems of classification that may be used in the study of plant diseases. The most widely used system for introductory study is to classify diseases by their major causal agents. This system's major advantage is the stark simplicity of its organization. This is the system presented here.

I. Infectious diseases incited by:

1. Bacteria
2. Mollicutes and/or related forms
3. Fungi
   - Phycomycetes
   - Ascomycetes
   - Basidiomycetes
   - Fungi Imperfecti

4. Viruses
5. Parasitic seed plants
6. Nematodes

II. Noninfectious diseases incited by:

1. Low temperatures (plant injury)
2. High temperatures
3. Unfavorable oxygen relations
4. Unfavorable soil-moisture relations
5. Lightning, hail, and wind (plant injury)
6. Mineral excesses
7. Mineral deficiencies
8. Naturally occurring toxic chemicals
9. Pesticide chemicals
10. Air pollution toxicity

Insects will be considered only in their role as vectors, or carriers, of infectious pathogens. This is because the study and control of insects belongs to the field of entomology.

Selected References

Heald, F. D. 1933. Manual of plant diseases. Ed. 2 McGraw-Hill Book Co., New York. pp. 1-7.

Horsfall, J. G., and A. E. Dimond. 1959. The diseased plant. pp. 1-17, In J. G. Horsfall, and A. E. Dimond, (ed.), Plant pathology. Vol. I. Academic Press, New York.

Melhus, I. E., and G. C. Kent. 1948. Elements of plant pathology. The Macmillan Co., New York. pp. 22-30.

Roberts, D. A. and C. W. Boothroyd. 1972. Fundamentals of plant pathology. W. H. Freeman and Company, San Francisco. pp. 3-14.

Stakman, E. C., and J. G. Harrar. 1957. Principles of plant pathology. The Ronald Press Co., New York. pp. 35-47.

Walker, J. C. 1969. Plant pathology. Ed. 3. McGraw-Hill Book Co., New York. pp. 1-12.

CHAPTER 3

PART I: BACTERIA AS CAUSAL AGENTS OF PLANT DISEASE

3-1 THE BACTERIAL WORLD

Bacteria Everywhere

Indeed, it is difficult to find a situation devoid of bacteria. Objects exposed to air, such as this page, your desk top and the floor, are generally covered with bacteria. Your skin may harbor millions of bacteria, and the digestive tracts of animals are home for billions of bacteria of many different kinds.

For most animals, bacteria are of enormous usefullness in aiding digestion. Plants would lack certain important compounds needed for their growth if it were not for bacteria in the soil. Much of the carbon necessary for the growth of all living things would be unavailable without bacteria in soil, water, and air to decompose dead plants and animals.

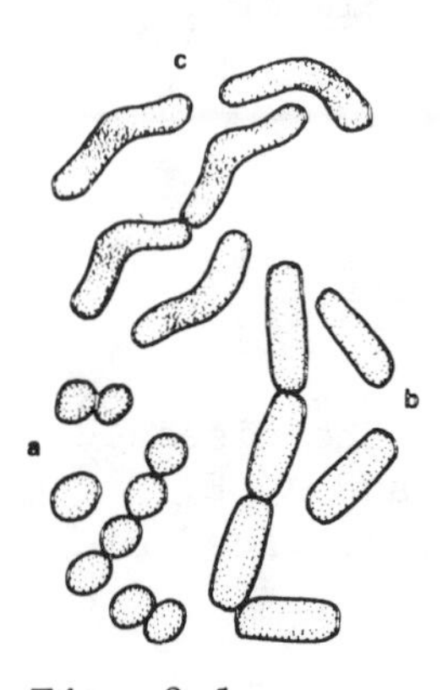

Fig. 3.1
Bacteria:
a. cocci
b. bacilli
c. spirilla

Three basic forms of bacteria are recognized; cocci (spherical), bacilli (rods), and spirilla (spiral or corkscrew). Some bacteria occur as single cells, others form colonies in pairs, chains, or irregular clusters. A bacterium does not have the well defined nucleus of the type found in cells of most higher plants and animals, it differs primarily by the lack of a nuclear membrane. Also, there is no nucleolus and no typical mitosis. The bacterial cell wall is a complex structure composed of carbohydrates and proteins. Most bacteria do not have cellulose in their walls as do higher plants.

Some Form Endospores

Some bacteria have the ability to form highly resistant endospores. These are produced by the formation of a thick cell wall that surrounds a single bacterium. Such endospores are capable of surviving exposure to high temperatures, or long periods of dryness.

Some Are Motile

Cocci are not motile, but bacilli and spirilla may or may not be. Motile bacteria have slender, whiplike threads called flagella that propel the bacteria through a film of water. The number and location of flagella are used to classify bacteria. The position of flagella is denoted by the following terms: monotrichous, one flagellum at one end; lophotrichous, more than one flagellum at one end; amphitrichous, one or more flagella at each end; and peritrichous, flagella scattered over the outer surface. The reaction of bacteria to Gram stain (a widely applicable technique testing the capacity of the fixed cells to retain crystal-violet stain in the presence of alcohol or an alcohol-acetone mixture) is a standard procedure used in classifying bacteria. Other characteristics used for bacterial classification are their abilities to grow on various carbon sources; gelatin liquefaction; nitrate reduction; production of indole, hydrogen sulfide and ammonia; growth in milk; and acid and gas formation with various carbon sources.

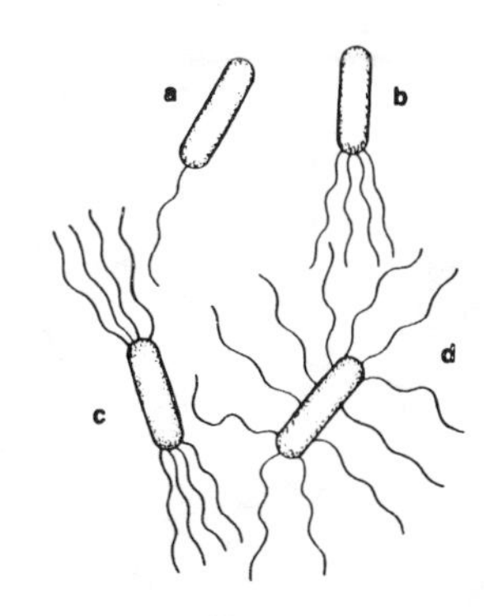

Fig. 3.2 Position of flagella:
a. monotrichous
b. lophotrichous
c. amphitrichous
d. peritrichous

Bacteria divide by binary fission, which is reproduction by means of splitting into two approximately equal parts. When given favorable temperature, moisture and nutrients, bacteria can divide every 20 minutes. With a theoretical maximum rate of growth, a single bacterium, after eight hours would increase to 16,777,216 cells, and after 16 hours there would be over 281 trillion cells. This growth rate is a geometric progression. After 7 days, there would be a bulk as large as the earth. Why, then, isn't the world overrun with bacteria? The answer is simply that they soon run out of nutrients and moisture. Also, as their population grows, the bacteria usually produce substances that accumulate and build up to a toxic threshold level.

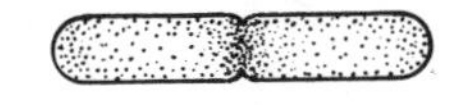
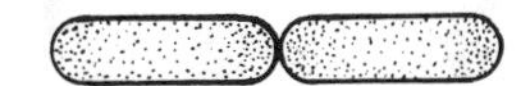

Fig. 3.3 Binary fission

3-2 GENERA OF BACTERIAL PATHOGENS

Rod Forms From Six Genera

There are about 200 different species of rigid bacteria that cause plant diseases. These species are primarily rod forms and come from only six genera: *Agrobacterium*, *Corynebacterium*, *Erwinia*, *Pseudomonas*, *Streptomyces*, and *Xanthomonas*. Only five species in the genus *Agrobacterium* are plant pathogens. The best known of these, *A. tumefaciens*, which causes crown gall, has received considerable attention because the growth of the gall resembles cancerous tissue in humans. The genus *Corynebacterium* has about 11 species that are plant pathogens. They are usually gram-positive; all other plant pathogenic bacteria are gram-negative. "Bergey's Manual" (Breed et al. 1957) lists 22 species of the genus *Erwinia*, all of which are plant pathogens and some of which are among the most difficult diseases to control. A foremost example is fire blight of apples and pears, caused by *E. amylovora*.

Pseudomonas contains about one-half of the known species of bacterial pathogens, and they cause a wide variety of rots, leaf spots, blights, and wilts. A closely related genus is *Xanthomonas*, which is the second largest genus of plant pathogens. It contains over 60 such species. They also cause a wide range of diseases including rots, blights, and leaf spots. These species characteristically produce a water-insoluble, yellow pigment in culture.

Only two plant pathogenic species are listed in "Bergey's Manual" as belonging to the genus *Streptomyces*. This genus is placed in the order Actinomycetales, which is considered by most authorities to be a distinct and separate group within the bacteria. They characteristically have a well-branched mycelium with conidia borne in chains. Endospores are commonly formed in this genus.

Rule -- Bacterial Parasitism

In respect to their general nature of pathogenesis, bacteria that cause plant disease are all facultative parasites; none are obligate parasites.

3-2.1 DISEASE TYPES

Bacteria generally cause one or more of the following categories of plant disease:

1. Soft rots involve primarily parenchyma tissue of various plant organs, resulting in a slimy softening and decay. The bacteria advance by the secretion of enzymes that hydrolyze the middle lamellae of the host cells and components of the cell walls. Affected host cells undergo plasmolysis and the tissue softens. As additional water exudes from the cells into the intercellular spaces, the tissue becomes necrotic, watery, and mushy. Entry

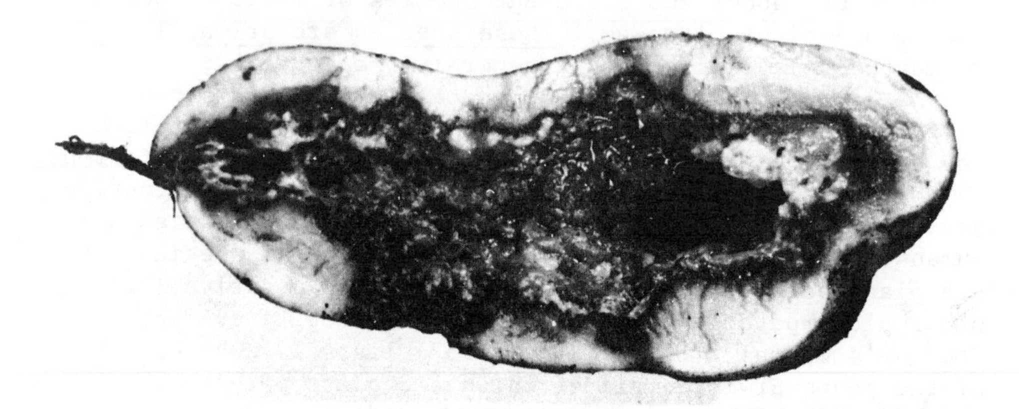

Fig. 3.4 Bacterial soft rot of potato, caused by *Erwinia carotovora*. *(Courtesy of R. W. Samson)*

of these organisms is generally through wounds.

Examples include: *Erwinia carotovora*, the incitant of bacterial soft rot of vegetables; and *E. atroseptica*, the causal agent of potato blackleg.

Leaf Spots

2. Leaf spots involve primarily leaf and stem parenchyma cells. Bacteria invade and multiply in substomatal cavities and intercellular spaces and eventually cause localized necrotic areas.

   Examples include: *Xanthomonas phaseoli*, the incitant of common blight of beans; *Pseudomonas phaseolicola*, the incitant of halo blight of beans; and *X. malvacearum*, the incitant of angular leaf spot of cotton.

Fig. 3.5 Bacterial blight of beans, caused by *Xanthomonas phaseoli*. Although the common name is "bacterial blight", the lesions are localized and appropriately fit the category of "leaf spots". *(Courtesy of the Department of Botany and Plant Pathology, Purdue University)*

Blights

3. Blights involve primarily leaf and stem, and also phloem tissue. There is generally a rapid progress of the bacteria through the host tissue. Dead leaves often remain attached, resulting in a scorched appearance of affected tissue.

Fig. 3.6A Fire blight of apple, incited by *Erwinia amylovora*. There is extensive necrosis of all branches. Some leaves and fruit still remain attached to the tree even though they are dead.
(Courtesy of E. G. Sharvelle)

Examples include: *Erwinia amylovora*, the causal agent of fire blight of apple and pear; and *Pseudomonas tabaci*, the incitant of wildfire of tobacco.

Fig. 3.6B Fruit blight phase of fire blight. The drops of exudate on the fruit contain millions of bacteria. *(Courtesy of the Department of Botany and Plant Pathology, Purdue University)*

Wilting

4. Wilting due to vascular disorders involves primarily the xylem vessels. Bacteria multiply in xylem tissue and become systemic. In later phases of the disease, the organisms may invade xylem parenchyma. The bacteria multiply and concentrate in the vascular system and may form a slime that reduces the water flow in the plant, resulting in wilting and often death of the host.

Examples include: *Erwinia tracheiphila*, the incitant of bacterial wilt of cucurbits; *E. stewartii*, the incitant of bacterial wilt of corn; and *Pseudomonas solanacearum*, the incitant of Southern bacterial (Granville) wilt of tomato, tobacco, potato, etc.

Fig. 3.7A The stunted corn plants are afflicted with the systemic phase of bacterial (Stewart's) wilt of corn, caused by *Erwinia stewartii*.
(Courtesy of A. J. Ullstrup)

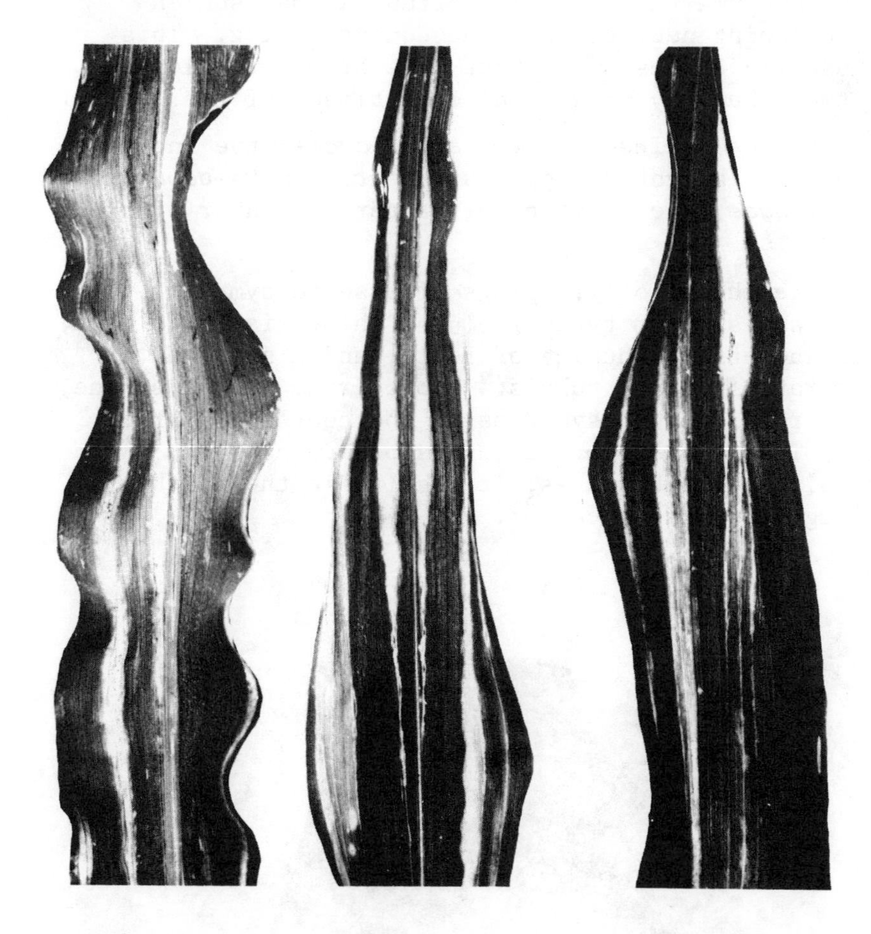

Fig. 3.7B The leaf blight, or nonsystemic phase of bacterial (Stewart's) wilt of corn.
(Courtesy of A. J. Ullstrup)

5. Galls involve hyperplasia and hypertrophy of host meristematic and parenchyma tissue.

   Examples include: *Agrobacterium tumefaciens*, the incitant of crown gall that attacks all fruit crops except strawberry; *A. rubi*, causes cane gall on brambles; and *Corynebacterium fascians*, which causes leafy gall on a number of host plants.

Fig. 3.8
Crown gall

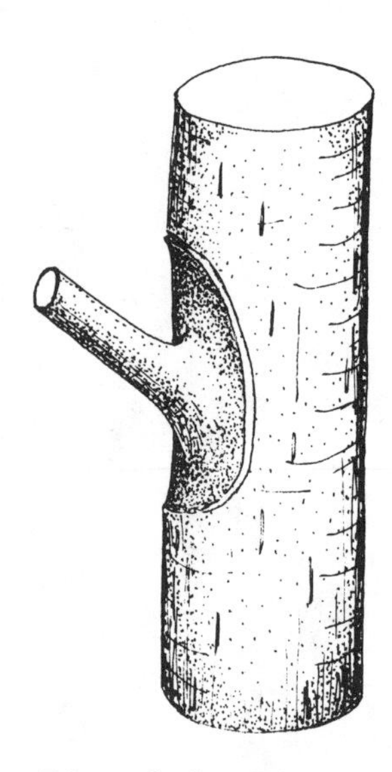

Fig. 3.9 Fire blight canker

6. Cankers involve primarily phloem and related parenchyma tissue, which often becomes sunken following necrosis and subsequent drying. This usually causes a distinct line of demarcation between healthy tissue and dead tissue of the canker.

   Examples include: *Erwinia amylovora*, the incitant of fire blight of apple and pear; and *Pseudomonas syringae*, the incitant of bacterial canker of stone fruits.

These are the major groupings of disease symptom types. Other disease symptom types include: chlorosis of foliage, abnormal increase branching of roots, and a scab and a vascular rot of potato tubers. A bacterium may cause one, two, or more different symptoms in the course of pathogenesis. Thus, *E. amylovora* causes branch cankers. When cankers girdle the branches, it results in the death of all foliage beyond the canker.

Fig. 3.10 Scab of potato, caused by *Streptomyces scabies*. *(Courtesy of R. W. Samson)*

PART II: MOLLICUTES AND RELATED ORGANISMS AS PROBABLE CAUSAL AGENTS OF PLANT DISEASE

3-3 INTRODUCTION AND HISTORY OF MYCOPLASMA AND L-PHASE BACTERIA

In the 19th century, Pasteur discovered the pleuropneumonia disease of cows was caused by an infectious agent. However, he was unable to isolate the agent, culture it, or observe it under the microscope. Then in 1892, Ivanowski showed the infectious entity of tobacco mosaic could pass through filters that entrap bacteria and proved the existence of disease agents that were smaller than bacteria. These disease agents were called viruses by Beijerinck in 1898. Consequently, it is not surprising that Pasteur's pleuropneumonia agent was also called a virus. Also in 1898, two Frenchmen, Nocard and Roux, isolated and grew the pleuropneumonia agent in a culture medium, proving that it was composed of true cells and was not a virus, which multiplies only in living organisms.

Pass Thru Bacteria-proof Filters

Other organisms similar to the pleuropneumonia disease agent were found years later, and they were simply referred to as "pleuropneumonia-like organisms", and abbreviated as PPLO. The PPLO were shown to exist in two sizes, the so-called "large cells", and the much smaller "elementary bodies". The elementary bodies may be as small as 125 millimicrons. The diameter of the large cells may be as much as several microns. Both forms of PPLO lack a rigid cell wall and are extremely delicate since they are bounded only by a thin membrane.

PPLO

At the Lister Institute in London, Klieneberger, in 1935, reported the discovery of an L-form of a bacterium. The letter "L" stands for Lister Institute. The term L-form has since given way to the terms L-phase or L-phase variant (Edward, 1960). Klieneberger's L-phase variant was an aberrant, soft, protoplasmic form of the bacillus *Streptobacillus moniliformis*, and was similar in size, shape and growth habit to bovine PPLO. Klieneberger successfully separated the L-phase from the bacilli and maintained it in pure culture thinking at first it was a PPLO that lived in symbiosis with the bacilli of *S. moniliformis*. It was later shown that the soft bodies were derived from the *S. moniliformis*.

L-phase -- A Bacterial Variant

Production of L-phase Variants

Dienes and his co-workers showed that under the influence of low, sublethal concentrations of penicillin, many bacteria produce soft bodies from which cultures of L-phase variants can be obtained. Usually there is a transition period when cultures must contain penicillin to prevent reversion to the parent rigid-bacterium. After several successive subcultures in contact with the penicillin, the soft forms often become stable L-phase variants and may not revert to the walled-bacteria form. In contrast to the parent bacterium, stable L-phase variants are insensitive to the action of penicillin. We now know that penicillin causes the lysis (dissolving) of bacterial cell walls. In light of this knowledge, it is understandable that this antibiotic does not inhibit organisms lacking cell walls. A number of chemical agents are capable of causing the lysis of walls of bacteria and thereby destroy them. Lysozyme is probably the best known of these "lytic" agents. If such a compound is present at low concentrations in certain bacterial cultures, it may cause the formation of large bodies which become granular in appearance. The membranes of the large bodies rupture and liberate many of the much smaller L-phase variants.

Smith, 1971, lists a number of agents including antibiotics, amino acids, enzymes and ultra-violet irradiation that have been used to transform various bacteria to the L-phase. However, reversion to the parent bacterial form frequently occurs when the lytic agents are removed.

Some L-phase Variants Arise Spontaneously

L-phase variants have also been shown to arise spontaneously in some bacterial cultures (Dienes, 1941), probably resulting from accumulation of metabolic by-products liberated by the parent bacteria.

L-phase Formation Is Common Bacterial Property

Within the last two decades, numerous bacteria have been shown to be capable of forming L-phase variants. These have been summarized by Hijmans et al., 1969. The ability to produce the L-phase is now a generally accepted property of bacteria. To plant pathology, this poses interesting questions concerning whether or not bacterial plant pathogens form the L-phase *in vivo* (in living organisms) as well as *in vitro* (non-living culture media). Also, (1) is it possible for a bacterial plant pathogen to reside inside a plant in the L-phase without causing plant disease, then revert to the parent "walled" form and initiate disease? (2) Are L-phase variants of plant pathogenic bacteria pathogenic? Such questions as these will surely come under investigation in years ahead.

Definition Of L-phase

L-phase bacteria have been defined by Hijmans et al., (1969) as "independent growth variants of bacteria that lack a rigid cell wall and have a potential reversibility to the parent strain".

PPLO were assigned to the bacterial class Schizomycetes, order Mycoplasmatales by Edward and Freundt, (1956). Within this order there was one family Mycoplasmataceae and one genus *Mycoplasma*. More recently, it was recommended that the order Mycoplasmatales be removed from the bacterial class and placed in a new class termed Mollicutes (Edward and Freundt, 1967).

Class Mollicutes

Edward and Freundt (1969) list 35 species of *Mycoplasma*, some of which are pathogens of man, large and small animals, and birds; others live saprophytically in sewage and soil.

Utilizing the classification system as given in Bergey's manual of determinative bacteriology, 7th ed. 1957, the relative positions of the mycoplasmas to bacteria, rickettsias and bedsonias are as follows:

Classification System

Kingdom: Plantae (Plant Kingdom)
 Division I: Protophyta (Primitive plants)
  Class I: Schizophyceae
   (Blue-green algae)
  Class II: Schizomycetes
   (Bacteria and related forms)
  Class III: Microtatobiotes
   Family Rickettsiaceae
   Family Chlamydiaceae (Bedsonia)
  Class IV: Mollicutes
   Order Mycoplasmatales
    Family Mycoplasmataceae
     Genus *Mycoplasma*

This classification system will undergo considerable change as relationships between groups become more defined.

The L-phase variants of bacteria resemble the mycoplasmas and this has been the subject of much discussion and controversy. Now it is generally agreed that despite many similarities, the L-phase variants are distinct and different from the mycoplasmas.

Two features which are among the most important characters separating the L-phase variants from mycoplasmas are: (1) L-phase variants are known to arise from parent bacteria while mycoplasmas have not been shown to arise from bacteria. (2) There are few or no genetic similarities between mycoplasmas and L-phase variants (McGee et al., 1967; Somerson et al., 1967). Also, the short filamentous projections produced by L-phase variants are unlike the filaments produced by mycoplasmas. An excellent discussion of the similarities and differences between L-phase variants and mycoplasmas is given by Edward and Freundt (1969) and Smith (1971).

3-4 CHARACTERISTICS OF MYCOPLASMA

The word mycoplasma means "fungus form". This is descriptive of the elongated, branching filaments that resemble fungal hyphae, which these unicellular organisms may produce. All mycoplasmas form numerous elementary bodies which are the smallest viable (able to keep alive) units. It is these elementary bodies ranging in size from 125 to 250 millimicrons that pass through bacterial filters. Starting with the elementary bodies, mycoplasma grow and complete their life cycle either forming the branching mycelioid structures or the "large cells" that approach the dimensions of bacteria. The position of the large cells in the life-cycle of mycoplasmas is uncertain. Some authors have theorized the large cells reproduce by buds in a manner similar to the budding of yeasts. Other workers state that granules may form in the large bodies and sometime thereafter the membrane ruptures and elementary bodies are released. Reproduction by means of binary fission is thought by some workers to occur in at least some known mycoplasmas. Thus, reproduction of mycoplasmas is the subject of much discussion and controversy.

Method Of Mycoplasma Reproduction Is Uncertain

The electron microscope reveals that mycoplasmas are devoid of cell walls, but possess a three-layered (trilaminate) membrane. They are highly pleomorphic (capable of assuming many different shapes), soft, fragile, plastic, and are subject to distortion under slight shifts in osmotic pressure. These organisms are gram-negative and stain poorly with ordinary bacterial stains. However, they do take Giemsa stain fairly well.

Pleomorphic, Soft, Fragile, And Plastic

Cultures may be maintained in both liquid and solid media. Growth on solid media gives rise to a characteristic "fried-egg" form. There is a dense central core (the yoke of the fried egg), that penetrates downward into the agar. Surrounding this inner area is a circular growth zone that is lighter in color and spreads on the surface of the agar. With the exception of three species, most prominant of which is *M. laidlawii*, all other mycoplasmas thus far described require the presence of sterol in the culture media in order to grow. A number of mycoplasmas grow well under aerobic conditions, while others prefer anaerobic conditions. Some are best isolated under partial anaerobic conditions and then may be maintained aerobically. Mycoplasmas possess absolute resistance to the action of penicillin, but are inhibited by the tetracycline compounds.

Most Require Sterol In Culture Media

A most important feature of all mycoplasmas thus far described and known to be animal pathogens is the fact that they are all extracellular (outside host cells) in nature.

3-5 MYCOPLASMA-LIKE BODIES AS AGENTS OF PLANT DISEASE

Yellows Diseases Suspect

Most plant diseases presently suspected of being caused by mycoplasmas are in a group called "yellows", previously thought to be caused by viruses. Yellows diseases are typified by chlorosis of leaves and/or growth abnormalities, such as proliferation of axillary shoots and uneven distortion of flowers which appear greenish and usually sterile. These diseases are caused by agents that are principally spread and transmitted in the field by leafhoppers. Furthermore, no viruses have been observed by examination of diseased tissues from these infected plants with the electron microscope. Also, these diseased tissues have not yielded purified virus extracts. Based mainly on the filterability of the causal agents, all of the diseases of this group were thought to be caused by viruses. Also, these disease agents could be experimentally transmitted by grafting a portion of a diseased plant to a healthy plant. This is the universal method of transmitting plant disease viruses by experimental manipulation. The inability to purify viruses from these diseased plants was previously ascribed to their so-called "instability", and their probable low concentration in infected tissue.

Soft Bodies In Diseased Phloem Cells

Tetracycline Is Therapeutic

A team of Japanese workers (Doi, et al., 1967, and Ishiie, et al., 1967), proposed that mulberry dwarf disease may be caused by mycoplasma or similar organisms. Their evidence was based upon consistently finding "mycoplasma-like" organisms within phloem cells and occasionally in phloem-parenchyma cells of diseased plants as seen with the aid of the electron microscope. Similar bodies were not found in healthy plants. Furthermore, treatment of diseased plants with tetracycline provided therapeutic effectiveness for diseased plants, and no mycoplasma-like bodies were observed in treated plants. Plant diseases caused by known viral agents usually appear to be unaffected by antibiotics such as the broad-spectrum tetracycline compounds.

Dio, et al., (1967), also detected mycoplasma-like bodies in infected tissues of three other yellows diseases, including aster yellows of petunia. At no time did these workers find mycoplasma-like bodies in healthy plants.

Soft Bodies In Infective Leafhoppers

By use of the electron microscope, Granados, Maramorosch, and Shikata (1968), found mycoplasma-like bodies in corn plants afflicted with corn stunt disease. Similar bodies were absent in healthy plants. They also found mycoplasma-like bodies in 4 out of 11 leafhoppers that were known to possess and were able to transmit the corn-stunt agent. They were unable to find such bodies in 10 leafhoppers that

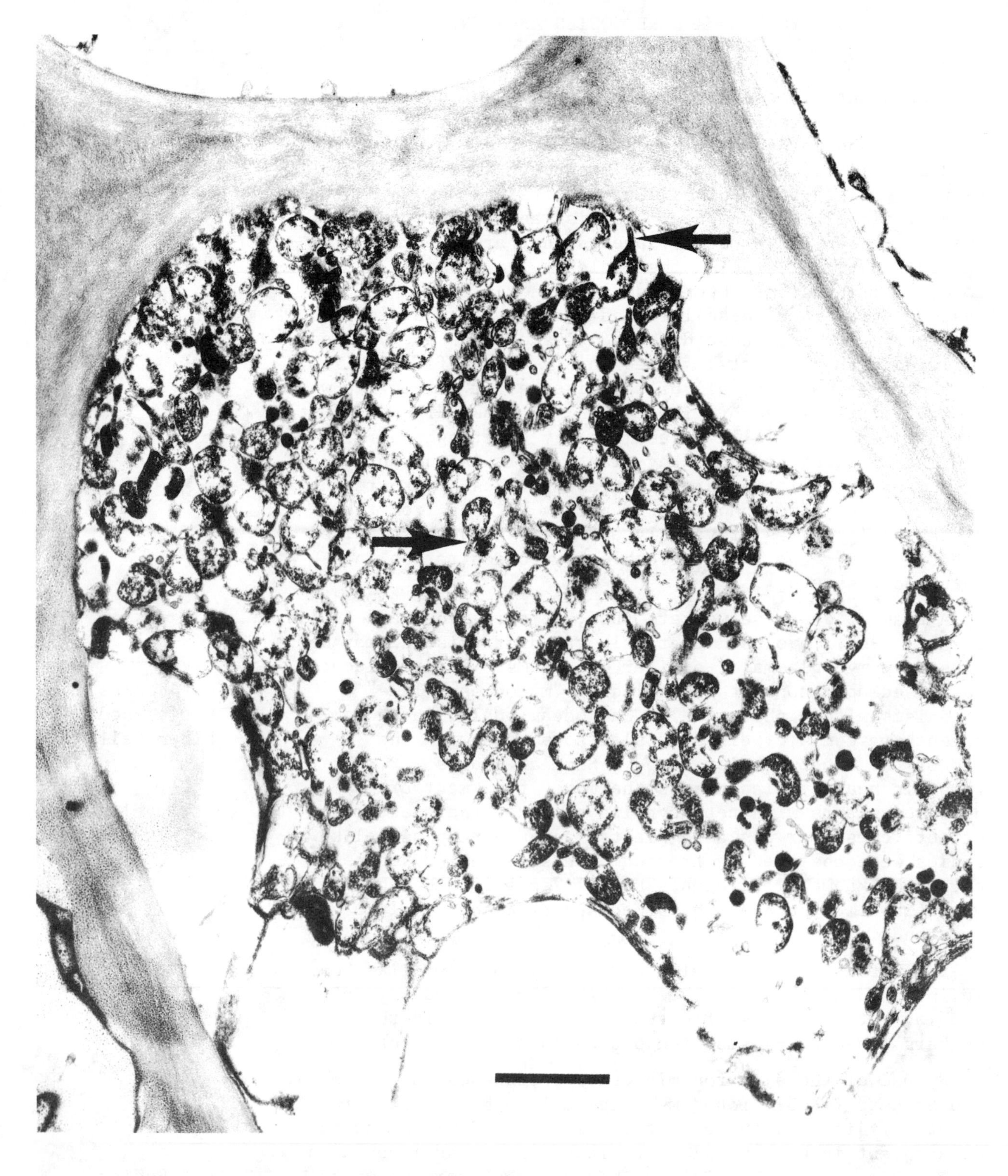

Fig. 3.11 Section of phloem sieve element from symptomless leaf of aster yellows diseased tobacco plant showing variety of mycoplasma-like forms enclosed within the plasmalemma. Filamentous growth (arrows) is associated with several of the larger spherical organisms. Bar equals 1 micron. *(Courtesy of J. F. Worley)*

were reared in an environment to assure that they were free from the corn-stunt infectious agent. These workers also examined earlier photographs, taken in 1964 through the electron microscope, and found mycoplasma-like bodies in two different host plants afflicted with American aster yellows disease, (Maramorosch, Shikata, and Granados, 1968).

Another group of workers, Davis, Whitcomb, and Steere, (1968a), worked with American aster yellows. They inoculated chrysanthemum plants with the aster yellows disease agent, and after several days the plants were treated by immersing the roots in 1,000 ppm chlorotetracycline (Aureomycin). The treated plants did not show disease symptoms. Leafhoppers that were injected simultaneously with the aster yellows agent and 1,000 ppm chlorotetracycline also failed to transmit the infectious agent to chrysanthemum, (Davis, Whitcomb, and Steere, 1968b). Aster yellows symptoms were surpressed by chlorotetracycline, tetracycline, and chloramphenicol. However, penicillin had no affect upon the symptom expression of diseased plants. This is consistant with the knowledge that penicillin has no affect on mycoplasmas.

Penicillin Is Not Therapeutic

In 1969, Hampton, Stevens, and Allen reported finding a mycoplasma-like body in association with alfalfa mosaic virus in diseased pea plants. The mycoplasma alone caused disease in both peas and cowpeas. However, the disease was noticeably more severe when these plants were infected with a combination of the mycoplasma and alfalfa mosaic virus. This was the first report of a mycoplasma-like body causing a plant disease which is not in the "yellows" category.

Hull (1971), lists over 40 plant diseases that have been reported to have mycoplasma-like bodies as their causal agents. Most of these diseases are of the "yellows" type.

Mycoplasma-like bodies have been frequently compared with known, described *Mycoplasma* spp. They have also been compared to L-phase bacteria, and *Chlamydia* spp.

At the time of this writing, several authors claim to have fulfilled Koch's postulates for mycoplasma-like organisms (Chen and Granados, 1970; Hampton et al., 1969; Nayar, 1971; and Giannotti, et al., 1971). However, although limited maintenance in culture has been reported for several mycoplasma-like organisms associated with plant disease, none as yet have been deposited in type-culture collections, and thus none have been compared with described species of *Mycoplasma*.

One significant difference between the described species of *Mycoplasma* and the mycoplasma-like bodies causing plant diseases is that the former are all extracellular (outside host cells), while the latter are intracellular (inside host cells) in both plant and insect hosts. The significance of the extra- versus intracellular growth habit prompted Davis and Whitcomb (1971) to suggest there may be important differences between mycoplasma-like agents of plants and known *Mycoplasma* spp.

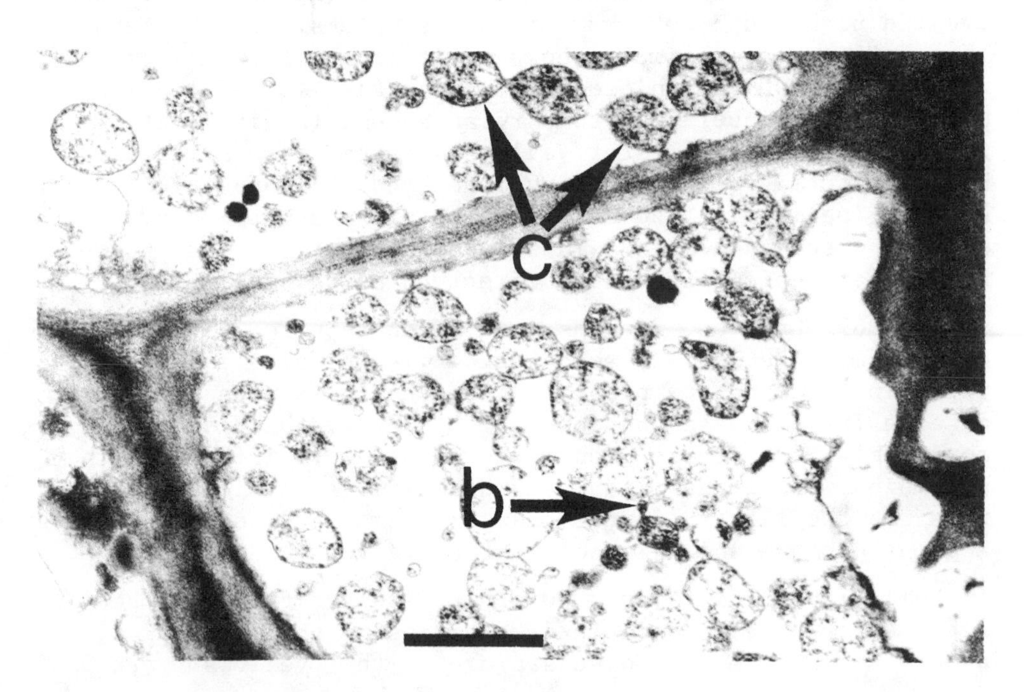

Fig. 3.12 Mycoplasma-like bodies showing chainlike configuration (c) and budding (b) in sieve elements of aster plant infected with aster yellows disease. Bar equals 1 micron. *(Courtesy of J. F. Worley)*

Fig. 3.13 Right. (A) Section of sieve element from chlorotic leaf of aster yellows diseased tobacco plant showing presence of irregular elongated forms (arrows) intermixed with other pleomorphic forms. (B) Structure indicative of binary fission of a mycoplasma-like organism. (C) Structure resembling filamentous growth of mycoplasma-like organism. Spherical bodies appear connected by thin filaments (arrows). Bar equals 1 micron. *(Courtesy of J. F. Worley)*

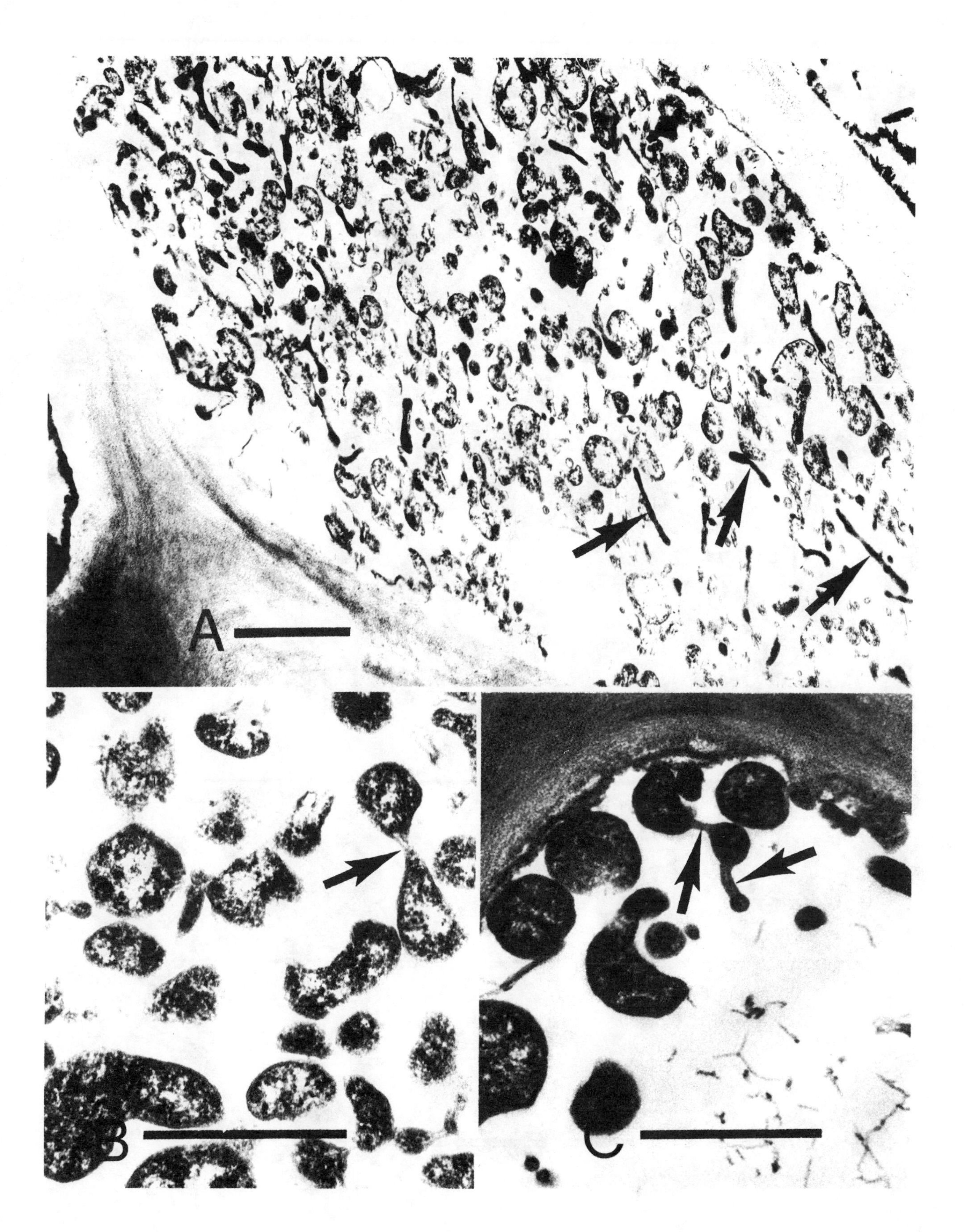
A
B
C

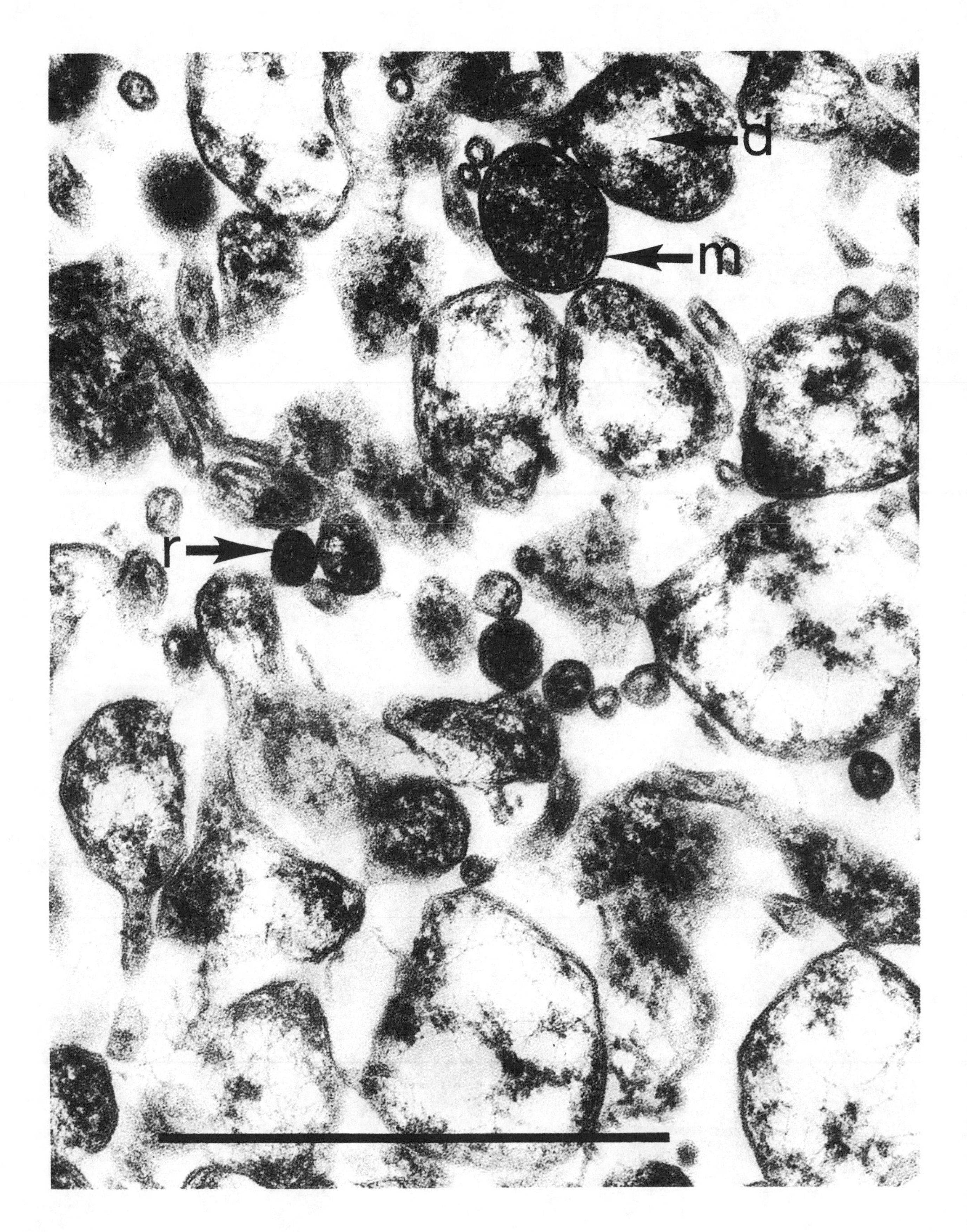
d
m
r

3-6 THE RICKETTSIAS

The family Rickettsiaceae is composed of a group of small gram negative organisms that appear coccus-shaped, rod-shaped, or pleomorphic. Their sizes are usually 0.5 to 0.8 by 1 to 2 microns. Thus their size is much larger than the mycoplasmas. They are retained by filters through which mycoplasmas, L-phase variants and bedsonias easily pass. Rickettsias possess a cell wall of varying thickness so as to often appear wavy.

Wavy Cell Wall

Most rickettsias are parasites of arthropods, but a few are pathogenic to man and other mammals. With one exception (the trench fever agent), the rickettsias that infect mammals are obligate intracellular parasites. Davis and Whitcomb (1971) state: "Their widespread occurrence among arthropods, and especially their presence in homopterous insects, suggests they may eventually be found in plants." For example, infected leafhoppers could inject rickettsia parasites into plants during feeding. In fact, at this time there are two reports of rickettsiae-like organisms associated with plant diseases previously thought to be caused by viruses; Pierce's disease of grapevines (Goheen et al., 1973), and phony peach disease (Hopkins et al., 1973).

Intracellular

3-7 THE CHLAMYDIACEAE (BEDSONIA)

This group of organisms are all obligate intracellular parasites that possess a distinctive life cycle whereby the organisms occur in two forms. There is a small form (elementary body) and a large form (initial body). The elementary body is about 300 millimicrons in diameter. It is composed of a dense core surrounded by a loose membranous cell wall. Elementary bodies are the infective entities that enlarge into the larger forms after entering a susceptible host cell. The initial bodies are spherical, about 1 or more microns in diameter, and they multiply by fission.

Intracellular

Chlamydia spp. are distinguished from mycoplasmas and L-phase variants by the presence of cell walls and subsequent sensitivity to penicillin. Also, they are not conspicuously pleomorphic and do not form filaments.

Fig. 3.14 Left. High magnification micrograph showing mycoplasma-like bodies bound by a trilaminar membrane (m). Small electron-dense bodies contain ribosomes (r) and larger bodies contain strands resembling DNA (d). Bar equals 1 micron. *(Courtesy of J. F. Worley)*

3-8 RELATED FORMS

Plant parasitic Mollicutes and related organisms are currently the subject of intensive research. As a result, a great amount of new information will shortly be added to our present meager knowledge. An illustration of an exciting development in this area is the report of Davis and Worley, 1973, in which a newly discovered type of organism has been found in association with the Rio Grande strain of corn stunt disease. Previous reports indicated a mycoplasma-like organism to be the presumed agent of corn stunt disease. These authors found a minute coiled or helical filamentous organism associated with corn stunt. Furthermore, the helical filaments were motile, intracellular and without cell walls or flagellae. The authors proposed the trivial name "spiroplasma" for these organisms. The filaments are 0.2 to 0.25 microns wide by 3 to 15 microns long. Some filaments have spherical bodies (0.4 to 0.6 microns in diameter) attached. Motility is achieved by spinning of the coil about its long axis and a bending or flexing of the coil. Clearly this "spiroplasma" does not fit any previously described category of life forms.

Spiroplasma

At least two strains of corn stunt exist, but at the time of this writing it is not known if "spiroplasmas" are associated with the other strain or not.

Selected References

Part I

Burkholder, W. H. 1948. Bacteria as plant pathogens. Annual Review of Microbiology. 2:389-412.

Carter, W. 1962. Insects in relation to plant disease. Interscience Publishers, New York. pp. 9-49.

Dowson, W. J. 1959. Plant diseases due to bacteria. Ed. 2. University Press, Cambridge.

Elliott, C. 1951. Manual of bacterial plant pathogens. Ed. 2. Chronical Botanica Co., Waltham, Mass.

Graham, D. C. 1964. Taxonomy of the soft rot coliform bacteria. Annual Review of Phytopathology. 2:13-42.

Riker, A. J., and A. C. Hildebrandt. 1953. Bacteria-small and mighty. pp. 10-15, In Plant diseases. U. S. Dept. Agr. Yearbook.

Starr, M. P. 1959. Bacteria as plant pathogens. Annual Review of Microbiology. 13:211-238.

Walker, J. C. 1969. Plant pathology. Ed. 3. McGraw-Hill Book Co., New York. pp. 104-174.

Part II

Davis, R. E., and R. F. Whitcomb. 1971. Mycoplasmas, Rickettsiae and Chlamydiae: possible relation to yellows disease and other disorders of plants and insects. Annual Review of Phytopathology. 9:110-154.

Hampton, R. O. 1972. Mycoplasmas as plant pathogens. Annual Review of Plant Physiology. 23:389-464.

Hayflick, L. (ed.). 1969. The mycoplasmatales and the L-phase of bacteria. Appleton-Century-Crofts, New York.

Hull, R. 1971. Mycoplasma-like organisms in plants. Review of Plant Pathology. 50:121-130.

Maramorosch, K., et al. 1970. Mycoplasma diseases of plants and insects. Advances In Virus Research. 16:135-193.

Smith, P. F. 1971. The Biology of Mycoplasmas. Academic Press, New York.

Whitcomb, R. F., and R. E. Davis. 1970. Mycoplasma and phytarboviruses as plant pathogens persistently transmitted by insects. Annual Review of Entomology. 15:405-464.

CHAPTER 4

FUNGI AS CAUSAL AGENTS OF PLANT DISEASE

4-1 THE FUNGAL WORLD

Fungi, Vital Segment Of Biological Cycle Of Life

There are more than fifty thousand species of fungi recorded. The great majority of these are obligate saprophytes. By their "obligation" to live on dead organic matter they provide a great service to green plants as well as to mankind. The decay caused by scavenging fungi, which transposes dead plant and animal matter into fertile soil, is one of the vital segments of the biological cycle of life. As men are dependent upon green plants for their food, so are the molds. However, with all equality, this also may be turned around. Green plants are sustained by the humus of the soil, which is formed by the decomposition of once-living tissue. Suppose for a moment what this world would be like if the soil could support green plants indefinitely without the need for replenishment with the products of decomposition; and further, suppose that there were no fungi! The face of the world, both land and sea, would be choked and smothered with garbage! So, surely man is both directly and indirectly dependent upon fungi for his food, comfort and well-being. Of course, this is not to minimize the role played by bacteria in the decay and decomposition of organic matter, for this is also essential and vital.

Some Fungi Pesky

Not all saprophytic fungi are totally beneficial. Some possess the pesky ability to grow on fabrics, shoes and other leather goods, woodwork of all kinds, and food products. This really shouldn't surprise us too much, for these fungi were here since the very beginning of earth's biological cycle, and they convert man's handywork into humus with as much relish as the fallen trees of the forest.

Human Parasites

Some fungi are parasites of humans, causing diseases which are all too often exceptionally difficult to treat. The most serious of these prevail in tropical climates where seasonal high temperatures and humidity offer an ideal environment for the year-round growth of fungi.

However, man is able to exploit some fungi to his own benefit. Typical of these is *Penicillium* sp. The antibiotic penicillin is extracted from the media upon which the fungus is grown. A large number of other antibiotics and drugs, along with enzymes and organic acids, are also obtained from various fungi. Fungi are also the basis of a number of industrial processes which involve fermentation to provide bread, beer, wine and certain cheeses. Also, we should not overlook the fact that some fungi are considered to be among that select circle of foods called delicacies, such as the common cultivated mushroom, and the tasty morels of the woodlands.

Fungi Exploited

Fig. 4.1
Mushrooms

Of necessity, any introductory or general course in plant pathology must spend a major share of its time and effort exploring plant pathogenic fungi, because the majority of important plant diseases are incited by fungi. It is fortunate indeed that just a portion of the described fungi are capable of causing plant disease, for the task of controlling many of the fungal parasites is well beyond our present capabilities.

Many Fungi Are Easily Visible, Most Plant Parasites Are Not

Many fungi are easily visible. The fleshy mushrooms of field and forest are large and conspicuous, as are the woody bracket fungi on tree trunks and rotting logs. Plant pathogenic fungi generally are much less discernible. The disease symptoms these fungi cause are, however, strikingly apparent. For centuries the obvious nature of most fungal disease symptoms ruled against man's seeking out their true cause, the generally not obvious fungi. This fact is repeatedly brought to light when new students of plant pathology fail to understand that symptoms are only the visible effects of disease and not the disease itself. As a general rule, when compared to the conspicuous fungi, plant pathogenic fungi are more closely related to the molds that invade bread, rot fabrics, and overrun oranges with a powdery, greenish spore mass.

4-2 FUNGI LIVE AND GROW

Fungi: Living Organisms, Have No Chlorophyll, Live As Saprophytes Or Parasites

The most important characteristic to remember about fungi is that they are living organisms. The second most important feature of fungi is that they do not have chlorophyll and cannot utilize the energy of the sun and thus manufacture their own food. Therefore, they must obtain their food from some other source. The substance or object on or in which a fungus grows and utilizes for food is called the substrate. This may be the humus of the soil, non-living remains of animals or plants, or sometimes our cultivated and prized green plants. Fungi that live on green plants do so simply because they are able to, and some are incapable of growth anywhere else. A few fungi are capable of living on only one particular host species; others are ubiquitous in their needs and may live on any or all of a hundred or so green plants. We speak of the various species of host plants upon which a parasite will grow as being its host range. Some species of fungal pathogens have an extreme specialization as to the host on which they grow. These species of fungi have sub-groups which look alike, but are physiologically different. Each sub-group is called a race. A race is able to attack only particular and special hosts. A race has its own narrow host range, which may overlap that of other races, but does not coincide with that of any other race. The fungus causing stem rust of wheat has many such races that are able to grow only on certain varieties of wheat.

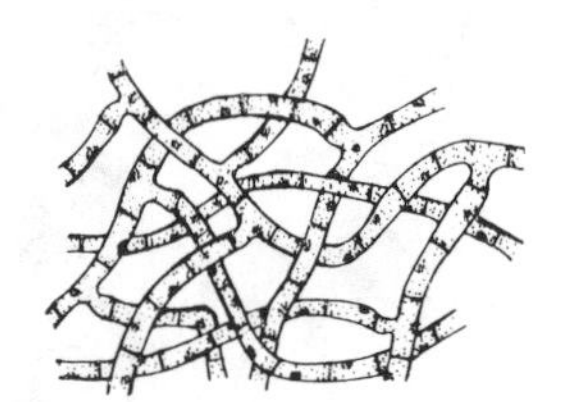

Fig. 4.2
Septate hyphae

Thallus (pl. thalli), is a term for a simple plant body that does not have stems, roots, or leaves; this term then well describes the vegetative body of a fungus. The vegetative (assimilative) phase, of the typical fungus consists of microscopic threads or filaments which branch intricately in all directions, spreading over or throughout the substrate. Each of these threads or filaments is called a hypha (pl. hyphae). A hypha is tublar in form with a thin transparent wall. Protoplasm may fill the hypha or form a thin layer inside of the hyphal wall. Depending upon the species, the protoplasm may be continuous or it may be divided by cross-walls, cutting the hypha into cells. The cross-walls are called septa (sing. septum), and a hypha containing septa is referred to as septate. A hypha without cross-walls is referred to as non-septate. Hyphal segments bounded by septa are called cells.

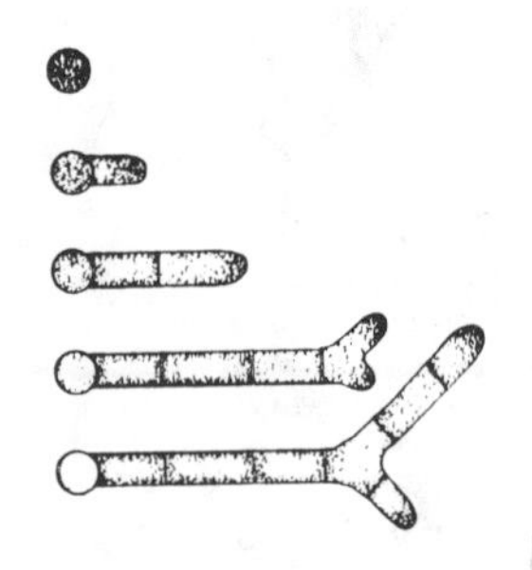

Fig. 4.3
Germinating fungal spore illustrating apical growth and septate hyphae.

Fungi possess nuclei that have a nuclear membrane, a nucleolus, and chromatin strands. Most fungal nuclei are extremely minute, making their study quite difficult.

Hyphal cells may contain one, two, or many nuclei, and are referred to as being uninucleate, binucleate, or multinucleate, respectively. When hyphae have no septa, the nuclei are scattered uniformly throughout the hyphae, and the condition is called coenocytic. Hyphal filaments elongate by apical growth, but most parts, or fragments of hypha, are capable of growth.

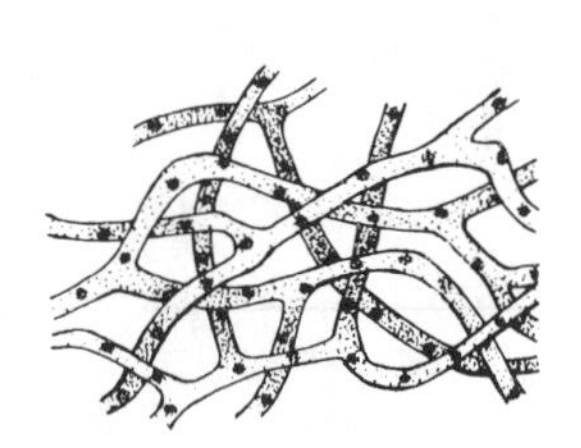

Fig. 4.4
Coenocytic and non-septate hyphae

A mass of hyphae is called a mycelium (pl. mycelia). Whereas the typical fungus is myceloid in nature, the thallus of certain fungi may consist of only a single globular cell no larger than individual cells of a hypha.

Fungi Secrete Enzymes

The protoplasm that fills the cells at the growing tips of the hyphae manufacture a great number of enzymes and organic acids that diffuse out of the hyphal cells into the substrate. The enzymes and acids break down cellulose, starches, sugars, proteins, fats, and other constituents of the substrate, which in turn, diffuse into the fungal cells and are utilized for food and energy for more growth.

The older hyphae die as the mycelium grows and branches. The dead cells quickly disintegrate, generally due to bacterial decay.

Mycelium External Or Internal

The mycelium of a parasitic fungus may be either external or internal in reference to the host. External mycelia are usually whitish or dark brown in color and cobweb-like, forming dense tangled threads on the surface of leaves, stems, or fruits. Internal mycelia either penetrate between the host cells, a condition called intercellular, or penetrate into the host cells, a condition called intracellular. Since they come in contact with the host protoplasm, intracellular hyphae absorb food directly. Intercellular hyphae absorb food as it diffuses out through the host cell wall, or send special absorbing organs, called haustoria (sing. haustorium), into the host cells. Haustoria are outgrowths of hyphal cells. They enter a host cell through a minute puncture in the cell wall. Haustoria may be a simple knob, long tube, or variously branched.

Fungi tend to grow radially in all directions. However, the character of the host substrate limits such growth. For example, fungal lesions on leaves of broad leaved plants are usually circular, but may be V-shaped or angular because of leaf veins. Lesions on plant stems and monocotyledonous leaves may also be restricted by veins and appear eliptical or elongated.

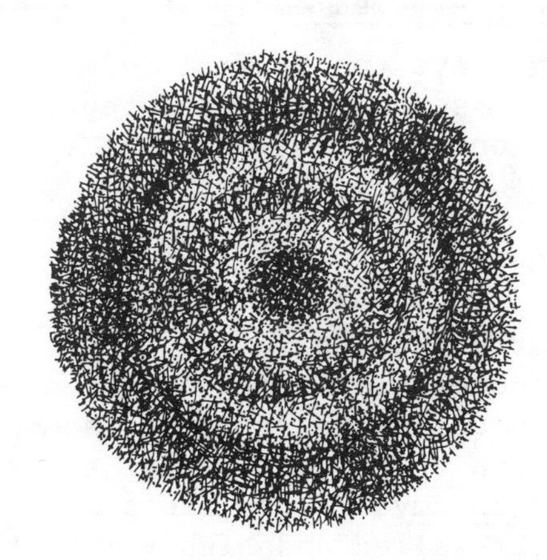

Fig. 4.5 Radial growth of a fungus that occurs in agar culture.

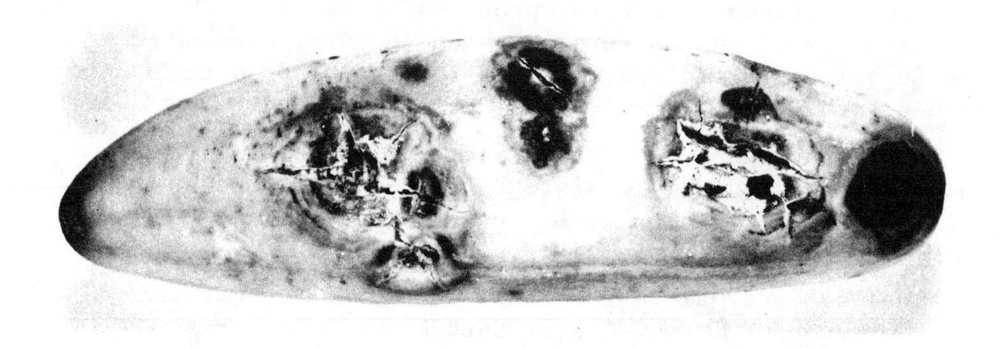

Fig. 4.6 Anthracnose of cucumber, caused by *Colletotrichum lagenarium*. Lesions are sunken and generally round. *(Courtesy of R. W. Samson)*

Mycelium Modifications

Mycelium may be tightly woven into thick strands called rhizomorphs. Rhizomorphs are hard-packed, with the hyphae losing their individuality. They are capable of remaining dormant in unfavorable conditions; when favorable conditions return, growth is resumed.

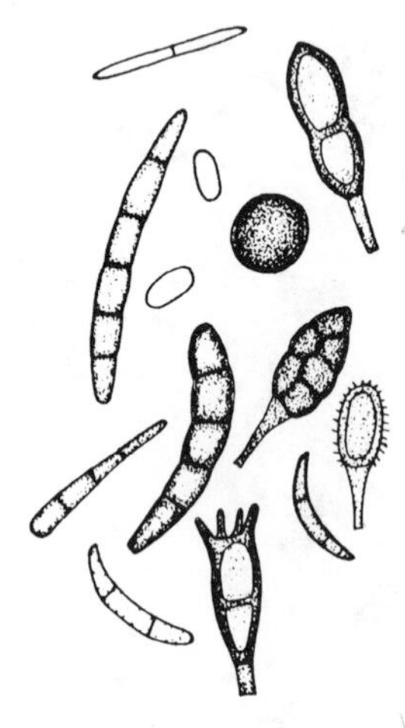

Fig. 4.7 Various shapes of fungal spores

Some fungi produce dense compacted aggregates of hyphae, called sclerotia (sing. sclerotium), which are resistant to unfavorable conditions, and may remain dormant for long periods of time. They may germinate upon the return of favorable conditions.

4-3 REPRODUCTIVE ENTITIES

The common reproductive entities of fungi are microsopic bodies called spores. A spore is a propagative unit which functions as a seed, but differs from it in that a spore does not contain a preformed embryo.

Typically, fungi have two general forms of reproduction, asexual and sexual. Asexual reproduction does not involve the union of nuclei. Sexual reproduction involves the union of nuclei.

4-3.1 ASEXUAL SPORES

Spores Variable

There is a great variation in the structure and origin of spores. Some spores are formed by the production of specialized spore-bearing branches, and some are formed by direct transformation of certain hyphal cells. The asexually produced chlamydospores are produced in this latter manner and are thick-walled and highly resistant spores. Only scattered cells may be organized as spores, or all of the cells of a hypha may be organized as spores. An example of the latter is the smut fungi.

4-3.2 SEXUAL SPORES

Since sexual reproduction involves the union of compatible nuclei followed by a process called meiosis, it is necessary to discuss this process before we examine the types of sexual spore forms. The normal chromosome content of the vegetative phase of a fungus is termed haploid, and is often written as (n). The union of nuclei doubles the chromosome content, and this phase is called diploid and is often written as (2n). The diploid nucleus quickly undergoes meiosis; this is a pair of nuclear divisions occurring in quick succession, one of which is reductional, resulting in four haploid nuclei. The student should not confuse meiosis with mitosis. Mitosis is the usual process of nuclear division in which each chromosome duplicates and pulls apart resulting in each daughter nucleus containing an identical complement of chromosomes.

Meiosis

Mitosis

The sex organs of fungi are termed gametangia (sing. gametangium). In cases where the male and female organs do not look alike, we call the female gametangia oogonia (sing. oogonium), and the male gametangia antheridia (sing. antheridium). The sex cells or sex nuclei that fuse in sexual reproduction are called gametes. When both sex organs look alike, we can make no distinction as to which is male or female. We simply say they are of opposite sex, call them compatible, or one is termed minus and the other plus.

In the higher fungi, there is a situation in which hyphal cells contain two nuclei, each from a separate, compatible parent. Such a condition is called dikaryotic. These nuclei fuse later in the life cycle of the fungus to produce the diploid cell during the sexual reproductive stage. However, many fungi may have binucleate cells that are not necessarily dikaryotic, because their nuclei do not fuse.

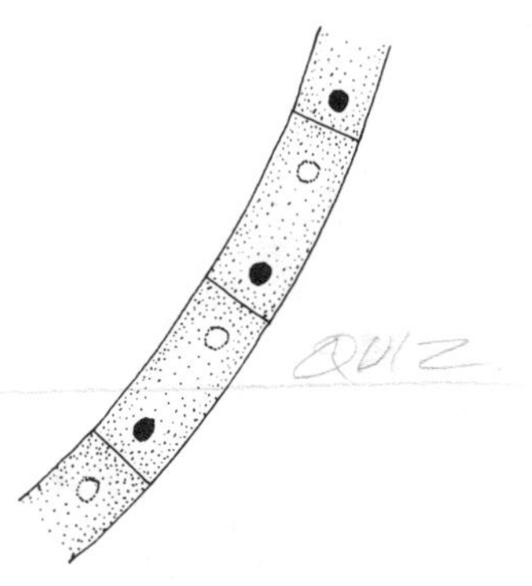

Fig. 4.8 Dikaryotic hyhal cells

A homothallic fungus is one in which sexual reproduction takes place in a single thallus, meaning that the fungus is self-compatible. The term heterothallic refers to fungi that require the union of two compatible thalli for sexual reproduction.

Homothallic

Heterothallic

An overwhelming majority of fungi produce spores, and most of these produce both sexual and asexual spores. Fungi never produce more than one type of sexual spore, but may produce up to three types of asexual spores.

Both sexual and asexual spores may occur at random over the surface of the relatively undifferentiated mycelium, or they may be enclosed in a fruiting body. There is a wide variety of fruiting bodies that may be formed. The type of fruiting body, and the manner in which spores are borne, are used in classifying fungi.

4-4 TYPES OF PLANT PATHOGENS

Obligate Parasites

The level of parasitism is often considered in discussing types of fungal plant pathogens. These are obligate parasites, faculative saprophytes, or facultative parasites. The obligate parasites include: rusts; powdery mildews; downy mildews; the white rusts (caused by the genus *Albugo*); and several lower Phycomycetes, such as *Synchytrium endobioticum*, which causes black wart of potato, and *Plasmodiophora brassicae*, which causes clubroot of cabbage.

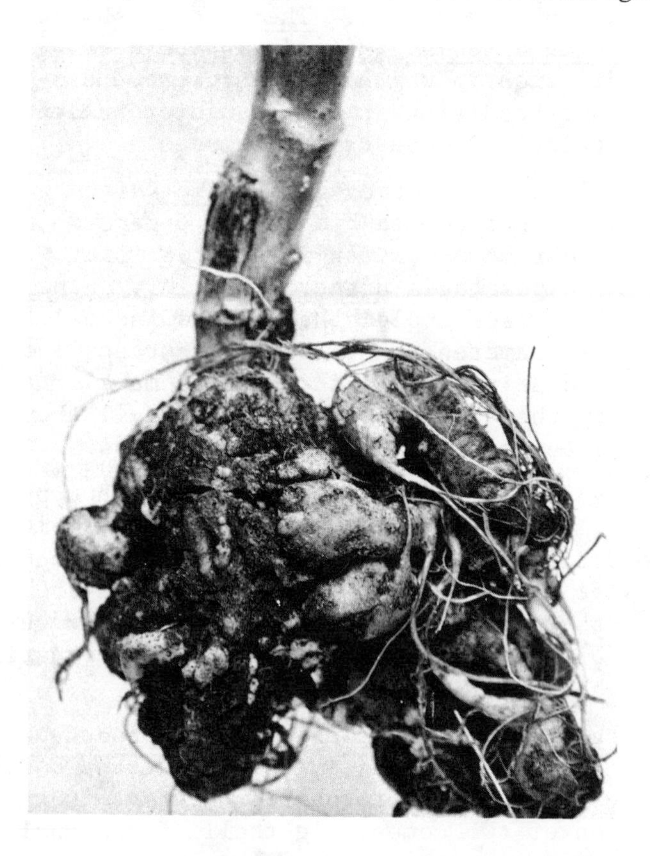

Fig. 4.9 Clubroot of cabbage, caused by *Plasmodiophora brassicae*. The heavily galled or clubbed root system is the result of multiple infections. *(Courtesy of the Department of Botany and Plant Pathology, Purdue University)*

Facultative Saprophytes

Many fungal pathogens are facultative saprophytes whose highest level of parasitism is exhibited by fungi that, when in the field, grow only on living hosts and do not multiply as saprophytes. However, they may be grown in culture in the laboratory. At the lower range of this category are many fungi that, when in the field, ordinarily grow as parasites, but may also multiply as saprophytes.

Fungi that are facultative parasites possess the lowest level of parasitism. They usually grow as saprophytes in the field, but become parasitic under conditions such as when a host is weak. These fungi are often called "weak" parasites. Other fungi of this category usually enter their hosts through wounds and thereby are called "wound" parasites.

Facultative Parasites

4-5 TAXONOMIC CLASSIFICATION OF MAJOR GROUPS OF MOST FUNGAL PLANT PATHOGENS

The study of any assemblage of living organisms can best be done only when relationships within the major groups are brought to light and the relationships to the living community are known. Many variations of the classification of fungi exist. This is as it should be, because all classification schemes are man-made and are based upon our present knowledge. When new relevant information becomes known, changes are then made in the classification system to reflect the new knowledge.

Most present authorities no longer recognize the fungal class Phycomycetes. In its place, they recognize several classes whose fungi were formerly in the class Phycomycetes. However, for sake of simplicity, and since most references of plant pathology still utilize the class Phycomycetes, it is retained in the classification system presented here.

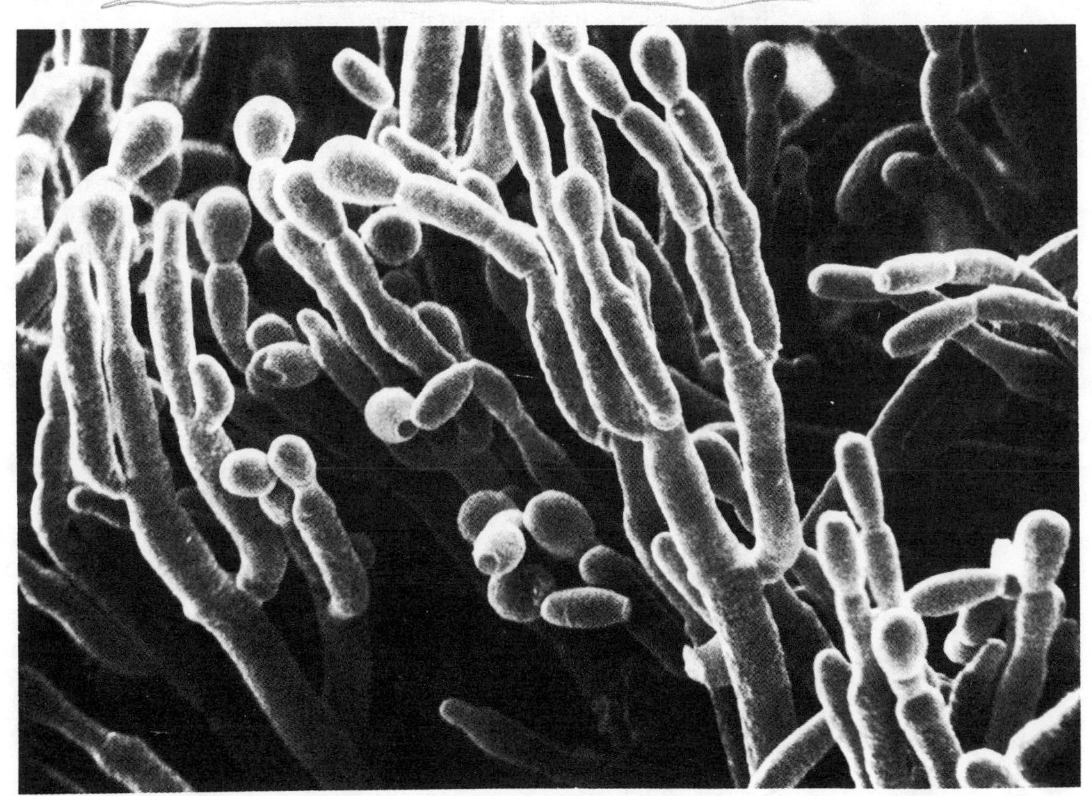

Fig. 4.10 Scanning electron micrograph of conidial chains of *Penicillium digitatum*, (x5,100). *(Courtesy of E. E. Butler)*

This classification includes the major groups of most fungal plant pathogens. Many groups of nonpathogenic fungi are purposely omitted.

Kingdom: Plantae (Plant Kingdom)
- Division: Mycota (the Fungi)
  - Subdivision: Myxomycotina (true slime molds)
  - Subdivision: Eumycotina (true fungi)
    - Class: Phycomycetes (lower fungi)
      - Sub-class: Hyphochytridiomycetes (n.p.i.)*
      - Sub-class: Chytridiomycetes
      - Sub-class: Plasmodiophoromycetes (club-root of crucifers)
      - Sub-class: Oomycetes (*Pythium*, late blight, downy mildews)
      - Sub-class: Zygomycetes (bread molds, pin molds)
    - Class: Ascomycetes (sac fungi)
      - Sub-class: Hemiascomycetes
        - Order: Endomycetales (yeasts)
        - Order: Taphrinales (peach leaf curl, etc.)
      - Sub-class: Euascomycetes
        - Series: Plectomycetes (powdery mildews)
        - Series: Discomycetes (brown rot, etc.)
        - Series: Pyrenomycetes (apple scab, etc.)
    - Class: Basidiomycetes
      - Sub-class: Heterobasidiomycetes
        - Order: Ustilaginales (smuts)
        - Order: Uredinales (rusts)
        - Order: Tremellales (jelly fungi, n.p.i.)
      - Sub-class: Homobasidiomycetes
        - Series: Hymenomycetes (mushrooms, bracket fungi, etc.)
        - Series: Gasteromycetes (puff-blass, earth stars, stink horns, birds' nests, etc., n.p.i.)
    - Class: Fungi Imperfecti (Deuteromycetes)
      - Form-order: Sphaeropsidales (*Diplodia*, *Septoria*, etc.)
      - Form-order: Melanconiales (anthracnose fungi, etc.)
      - Form-order: Moniliales (*Fusarium*, *Alternaria*, *Helminthosporium*, etc.)
      - Form-order: Mycelia Sterilia (*Rhizoctonia*, etc.)

*n.p.i. Indicates not phytopathologically important.

4-6 PHYCOMYCETES

The Lower Fungi

This class of fungi contains about 1,400 species, and includes many causal agents of plant disease. The parasitism exhibited by these pathogens varies from obligate through facultative. Many species are strict saprophytes.

The Phycomycetes are often referred to as the lower fungi since they are thought to occupy a lower evolutionary level than other classes of fungi. The reasoning for this is readily evident. The most primitive forms are those possessing the simplest structure, and members of this class are simple in form. The lowest members within the Phycomycetes are chiefly aquatic in habitat.

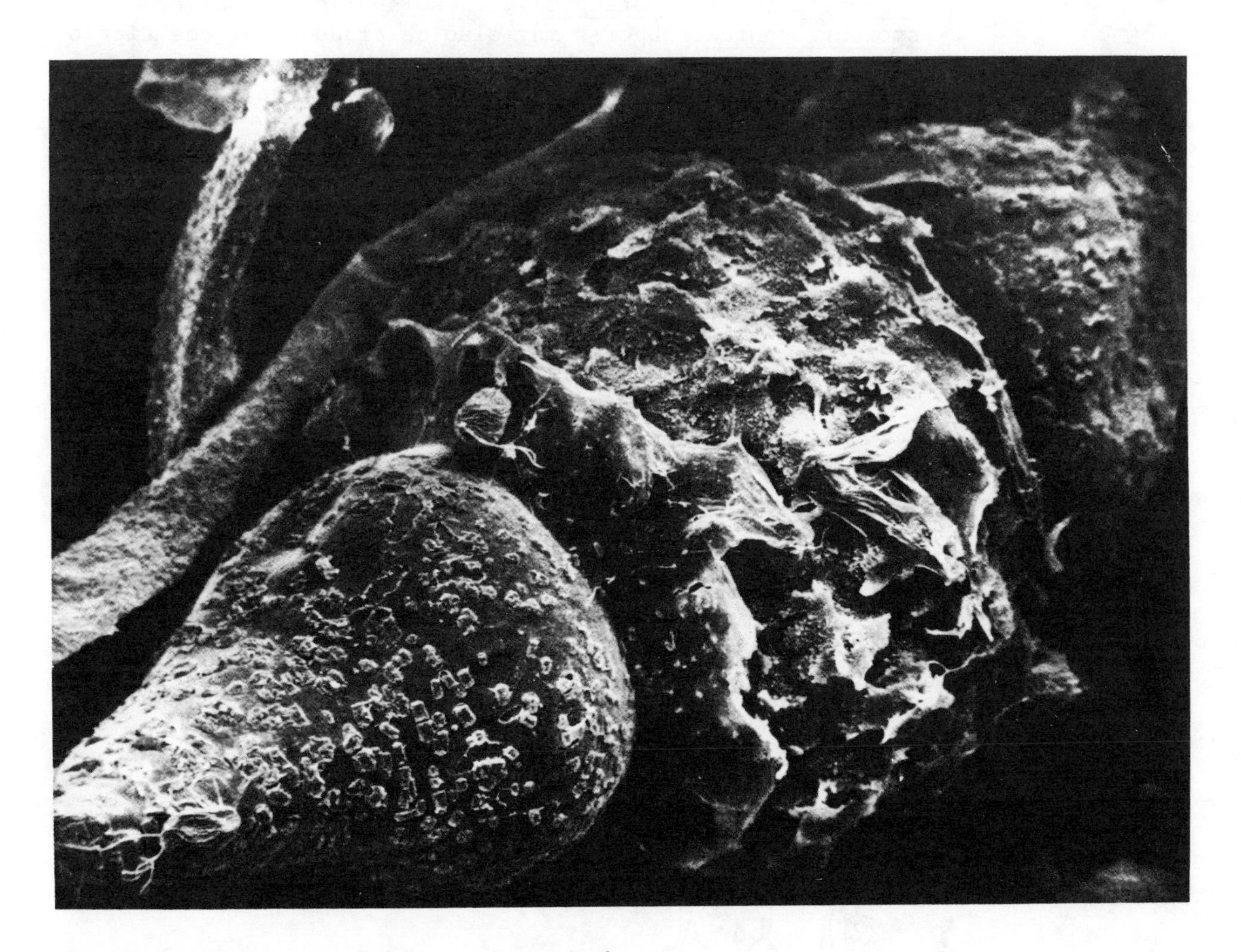

Fig. 4.11 Scanning electron micrograph of a zygospore of *Gilbertella persicaria*, (x3,000). *(Courtesy of E. E. Butler)*

4-6.1 VEGETATIVE FORMS

Some species are without mycelial organization, and the entire tallus is a single unbranched cell. Others are only slightly branched. These are the most primitive members. More advanced species do have a prolific, well-branched, coenocytic mycelium.

The simplest members of the Phycomycetes are holocarpic, meaning their entire protoplasm is used up in reproduction. This condition is also considered primitive. Specialized reproductive organs are borne on the mycelium of more advanced species.

4-6.2 ASEXUAL SPORES

Asexually produced spores may be borne in vessels called sporangia (sing. sporangium), these spores are called sporangiospores. Spores may also be produced at the tips or sides of hyphae in a number of ways and are then termed conidia (sing. conidium). A specialized hypha that bears conidia is called a conidiophore.

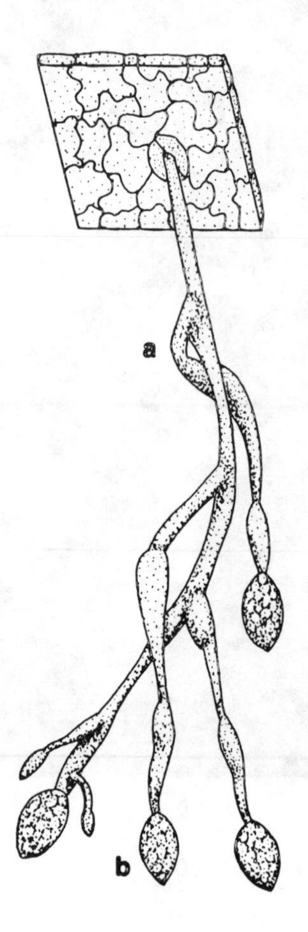

Fig. 4.12 A sporangiophore (a) and sporangia (b), characteristic of the late blight fungus, *Phytophthora infestans*, growing out of a stoma of an infected leaf.

The protoplasm of sporangia is converted into one or more, usually many, sporangiospores. These may be motile or non-motile. If motile, they are usually called zoospores. The majority of the Phycomycetes form zoospores. In holocarpic types, the entire thallus may function as a zoosporangium, or divide into a number of zoosporangia.

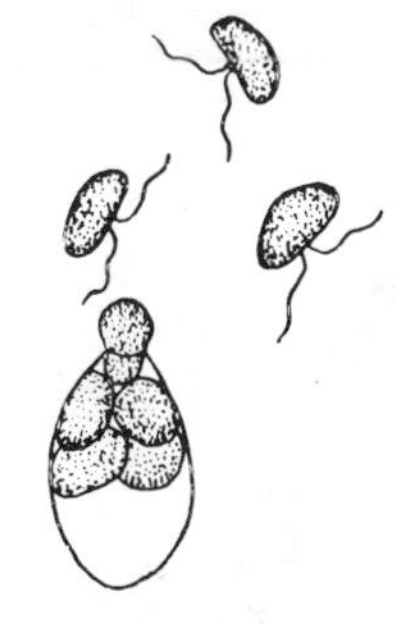

Fig. 4.13 Zoospores emerging from a sporangium, typical of the late blight fungus.

4-6.3 SEXUAL SPORES

The sexual spores are represented by zygospores or oospores. A zygospore develops from the union of gametes which look alike. An oospore develops from the union of unlike male and female gametes within the oogonia.

Both heterothallism and homothallism occurs. In holocarpic forms, two thalli may fuse to form a zygospore, or motile gametes may be formed that fuse to form a motile zygospore. In mycelial forms, such as in the order Mucorales, multinucleate gametangia of identical size and shape are formed on hypha of each of two mating strains that are positioned close together. The gametangia grow together until they touch. Protoplasmic fusion occurs and a thick-walled zygospore is formed. The genus *Rhizopus* is an example of a fungus that forms zygospores. Fungi that produce zygospores are placed in the sub-class Zygomycetes, (see Fig. 4.11 and 4.14).

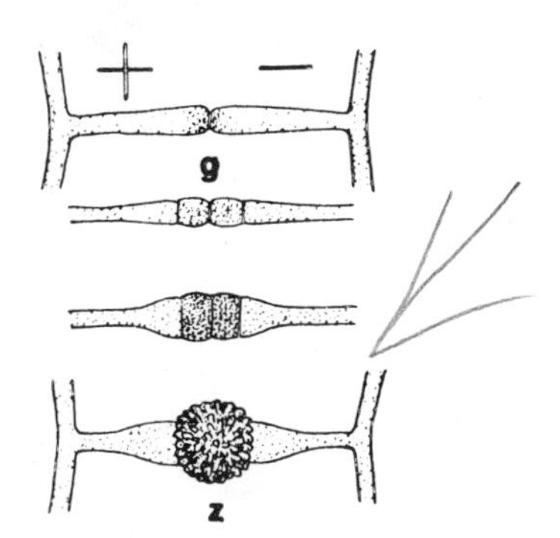

Fig. 4.14 *Rhizopus* spp., showing the fusion of like gametes (g) to form a zygospore (z).

In the order Peronosporales, unlike sex organs are produced on the same mycelium. A fertilization tube grows out from the antheridium and contacts the oogonium. One or more oospheres (female gametes), are differentiated in each oogonium. One or more nuclei from the antheridium pass into one or more oospheres and nuclear fusion occurs. A thick-walled oospore develops from each oosphere in which nuclear fusion occurs. The genera *Pythium* and *Phytophthora* are examples of fungi that form antheridia and oogonia. Such fungi are placed in the sub-class Oomycetes.

4-7 ASCOMYCETES

Over 15,000 species of fungi are included in this class, and a great number of them are parasites of green plants. Some are obligate parasites, but most are facultative parasites. These fungi are all of a higher evolutionary scale than the Phycomycetes.

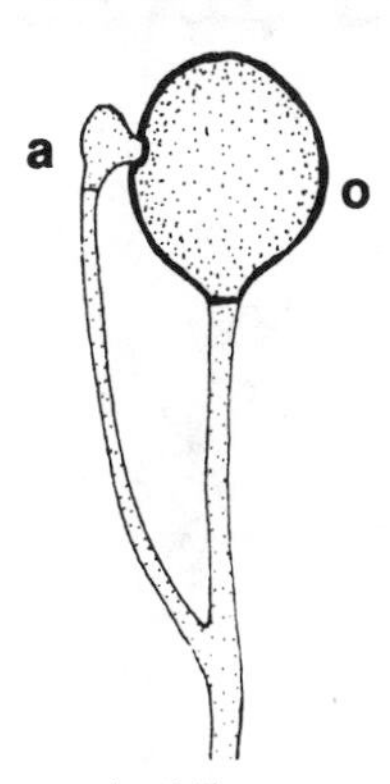

Fig. 4.15 *Pythium* spp., showing the fusion of an antheridium (a) with an oogonium (o).

4-7.1 VEGETATIVE FORMS

Except for organisms such as the yeasts, Ascomycetes have well-developed mycelia that are septate and generally uninucleate. Some forms are multinucleate. Sclerotia are formed by many species. All fruiting bodies, both sexual and asexual, arise from a stroma (pl. stromata), which is a tissue-like mass of hyphae in or from which the fructifications are produced.

4-7.2 ASEXUAL SPORES AND SPOROCARPS

Sporocarps

Most species of this class produce conidia. Some produce chlamydospores. Conidia may be produced on conidiophores that arise singly or in fruiting bodies collectively called sporocarps. The following asexual sporocarps are recognized:

> *Synnema* (pl. *synnemata*), a dense cemented fascicle of erect conidiophores, with a head of conidia-bearing branches.
>
> *Acervulus* (pl. *acervuli*), a depressed structure which forms conidia on conidiophores that are uniformly borne over the exposed surface.
>
> *Sporodochium* (pl. *sporodochia*), a cushion-like mass of fungal tissue that breaks through the host tissue and bears conidiophores over its surface.
>
> *Pycnidium* (pl. *pycnidia*), a hollow, flask-shaped or globular structure that is embedded in the substratum and open to the outside by a pore known as an *ostiole*. This sporocarp is generally blackish and is lined inside with conidiophores.

The asexual stages of the Ascomycetes generally serve to propagate, disseminate, and carry the species through the spring and summer months. The asexual (*imperfect*) stages of the Ascomycetes are similar to the asexual stages of the Fungi Imperfecti, in which the sexual (*perfect*) stages are unknown.

4-7.3 SEXUAL SPORES AND ASCOCARPS Fruiting Body containing Asci

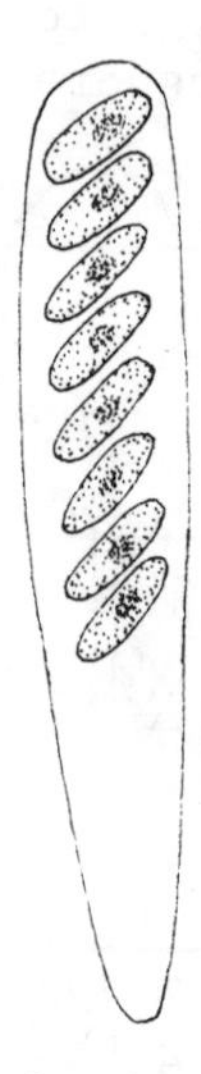

Fig. 4.16 Ascospores borne in an ascus

All members of this class produce sexual spores called *ascospores*. These spores are formed within a saclike structure, the *ascus* (pl. *asci*). The usual number of ascospores per ascus is eight, although there are many exceptions to this. In some of the Ascomycetes, the sexual organs (antheridia and ascogonia) are abortive. In others, they are functional, well-defined and differentiated; and in still others they are morphologically alike. The two compatible nuclei that are brought together in the same cell may fuse quickly or remain in close association in a dikaryotic condition. The nuclei of a dikaryon may undergo many conjugate divisions that result in a mass

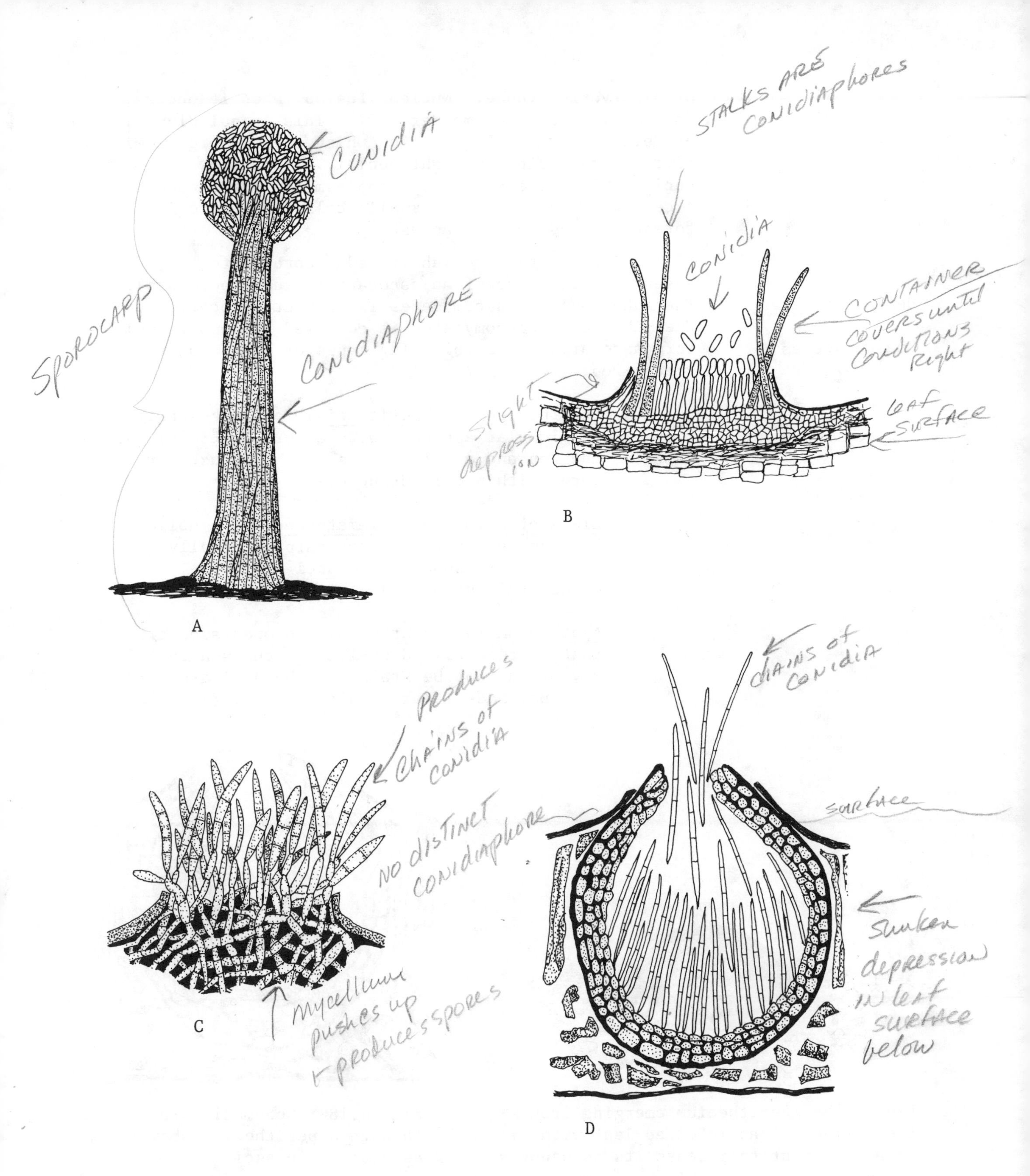

Fig. 4.17 A. Synnema B. Acervulus C. Sporodochium D. Pycnidium

Sporocarps that contain conidia

of dikaryotic hyphae. Nuclear fusion, when it occurs, does so in the ascus mother cell. This is quickly followed by meiosis forming four nuclei which then undergo a mitotic division resulting in eight nuclei. Typically, each nucleus with some adjacent cytoplasm, is surrounded by a cell wall in a process usually called "free cell formation" and becomes an ascospore.

Hemiascomycetes

Euascomycetes

The Ascomycetes are subdivided according to the absence or presence of an ascocarp (sexual sporocarp). The sub-class Hemiascomycetes is without ascocarps, and the sub-class Euascomycetes is composed of members which form ascocarps. The following types of ascocarps are recognized:

> Perithecium (pl. perithecia), a flask-shaped, or globular ascocarp with a pore (ostiole) at the top and a wall of its own. The asci are borne within this structure.
>
> Cleistothecium (pl. cleistothecia), a hollow, but completely closed ascocarp, generally spherical in shape. The asci are borne within this structure.
>
> Apothecium (pl. apothecia), an open ascocarp that is generally disk-like or cup-shaped. It may or may not be stalked. The asci are borne in a palisade layer within the cup.

Fig. 4.18A Perithecium emerging from the surface of an infected leaf with an ascus about to release its ascospores.

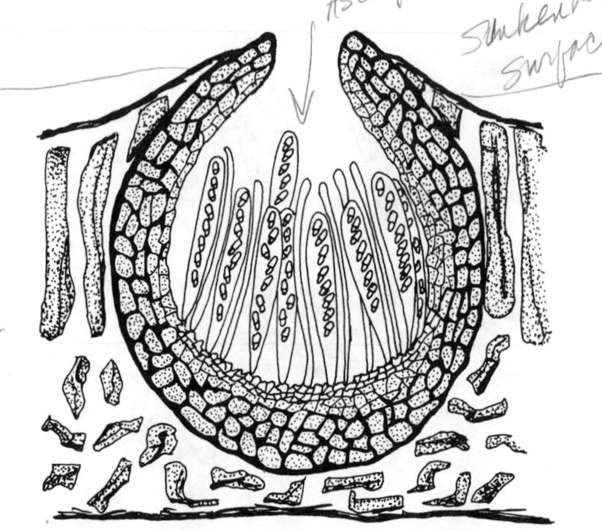

Fig. 4.18B Schematic drawing through a perithecium showing ascospores in asci.

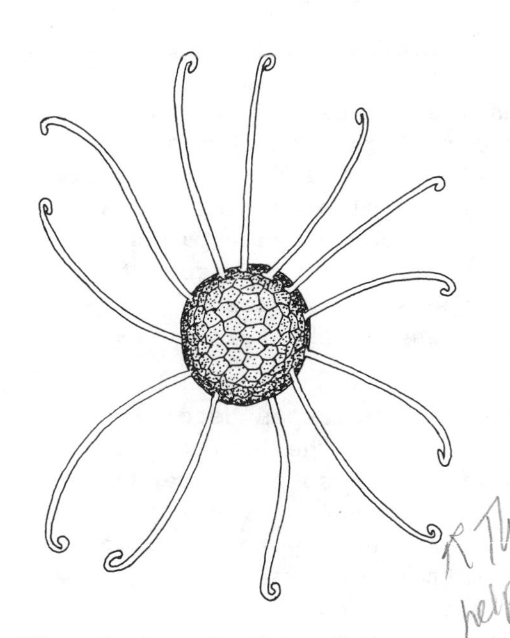

Fig. 4.19A A cleistothecium characteristic of the genus *Uncinula*.

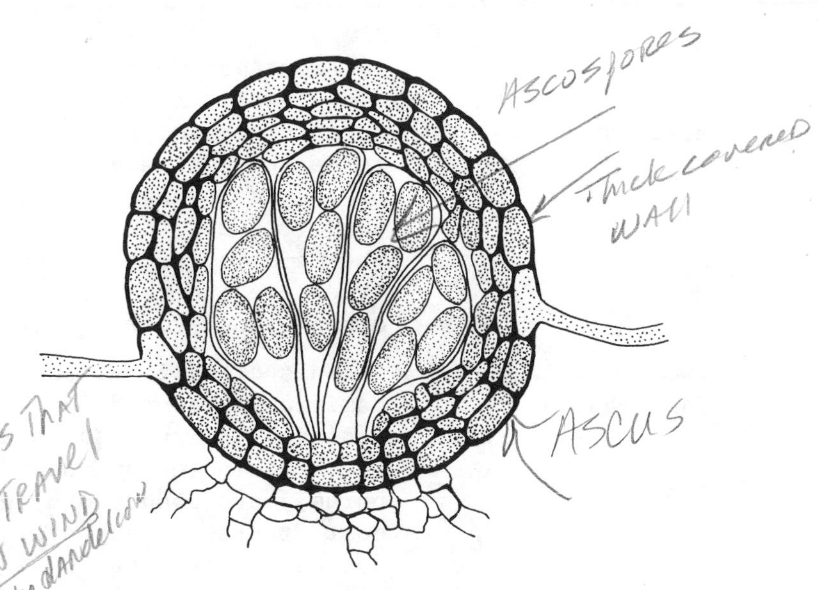

Fig. 4.19B Schematic drawing through a cleistothecium showing ascospores in asci.

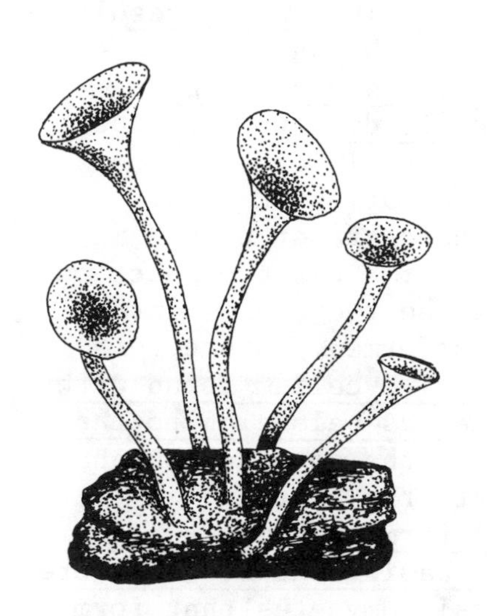

Fig. 4.20A Apothecia arising from a mummified fruit.

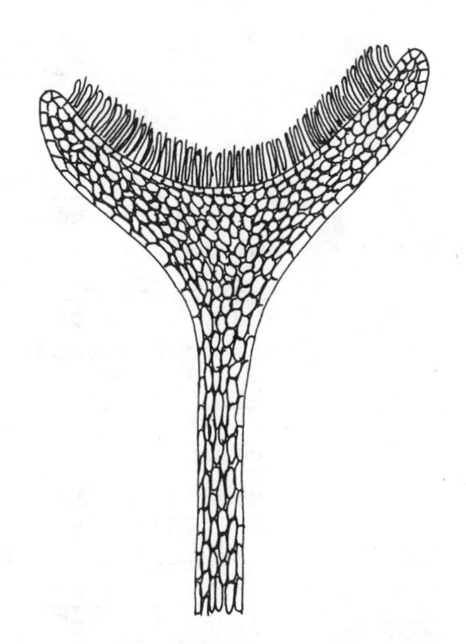

Fig. 4.20B Schematic drawing through an apothecium showing the palisade layer of asci.

Rule -- Overwintering Stage Of Ascomycetes

The same mycelium that bears asexual conidia later produces the sexual ascospores. Typically, some phase of the sexual cycle is involved in overwintering of the Ascomycetes. It then follows that ascospores are typically the primary inoculum, inciting the primary infections.

4-8 BASIDIOMYCETES

Most Advanced Group Of Fungi

About 15,000 species of fungi are included in this class, which is composed of the rust and smut fungi as well as most of the fleshy fungi. This class is considered to be the most advanced group of fungal organisms. Most of the rust fungi are considered to be obligate plant parasites and the smut fungi are facultative saprophytes. The majority of the highly destructive diseases of forest trees are incited by fungi belonging to this class, as are the wood-rotting fungi.

Two Sub-classes

This class is divided into the sub-classes Heterobasidiomycetes and Homobasidiomycetes. The former group contains the rust and smut fungi and is so-named because the basidia (organs upon which the sexual spores are borne), are produced by germinating teliospores, and in most orders are septate. The latter group has the typical club-shaped basidia, which are common to the mushrooms, puffballs, conks, and shelf or bracket fungi.

4-8.1 VEGETATIVE FORMS

Three Mycelial Forms

The hyphae are well-developed, septate, and of three forms. The first is the primary mycelium, which results from a germinating basidiospore. Its hyphal cells are haploid, as is the basidiospore. The second form is termed the secondary mycelium, and is dikaryotic. The dikaryotic condition results from a cellular fusion of primary mycelia from two compatible mating strains. Clamp connections are formed at the septa of hyphae in many Basidiomycetes. The function of a clamp connection is to allow the passage of one of the two nuclei into a new daughter cell as the septum is formed. Since this process is unique to this class, any mycelium found with clamp connections may be identified as belonging to the Basidiomycetes. In the smut fungi, the secondary hyphae is generally the stage that infects the host and initiates disease. The third mycelial form is composed of complex tissues that make various basidiocarps, and therefore, is limited to the higher Basidiomycetes that form such fruiting bodies.

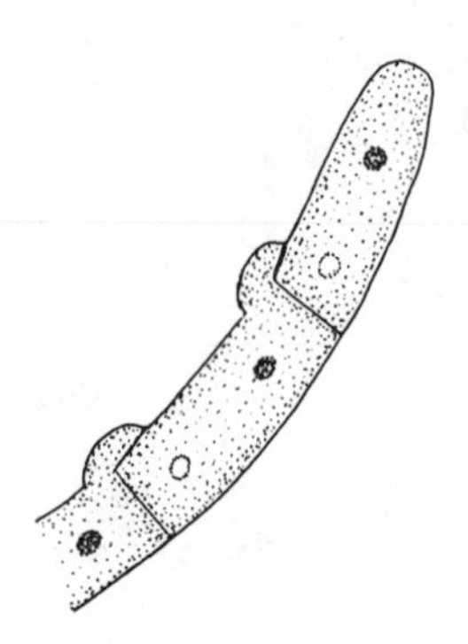

Fig. 4.21 Clamp connections of a dikaryotic hypha.

4-8.2 ASEXUAL SPORES

Asexual reproduction plays a highly significant role in the rust and smut fungi, and in a few other forms of the Basidiomycetes. However, it is probably of little significance in most of the higher forms. In fact, many of the fleshy fungi have not been shown to produce asexual spores.

The most common asexual spore of the smut fungi is the dikaryotic teliospore. It is formed from a dikaryotic hyphal cell and is generally dark colored with thick walls. For this reason some authors refer to them as chlamydospores. The teliospores are generally produced in large dusty masses called sori. Upon maturity, nuclear fusion occurs within these spores and initiates the sexual stage.

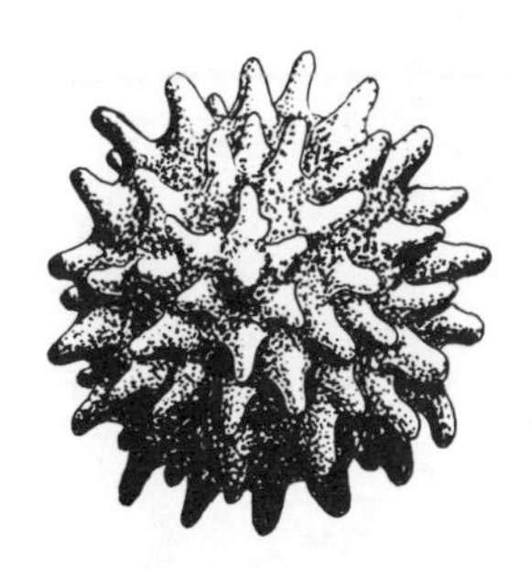

Fig. 4.22 A teliospore or chlamydospore of *Ustilago maydis*, the corn smut fungus.

In the rust fungi, three types of dikaryotic spores may be produced from dikaryotic hyphae; they are aeciospores, urediospores, and teliospores. Only some rust fungi produce all three forms. Aeciospores are produced in cup-shaped structures known as aecia (sing. aecium) and this phase of growth is called the aecial stage. Aeciospores germinate and infect host plants (often an alternate host), giving rise to either the uredial or telial stage. The uredial stage produces urediospores in

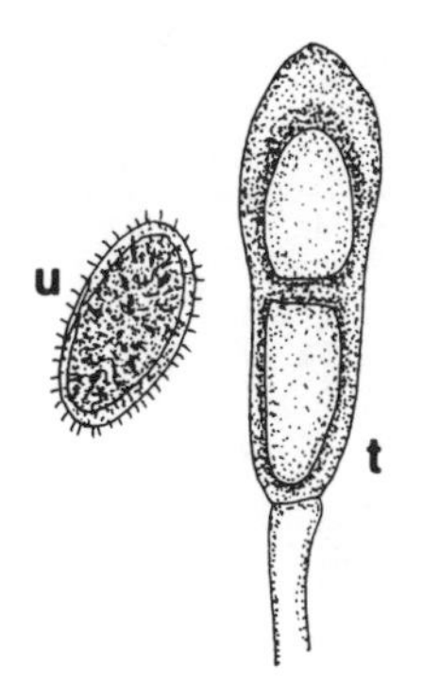

Fig. 4.23 A drawing of a urediospore (u) and teliospore (t) of a rust fungus.

Fig. 4.24A Wheat stems infected with stem rust, *Puccinia graminis* var. *tritici*. This is the telial stage in which masses of teliospores are borne in pustules or sori and rupture the epidermis layer of the stem. *(Courtesy of R. W. Samson)*

sub-epidermal structures called uredia (sing. uredium). Urediospores infect similar host plants and repeatedly give rise to the uredial stage and finally the telial stage. In stem rust, the telia are formed near the end of the growing season of the host. The old uredia may be converted into telia (sing. telium), which resemble the uredia in size and shape. Teliospores are borne in the telia and they are at first dikaryotic. As they mature, nuclear fusion occurs which initiates the sexual stage.

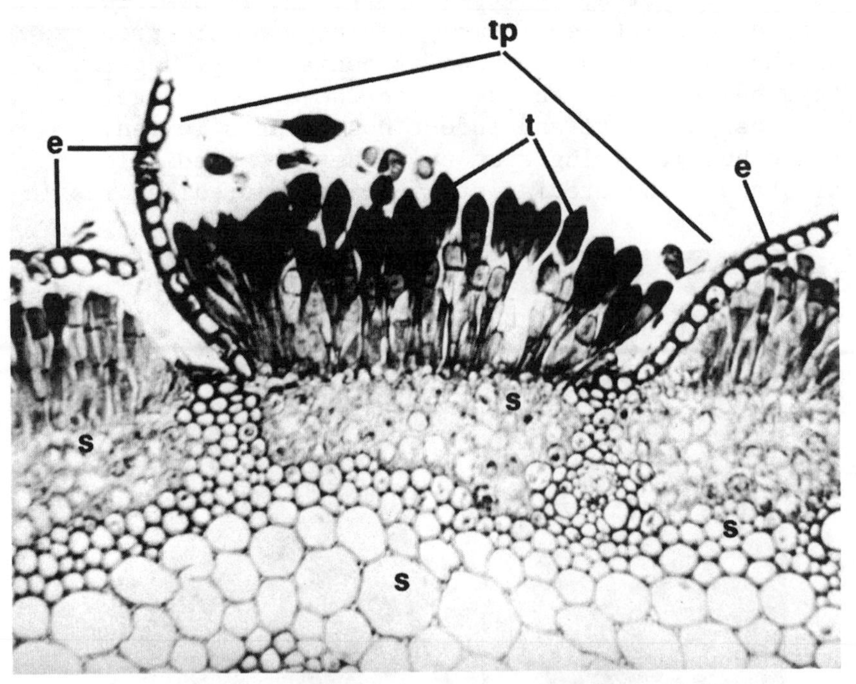

Fig. 4.24B A stained cross-section through a wheat stem (s) and telial pustule (tp) of stem rust of wheat. Note the broken epidermis (e) and dark stained teliospores (t). *(Courtesy of R. J. Green)*

4-8.3 SEXUAL REPRODUCTION

Nuclei of teliospores of the rust and smut fungi undergo a nuclear fusion, which is followed by meiosis, resulting in four haploid nuclei. Upon germination, a teliospore produces a basidium (pl. basidia), which is often called a promycelium. Basidiospores are formed on the outside of the basidium, and a single haploid nucleus migrates into each spore as it is formed. Rust fungi generally produce definite numbers of basidiospores, usually four. However, the smut fungi are generally inconsistant in the numbers of basidiospores produced.

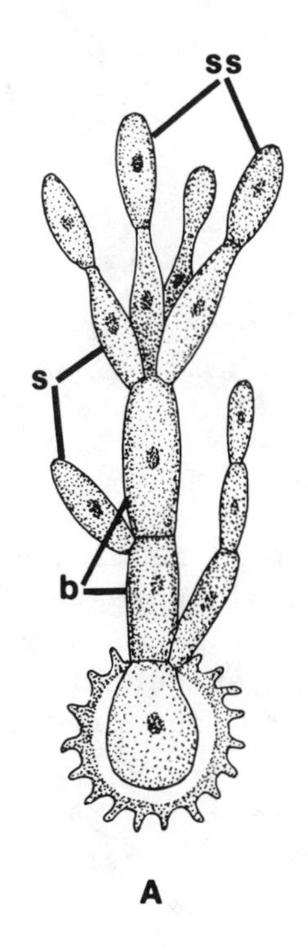

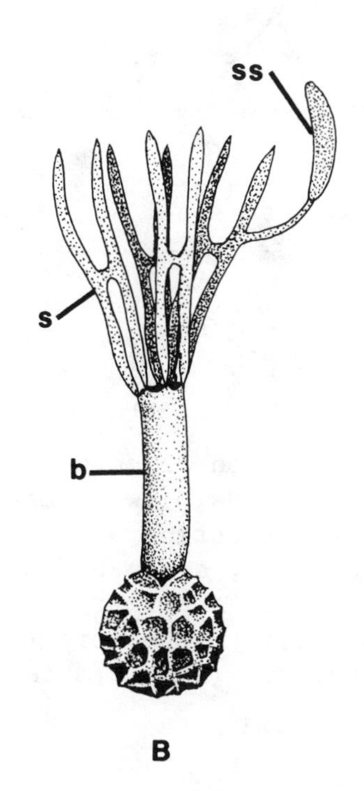

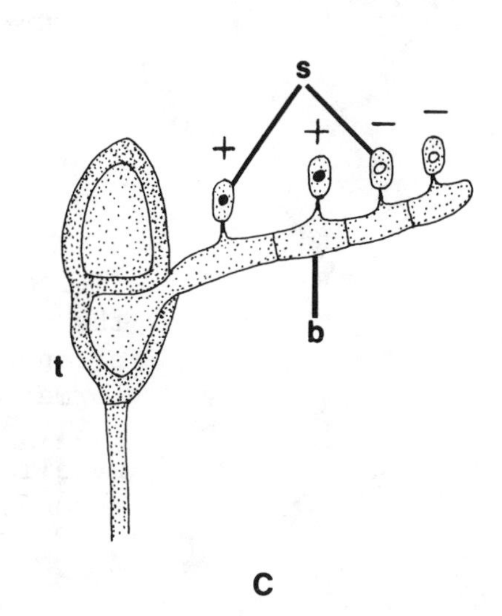

Fig. 4.25
A. A germinated teliospore of corn smut with a basidium, or promycelium (b), bearing haploid basidiospores, or sporidia (s), and secondary basidiospores, or sporidia (ss).
B. A germinated teliospore of *Tilletia caries*, the wheat bunt fungus. Haploid elongate basidiospores (s) fuse with compatible basidiospores to form characteristic H-shapes while still attached to the basidium or promycelium (b). The fused basidiospores produce dikaryotic secondary infection hyphae or secondary sporidia (ss), which may form infection hyphae.
C. A germinated rust teliospore (t) which has formed a basidium or promycelium (b) and four basidiospores (s).

Smut Fungi

Basidiospores of the smut fungi generally germinate to produce uninucleate primary mycelia that fuse in one of several ways with primary mycelia of compatible strains. This brings about dikaryotization and the formation of secondary mycelia.

Rust Fungi

Basidiospores of the rust fungi germinate to form the primary mycelia. This is capable of infecting a host plant and gives rise to the sex organs spermatia (male), and receptive hyphae (female). The spermatia are produced in flask-shaped or flattened spermagonia, which

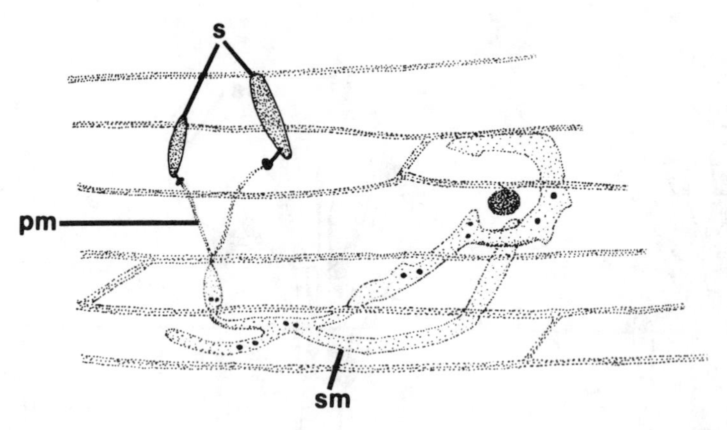

Fig. 4.26 Infection of corn leaf by *Ustilago maydis*, the corn smut fungus. Two compatible basidiospores, or sporidia (s), resting on a corn leaf have germinated, formed appressoria, and penetrated into an epidermal cell, each forming a primary mycelium (pm). The primary mycelia fuse to form the infectious dikaryotic, or secondary mycelium (sm). *(After Hanna, 1929)*

erupt through the host epidermis. The spermatia are exuded in a drop of nectar. The receptive hyphae are generally produced from the spermagonial wall and protrude out of the spermagonium. The receptive hyphae and spermatia produced by the same primary mycelium are not compatible. Dikaryotization is brought about by the fusion of spermatia with receptive hyphae of a compatible mating strain. Dikaryotization results in the production of secondary mycelia which then produce the aecial or telial stage depending upon the rust species.

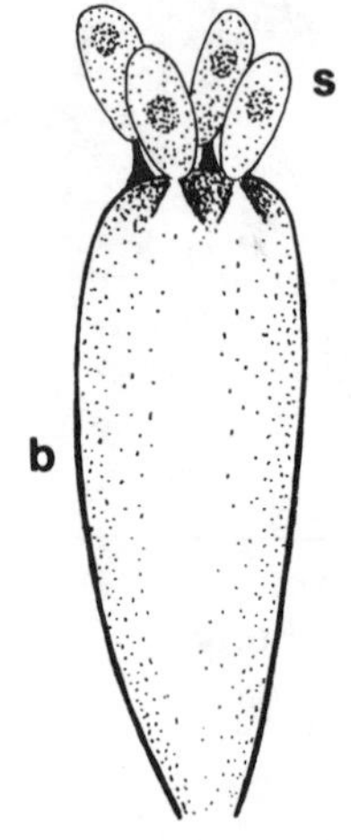

Fig. 4.27 A basidium (b) of the higher basidiomycetes bearing the typical four basidiospores (s).

The higher Basidiomycetes produce their basidia in the highly organized basidiocarps. The function of these fruiting bodies is the production, protection, and release of large numbers of spores. Basidiocarps may be thin or thick, soft or woody, papery or leathery, and stalked or sessile. They may be hoof-shaped conks on the sides of trees, the typical mushroom, the small or large round puffballs, or the delicate coral fungi. With some notable exceptions, basidia are typically formed in layers called hymenia (sing. hymenium). A basidium is a terminal cell that is at first dikaryotic. The nuclei fuse to form the diploid nucleus and then meiosis occurs. Four tiny hyphal branches called sterigmata (sing. sterigma) are formed at the end of the basidium, and their tips begin to enlarge forming the young basidiospores, one per sterigma. A single nucleus migrates into each spore.

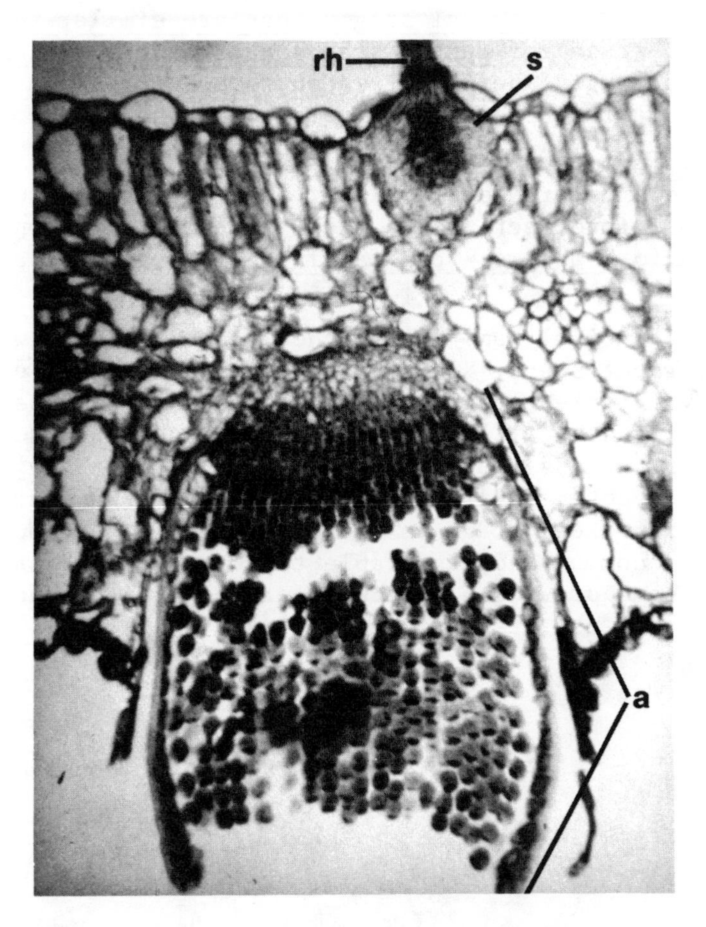

Fig. 4.28 A stained cross-section through an apple leaf infected with cedar-apple rust, *Gymnosporangium juniperi-virginianae*. A spermagonium (s) with receptive hyphae (rh) is seen breaking through the upper epidermis of the leaf. An aecium (a) bearing chains of aeciospores has ruptured the lower epidermis of the hypertrophied leaf. *(Courtesy of R. J. Green)*

4-9 FUNGI IMPERFECTI

This class contains about 15,000 fungi, a number of which are among the most serious causal agents of plant disease. The fungi in this class have septate mycelia and most of them possess an asexual (imperfect) reproductive stage. The sexual (perfect) stage is either lacking or unknown. One order contains fungi for which no reproductive spores are known.

Since our scheme of classifying fungi is based predominantly upon various characteristics of sexual reproduction, this class is an artificial group. The relationships between and within the sub-groups of this class are based on asexual stages that are similar to the asexual stages of fungi whose sexual stages are known. Most of the Fungi Imperfecti resemble those of the Ascomycetes and a few appear to be related to the Basidiomycetes.

The Phycomycetes, Ascomycetes, and Basidiomycetes are sub-divided into orders, families, genera, and species. In the Fungi Imperfecti these equivalent sub-divisions are called form orders, form families, form genera, and form species.

A Latin binomial name is assigned to each fungus with a known perfect stage. The same fungus may also have a distinctly different Latin binomial name for its imperfect stage. For instance, *Rhizoctonia solani*, which causes a root rot of various host plants is the imperfect stage of the Basidiomycete, *Pellicularia filamentosa*. However, since the imperfect stage is the one more often encountered, it is common and acceptable to refer to it by the imperfect name providing we are referring to its imperfect stage and not its perfect stage.

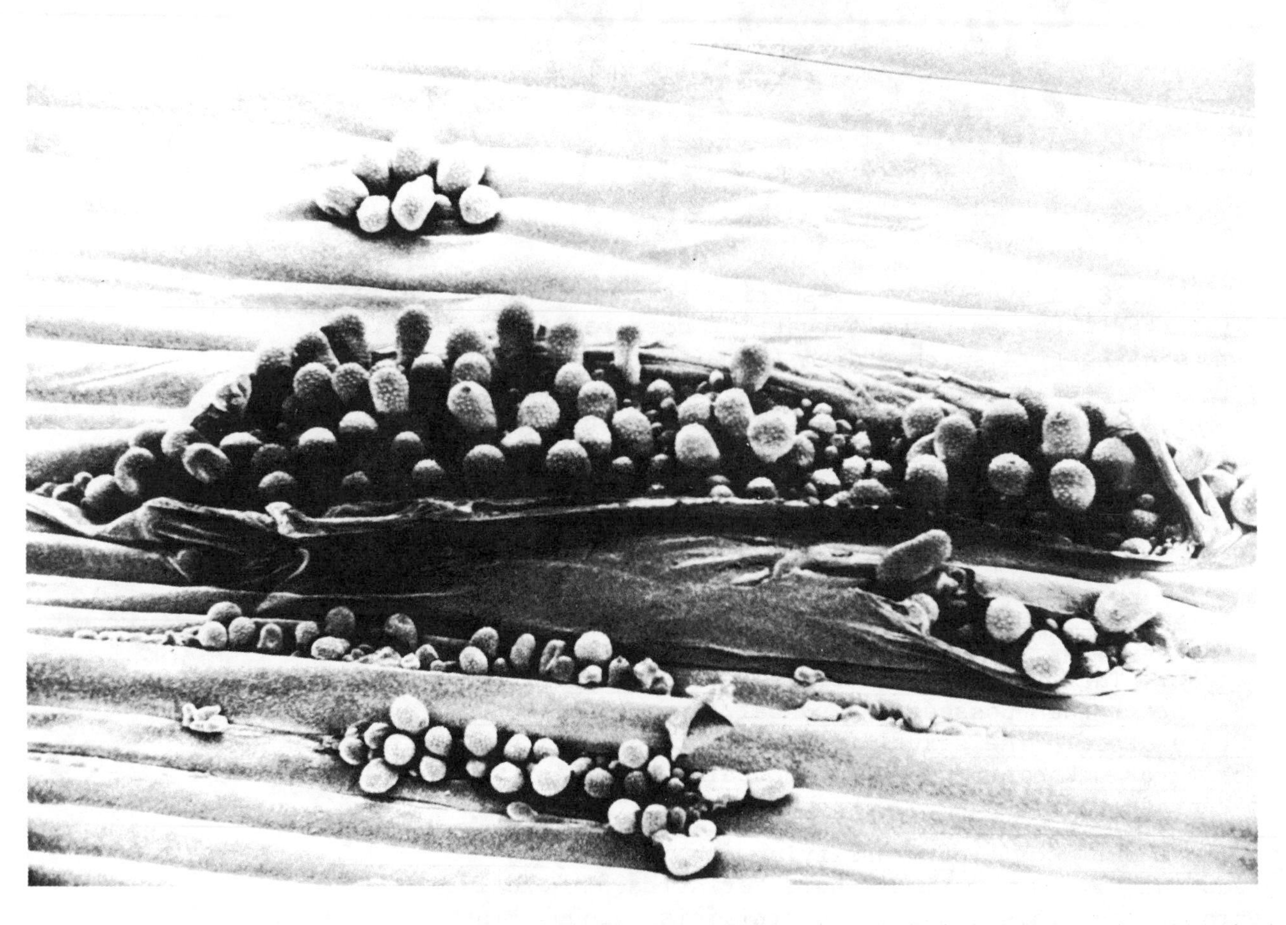

Fig. 4.29 Scanning electron micrograph of uredial pustules of *Puccinia graminis*. Notice the markings of the urediospores and how the pustules rupture the epidermis of the host stem, (x406). *(Courtesy of E. E. Butler)*

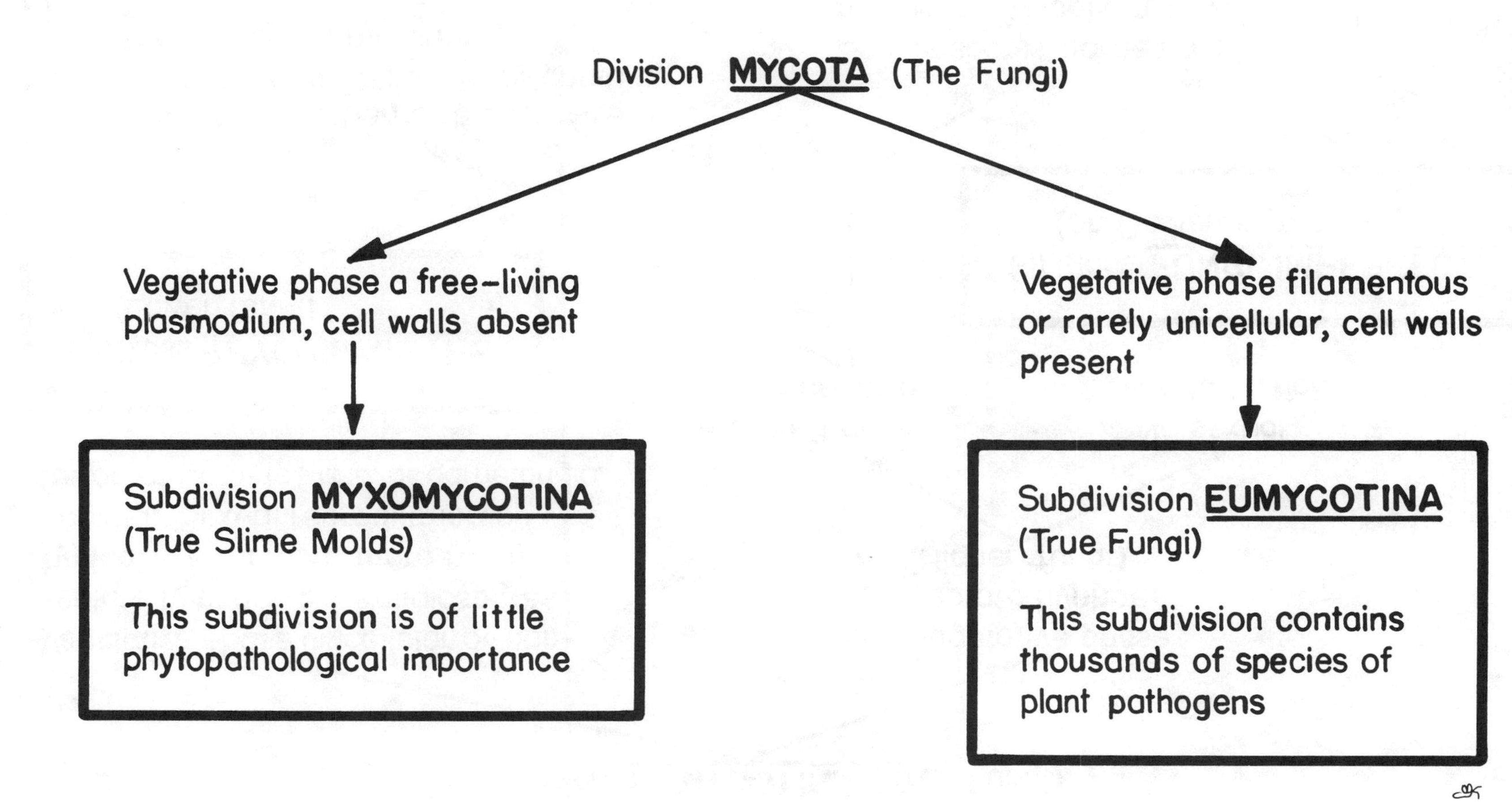
MAJOR GROUPS OF MOST FUNGAL PLANT PATHOGENS
(Many groups of non-pathogenic fungi are omitted)
Division MYCOTA (The Fungi)
Vegetative phase a free-living plasmodium, cell walls absent
Vegetative phase filamentous or rarely unicellular, cell walls present
Subdivision MYXOMYCOTINA
(True Slime Molds)
This subdivision is of little phytopathological importance
Subdivision EUMYCOTINA
(True Fungi)
This subdivision contains thousands of species of plant pathogens

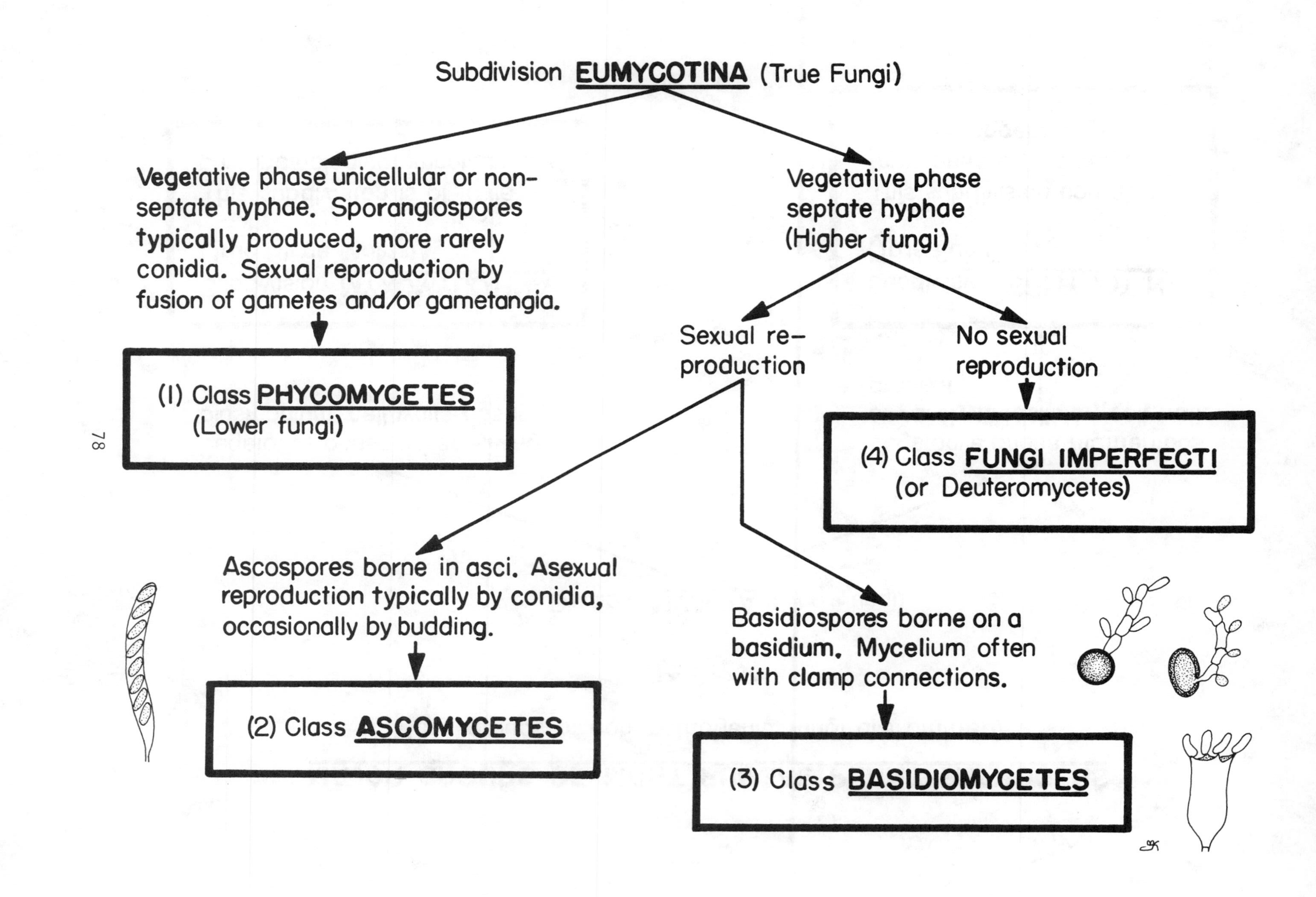
Subdivision EUMYCOTINA (True Fungi)
Vegetative phase unicellular or non-septate hyphae. Sporangiospores typically produced, more rarely conidia. Sexual reproduction by fusion of gametes and/or gametangia.
(1) Class PHYCOMYCETES
(Lower fungi)
Vegetative phase septate hyphae (Higher fungi)
Sexual re-production
No sexual reproduction
(4) Class FUNGI IMPERFECTI
(or Deuteromycetes)
Ascospores borne in asci. Asexual reproduction typically by conidia, occasionally by budding.
(2) Class ASCOMYCETES
Basidiospores borne on a basidium. Mycelium often with clamp connections.
(3) Class BASIDIOMYCETES

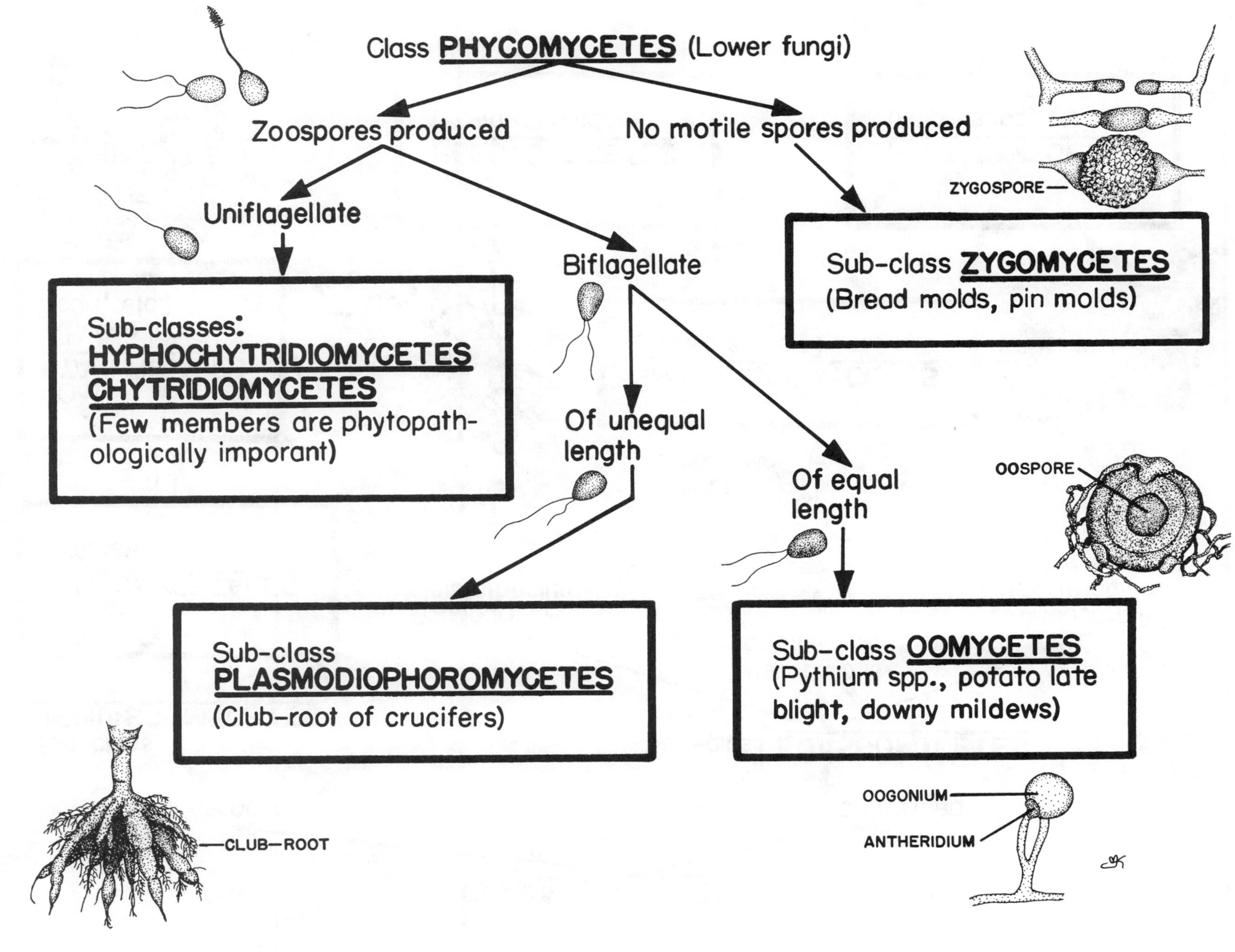
Class PHYCOMYCETES (Lower fungi)
Zoospores produced
No motile spores produced
ZYGOSPORE
Uniflagellate
Biflagellate
Sub-class ZYGOMYCETES
(Bread molds, pin molds)
Sub-classes:
HYPHOCHYTRIDIOMYCETES
CHYTRIDIOMYCETES
(Few members are phytopath-
ologically imporant)
Of unequal
length
Of equal
length
OOSPORE
Sub-class
PLASMODIOPHOROMYCETES
(Club-root of crucifers)
Sub-class OOMYCETES
(Pythium spp., potato late
blight, downy mildews)
OOGONIUM
ANTHERIDIUM
CLUB-ROOT

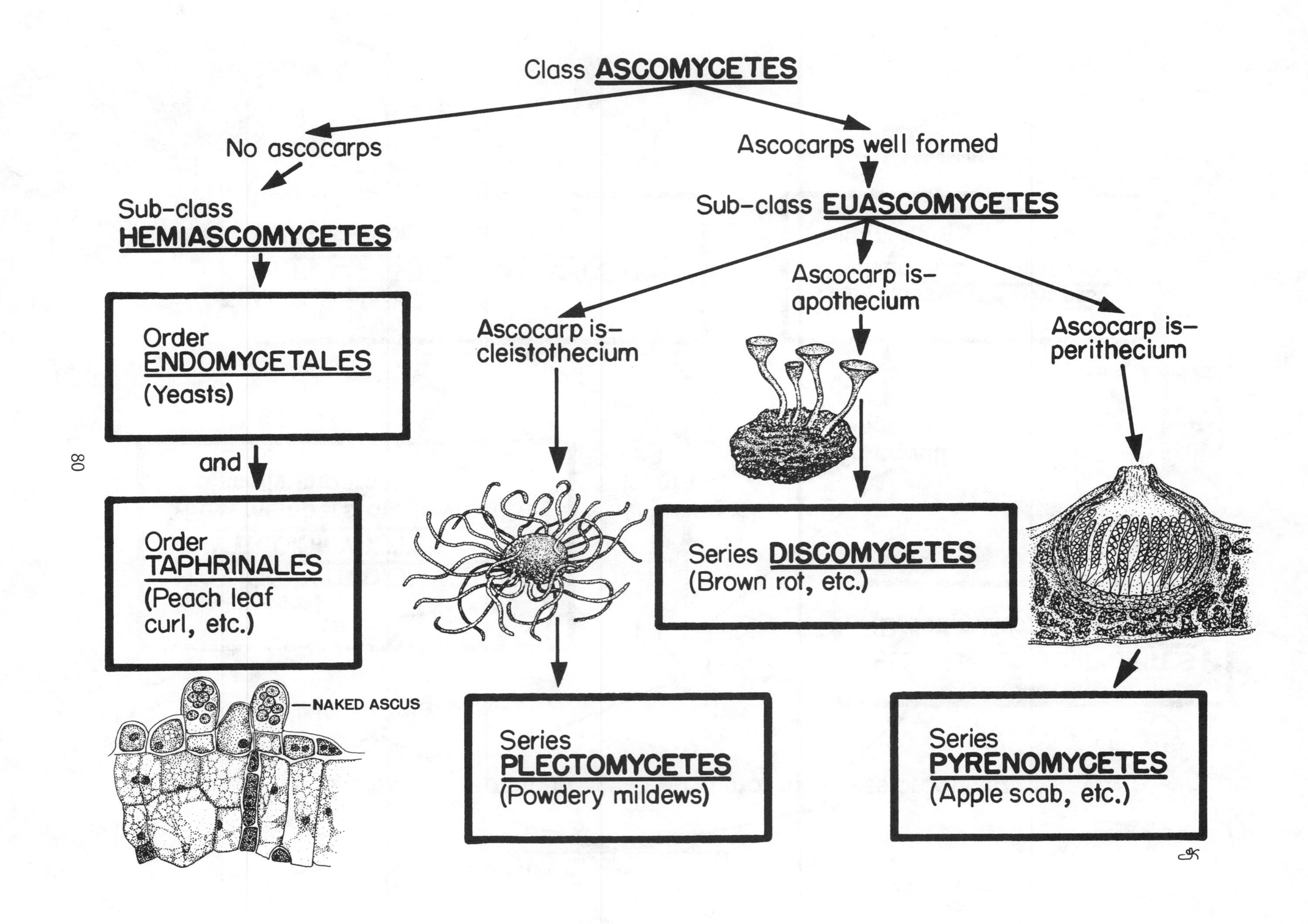
Class ASCOMYCETES
No ascocarps
Ascocarps well formed
Sub-class HEMIASCOMYCETES
Sub-class EUASCOMYCETES
Order ENDOMYCETALES (Yeasts)
and
Order TAPHRINALES (Peach leaf curl, etc.)
NAKED ASCUS
Ascocarp is- cleistothecium
Ascocarp is- apothecium
Ascocarp is- perithecium
Series DISCOMYCETES (Brown rot, etc.)
Series PLECTOMYCETES (Powdery mildews)
Series PYRENOMYCETES (Apple scab, etc.)

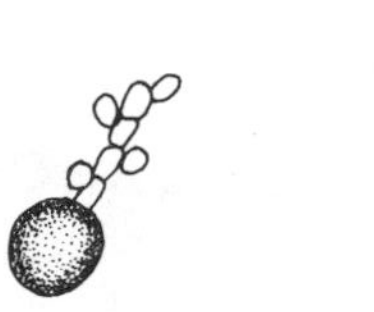

Class **BASIDIOMYCETES**

Basidia septate or deeply divided, often elongated. Basidiospores sometimes germinate by budding.

Basidia simple and unicellular. Basidiospores typically germinate to form a mycelium.

Sub-class **HETEROBASIDIOMYCETES**

Teliospores formed; no basidiocarp

- Basidiospores forcibly discharged → **Order UREDINALES** (Rusts)
- Basidiospores not forcibly discharged → **Order USTILAGINALES** (Smuts)

Basidiocarp formed; no teliospores → **Order TREMELLALES** (Jelly fungi, n.p.i.)

Sub-class **HOMOBASIDIOMYCETES**

Hymenium present, exposed before spores mature → **Series HYMENOMYCETES** (Mushrooms, bracket fungi etc.)

Hymenium absent or present, enclosed until spores are released from basidia → **Series GASTEROMYCETES** (Puff-balls, earth stars, stink horns, birds-nests etc., n.p.i.)

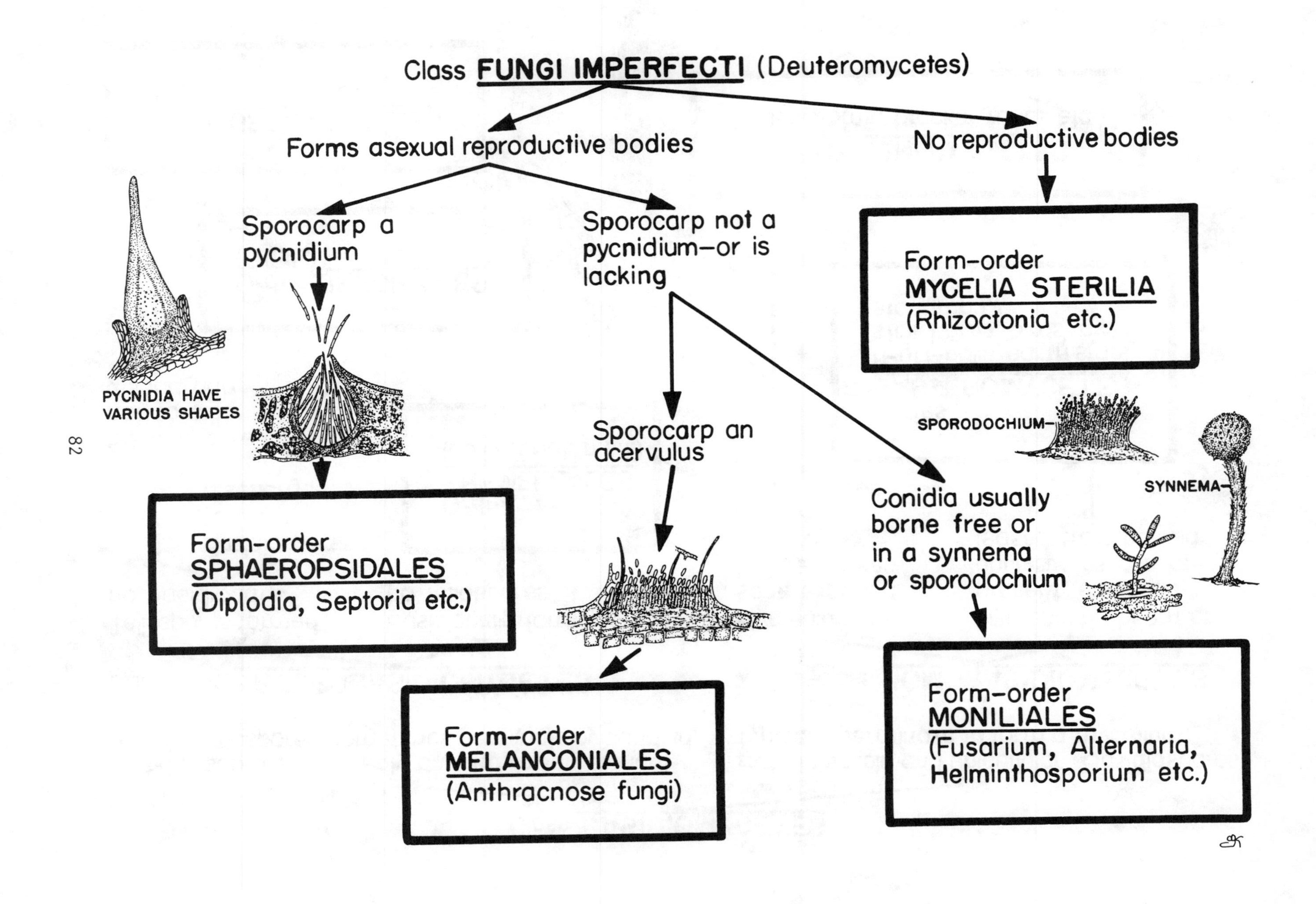

Class FUNGI IMPERFECTI (Deuteromycetes)
Forms asexual reproductive bodies
No reproductive bodies
Sporocarp a pycnidium
Sporocarp not a pycnidium—or is lacking
Form-order
MYCELIA STERILIA
(Rhizoctonia etc.)
PYCNIDIA HAVE VARIOUS SHAPES
Sporocarp an acervulus
SPORODOCHIUM
SYNNEMA
Conidia usually borne free or in a synnema or sporodochium
Form-order
SPHAEROPSIDALES
(Diplodia, Septoria etc.)
Form-order
MELANCONIALES
(Anthracnose fungi)
Form-order
MONILIALES
(Fusarium, Alternaria, Helminthosporium etc.)

Selected References

Ainsworth, G., and A. E. Sussman (ed.), 1965-1966. The fungi - an advanced treatise. Vols. 1 and 2. Academic Press Inc., New York.

Alexopoulos, C. J. 1962. Introductory mycology. Ed. 2. John Wiley and Sons, New York.

Alexopoulos, C. J., and H. C. Bold. 1967. Algae and fungi. The Macmillan Company, New York. pp. 79-116.

Burnett, J. H. 1968. Fundamentals of Mycology. St. Martin's Press, New York.

Christensen, C. M. 1961. The molds and man. Ed. 2. Lund Press, Inc., Minneapolis.

Gray, W. D. 1959. The relation of fungi to human affairs. Henry Holt and Company, Inc., New York.

Hawker, L. E. 1966. Fungi. Hutchinson and Co. Ltd., London. (Paperback edition by Anchor Press, and bound by Wm. Brendon, both of Tiptree, Essex).

Heald, F. D. 1933. Manual of plant diseases. Ed. 2. McGraw-Hill Book Co., New York. pp. 393-422.

Large, E. C. 1940. The advance of the fungi. Holt, New York. (Paperback edition by Dover Publications, New York).

Moore-Landecker, E. 1972. Fundamentals of the fungi. Prentice-Hall, Inc. Englewood Clifts, N. J.

Stevens, R. B. 1953. The fungi are living organisms. pp. 27-31, In Plant diseases. U. S. Dept. Agr. Yearbook.

Wolf, F. A., and F. T. Wolf. 1947. The fungi, Vols. 1 and 2. John Wiley and Sons, New York.

CHAPTER 5

VIRUSES AS CAUSAL AGENTS OF PLANT DISEASE

5-1 DEFINITION AND GENERAL STRUCTURE

Definition Of Viruses

Viruses are virulent pathogens of most forms of higher plant and animal life, as well as of bacteria. Viruses may be defined as ultramicroscopic infectious entities that contain only one of the two forms of nucleic acid, are synthesized only within suitable host cells by utilizing the synthetic mechanisms of the cells to produce the viral substances, and are unable to increase in size. Viruses are incapable of direct reproduction, which is characteristic of true living forms. In the scheme of classification of everything biological, the status of viruses is eloquently explained by McWhorter, (1965), "viruses are placed midway between the quick and the dead."

Protein Outside Nucleic Acid Inside

The plant viruses that have been purified to a state where their chemical compositions have been ascertained, have shown a marked chemical similarity. They each contain only a protective sheath of protein, sometimes called a capsid, and an inner core of nucleic acid, which is called a nucleocapsid. The nucleic acid of the vast majority of plant viruses is ribonucleic acid (RNA). Bacterial viruses (called bacteriophages), and most animal viruses contain deoxyribonucleic acid (DNA); however, some animal viruses contain a core of RNA. Viruses that contain DNA have considerably larger particle sizes than those with RNA.

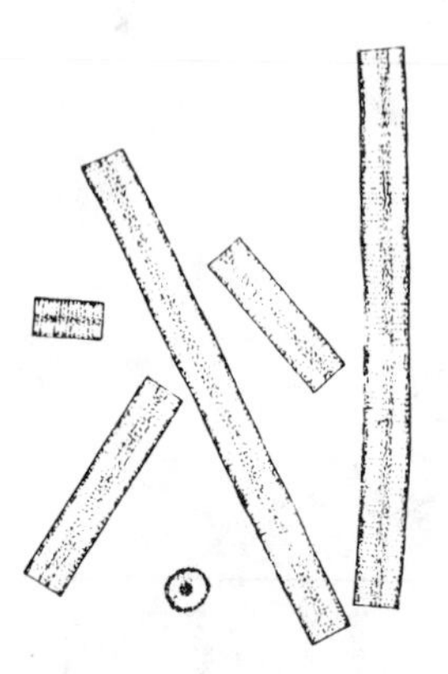

Fig. 5.1 Two rigid rod virus particles and fragments.

The shape of plant virus particles may be anisometric (unsymmetrical parts), or isometric (three equal axes at right angles to one another). Anisometric virus particles may be short structures, where the length is only several times the width. The most common of these are the short rigid rods, less common are the bullet-shaped viruses. Long rigid rods and flexible threads are also common.

Isometric virus particles usually have a rotational symmetry of an icosahedron (a polyhedron of twenty faces where each face is an equilateral triangle). These viruses are usually termed "spherical".

The capsids of rod-shaped viruses are single protein structure units, repeated up to several thousand times, and arranged in a helical manner to make cylindrical open-end tubes. The length of these capsids appears to be determined by the length of the embedded nucleic acid molecules. The capsids of spherical viruses are closed shells. The nucleic acid arrangement in the center of spherical viruses is unknown. However, the ratio of nucleic acid to protein is characteristically constant for each particular virus.

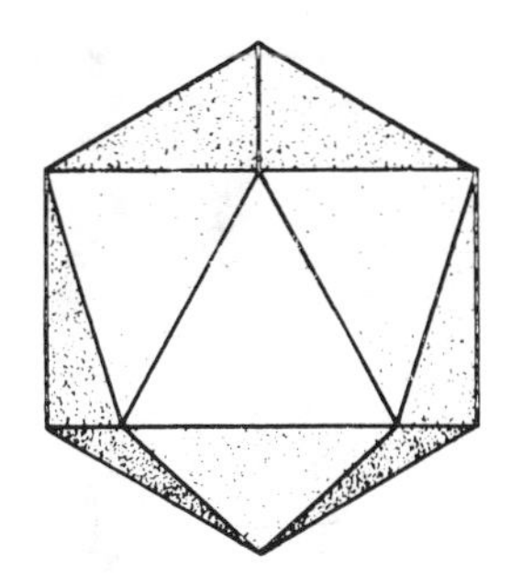

Fig. 5.2
An icosahedron

Unlike some bacterial viruses, plant viruses are not known to possess any enzymes. Therefore, these viruses occupy a unique niche among plant pathogens in that their parasitism is completely obligate in nature.

Viruses are so small they cannot be seen with a compound microscope and are termed ultramicroscopic. Only with the advent of the electron microscope has it been possible to see and photograph virus particles. Spherical

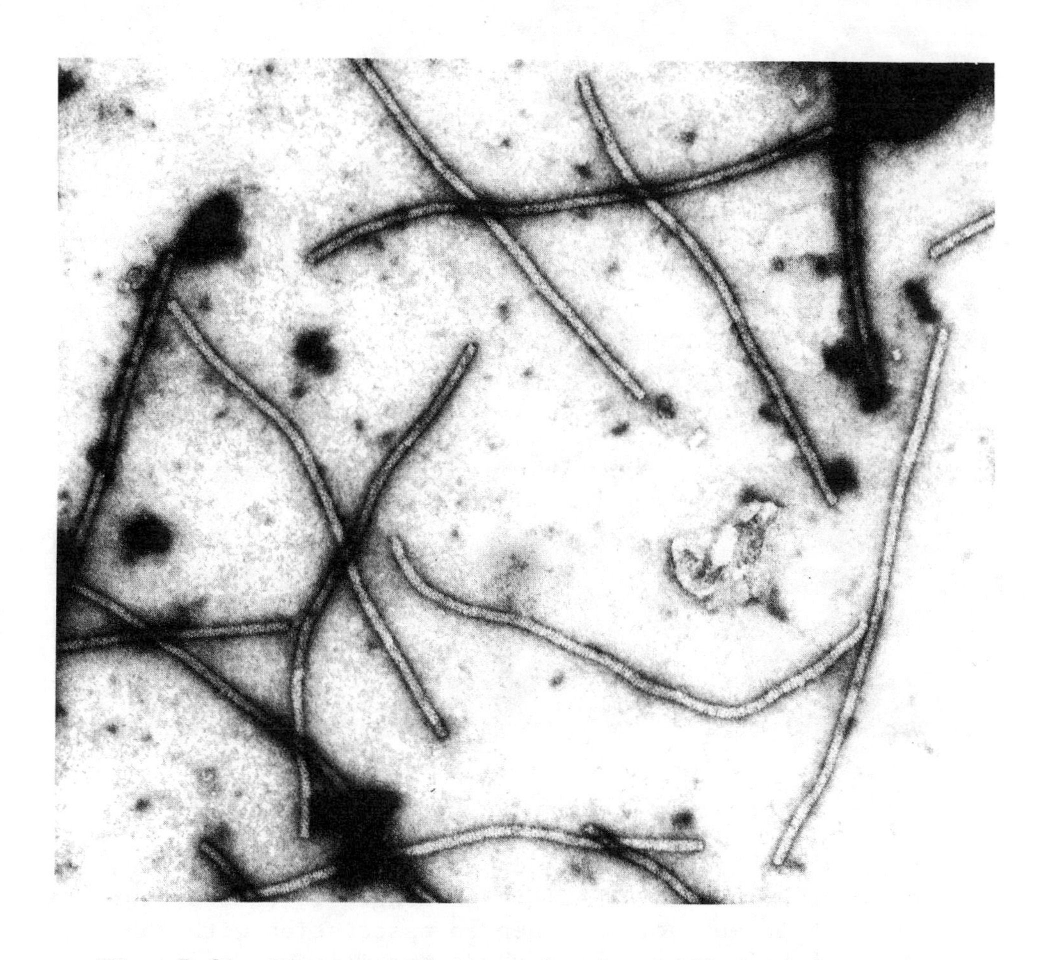

Fig. 5.3A Electron micrograph of purified apple stem grooving virus, (x106,000). *(Courtesy of R. M. Lister, C. E. Bracker, and A. F. El-Hadidi)*

plant viruses may have a diameter ranging from below 20 millimicrons (mμ) for the smallest to at least 60 mμ for the largest. The flexible rod shaped viruses may have lengths of 300 to 1,250 mμ. One mμ is about 1/25,000,000 of an inch. Therefore, it would take about 1,000,000 particles of a spherical virus with a mean diameter of 25 mμ laid side-by-side to measure one inch.

Fig. 5.3B Electron micrograph of purified tobacco rattle virus, a multicomponent virus. The longest particles (195 mμ) are infective but unstable. The short particles are 90 mμ (they are intermediate in length in this picture). They are not infective, but when in association with the long particles in a host, the short particles code for the protein coat for both particles. When the long particle has its protein coat, it is stable. The smallest particles in the picture apparently play no role in infection or viral replication, (x105,000). *(Courtesy of R. M. Lister and C. E. Bracker)*

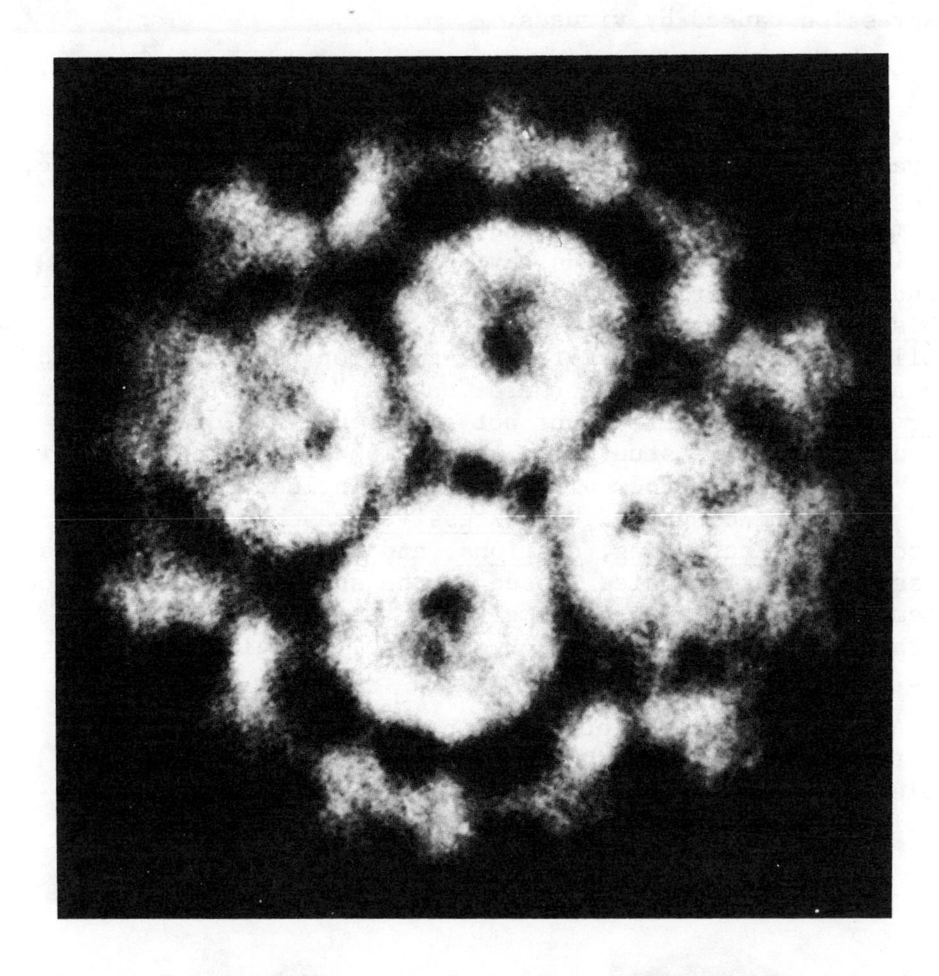

Fig. 5.4 Electron micrograph of a single cowpea chlorotic mottle virus particle, (approx. x4,000,000). *(Courtesy of J. B. Bancroft, G. J. Hills, and R. Markham)*

5-2 INFECTION AND DISEASE SYMPTOMS

There are two general types of infection, local and systemic. The term systemic refers to the spread from point of entry, usually a leaf or root, throughout the entire plant. Under natural conditions, plant diseases due to viruses are the result of systemic infection. Local infection occurs at the point of inoculation and fails to spread further. It is usually seen under conditions where the host is experimentally inoculated. Visible host responses at local infection sites are called primary symptoms. If the symptoms fail to spread, these are then termed local lesions. Symptoms resulting from systemic infection are termed systemic, also. Systemic symptoms usually are first seen on the young leaves around the top of the plant. Later, symptoms may also be seen on the systemically infected older leaves. There is great variation in respect to symptom

Rule --
Virus Disease,
Result Of
Systemic
Infection

expression caused by viruses.

Local lesions may be small or large, dry or watersoaked, and variously colored with shades of red, white, brown or tan. Major symptom differences exist in systemically invaded leaves and stems and they may be listed as: mosaic, ringspot, yellowing, and stunting. Mosaic is a spotted or mosaic pattern of green and yellowish (chlorotic) areas on leaves. Ringspot is an obvious pattern of brown, reddish brown, or yellow border surrounding a green center. This is usually a primary symptom, however, it is often followed by systemic symptoms of zonation that are arranged in an irregular fashion or are "oakleaf" in design. Yellowing and stunting often, but not always, occur together. When the plant is stunted, a shortening of the internodes is characteristic. Other virus symptoms include leaf-curling and crinkling of the edges, excessive branching (witches'-brooms), veinclearing, enations, and breaking. Enations are masses of hypertrophied tissue on the surface of the leaf or stem. Breaking is the term given to the striping of the petals or flowers. It is common on tulips, gladiolus, stock, etc.

A virus may cause local lesions on certain hosts and systemic symptoms on others. The host range is also variable. A virus may attack a wide number of plants, or only a single plant species.

Fig. 5.5A Symptoms of tobacco ringspot virus on tobacco, (also see Fig. 5.8). *(Courtesy of R. W. Samson)*

Fig. 5.5B A mosaic symptom pattern produced by tobacco mosaic virus on tobacco. *(Courtesy of R. W. Samson)*

Generally Reduce Yield And Quality

The general overall effect of a plant virus is a mild stunting with a reduced quality and yield. Usually, but not always, this is associated with other symptoms.

Two Conditions, Both With No Visible Symptoms

A plant may be infected by a virus, and yet there may be no symptoms. There are two situations where this phenomenon occurs. The first condition is one in which the symptoms are said to be "masked". This exists for a number of viruses that cause symptoms only under specific environmental conditions. For instance, a number of viruses will not produce symptoms at higher temperatures, while at low temperatures the leaves show a typical symptom pattern. The other situation occurs when a plant infected by a virus does not show symptoms regardless of what the environment is; in which case, the infected plant is said to be a symptomless carrier. This condition exists in many varieties of potatoes infected with the latent-mosaic virus. Tobacco commonly outgrows the symptoms of the tobacco ringspot virus. These plants are also referred to as symptomless carriers. Thus, a symptomless carrier is an infected plant that will not show symptoms regardless of environment, while a plant with a masked infection will show symptoms if the environment is favorable.

5-3 CROSS-PROTECTION AND SYNERGISM

A plant may be infected with several viruses at the same time. This is common in perennial plants. There are many ways that the presence of one virus in a host plant can affect the multiplication of, and symptom expression of another. When a plant infected with one virus is attacked by another, usually the symptoms of the second virus are merely superimposed on those produced by the first. However, plants infected first with an avirulent (non-pathogenic) strain may be protected from infection by a virulent strain. This is called cross-protection.

Cross-protection By Avirulent Virus

It has also been found that two unrelated viruses, for the most part, multiply independently of one another; each produces its characteristic symptom effects and reaches amounts comparable to when it is present alone. However, they may interact to produce more severe symptoms than either produces alone, and one or the other may also reach a higher concentration than when present alone. This symptom phenomenon has been called synergism. An example of this is double-virus streak in tomato caused by infection with both tobacco mosaic virus and potato virus X. When both viruses are present in a tomato plant, the resulting symptoms are an increase in stunting and extensive necrosis on leaves and stems that are not characteristic of either virus alone.

Synergism

5-4 VIRUS SYNTHESIS

Rule -- Viral Pathogenesis

As stated by Diener, (1963), "pathogenesis in viral infections is not a metabolic but a genetic phenomenon." Virus multiplication involves the interaction of the virus genome and that of a compatible host. The virus nucleic acid is the infective portion of the virus particle. It is the carrier of the viral genetic information. Since the nucleic acid is enveloped by a protective protein coat, the first step in the establishment of the virus is thought to be the removal of the protein. This process and its details are unknown at present, and we can only speculate as to whether or not it is similar in all plant viruses. However, it must be accomplished by a mechanism of the host cell.

Nucleic Acid Is Infective Portion

The naked nucleic acid, or the complete virus particle which is subsequently stripped of its protein, is transported by some means to a site of synthesis within the cell. At the site of synthesis, if the nucleic acid fraction is successful in gaining control of the metabolism of the host cell, synthesis of viral nucleic acid fractions takes place. The viral genetic information is "read", and the host cell "copies" this information and produces exact replicas of the nucleic acid and the protein coat. The virus nucleic acid is thought to be manufactured apart from the protein, and the separate components come together by a self-assembly process.

Nucleic Acid And Protein Manufactured Separately

A plant may be a favorable host for a virus while it is in one physiological state, but may not be receptive and support virus multiplication when it is in a different state. The environmental conditions that predispose plants to infection are not always those that favor optimum virus multiplication. Three environmental conditions that often favor virus infection are: (1) plants kept in darkness or shaded for one to two days prior to inoculation, (2) plants kept at a high temperature prior to inoculation, and (3) a high level of host plant nutrition. Two factors that generally favor virus multiplication are: (1) a normal plant temperature, and (2) a high level of fertilizer. Viruses are not alike in the environmental conditions that favor their infection and/or multiplication.

For many years it was thought that viruses multiplied only in the cells of their plant hosts and not in any of the insect vectors. However, just to prove that in biology there is an exception to almost every rule, a few years ago it was found that wound-tumor virus multiplies in the leafhopper insect vectors as well as in plants. Now, we have proof that a number of leafhopper, and a few aphid transmitted viruses are propagated inside their insect vectors. These viruses are termed propagative. Thus, our previous exception to a general rule has grown to become a general rule of its own, an excellent example of how our scientific knowledge grows.

Propagative Viruses Multiply In Insect Vectors

5-5 MULTICOMPONENT PLANT VIRUSES

It is now well established that a number of plant viruses have their total nucleoprotein divided into two, three, or more component particles. In other words, the genetic information (genome) required for virus production is divided between two or more nucleoprotein particles.

In a recent review, Lister (1969), tabulated 18 plant viruses with known divided genomes. Some of these viruses may be placed into rather well-defined categories. However, some multicomponent viruses have interacting components that are not well understood at this writing, or are too complex for presentation here. Therefore, this discussion is limited to three rather well-defined categories of multicomponent viruses. For a more complete review of this subject, the reader is referred to Lister, 1969, and van Kammen, 1972.

Viruses With Divided Genomes

5-5.1 SATELLITE VIRUSES

A satellite virus by itself is unable to infect a plant and become replicated. In order to become replicated, a satellite virus must be in association with another specific virus in a susceptible host. The latter virus is entirely autonomous, and is not dependent upon the satellite virus. However, the satellite virus is dependent upon the autonomous virus.

Perhaps the best example of this category is the satellite virus (SV) that is multiplied only in plants infected with tobacco necrosis virus (TNV). Replication of SV occurs only in association with the TNV. This relationship is said to be highly specific, because SV does not become replicated in association with any other virus. Tobacco necrosis virus is a fully autonomous virus that does not require the presence of the satellite virus to infect a host and become replicated. However, it appears that if tobacco necrosis virus and SV are replicated together within a host, the number of TNV particles is less than when SV is absent.

5-5.2 TWO COMPONENT VIRUSES--ONE COMPONENT IS INFECTIVE BUT UNSTABLE

Long Particle Infectious But Unstable

This group is perhaps best typified by the rod-shaped tobacco rattle virus (TRV). TRV has both a long and short particle (see Fig. 5.3B). The long particle is capable of infecting a host. However, only the ribonucleic acid (RNA) of the particle is formed and not the outer protein coat (capsid). These protein deficient RNA cores are not stable and cannot reinfect another host.

Short Particle Codes For Protein

The short tobacco rattle virus particles in association with the long particles can infect a host and both long and short complete particles are formed. The short particles of tobacco rattle virus carry the genome for the protein coat for both the long and short particles. The protein is the same for both particles but differs only in capsid length. The protein coat confers stability to the long particle.

5-5.3 TWO COMPONENT VIRUSES--BOTH REQUIRED FOR INFECTION AND REPLICATION

Both Particles Required For Infection

Several viruses belong to this group, one of which is cowpea mosaic virus (CPMV). The viruses of this group have their genomes divided between two nucleoprotein components. Neither can cause infection or become replicated without the presence of the other.

The two component particles of each virus within this group share the same protein capsid, only their nucleocapsids (nucleic acid cores) are different.

Cowpea mosaic virus, a spherical (icosahedral) virus, actually forms three types of particles, the third type is an empty protein capsid; a capsid that is devoid of a nucleic acid core. A number of plant viruses are known to have such empty protein capsid particles formed. Empty protein capsids do not appear to influence the infectivity or stability of complete virus particles.

5-6 FREE, STABLE RNA MOLECULES

Free Stable RNA

Several recent reports have brought to light facts of great significance in the field of plant virology. The most important of these may well be the discoveries that some plant diseases may be caused by free RNA molecules. These RNA molecules are without protein coats, yet they are stable when isolated in purified preparations. Three plant diseases are now known to be caused by free RNA molecules. They are: potato spindle tuber (Diener, 1971); citrus exocortis (Semanick and Weathers, 1972); and chrysanthemum stunt (Diener and Lawson, 1972).

Viroid

Diener (1971) proposed the name "viroid" for this group of infectious, free RNA molecules. Whether this name will become accepted is not clear at this time.

The infectious, free RNA molecules are of extremely small dimensions, much smaller than the smallest known infectious virus particle. They may be primative or advanced forms of viruses, or may have developed independently.

5-7 TRANSMISSION

Transmission In Nature

In nature, viruses are mainly transmitted by means of: (1) insects and other arthropods; (2) nematodes in the soil; (3) a group of soil-borne lower fungi; (4) infected seeds; (5) infected vegetative plant parts used for propagation; and (6) mechanical means. The first three groups are composed of natural vectors.

Rule -- Wound Pathogens

Plant viruses are all wound pathogens. Therefore, it is reasonable to think that all viruses lend themselves easily to experimental transmission by mechanical means. This is not the case. Some viruses are easily transmitted by this means and other have never been mechanically transmitted. Investigators also experimentally employ insects, parasitic seed plants (dodder), and grafting techniques to transmit viruses.

5-7.1 INSECTS AND OTHER ARTHROPODS

Rule -- Insect Transmission

In the field, insect transmission is overwhelmingly the most common, and therefore, the most important method of virus transmission. The insects that are vectors of viruses are: aphids, leafhoppers, treehoppers, whiteflies, mealybugs, thrips, earwigs, grasshoppers, and some beetles. Eriophyid mites of the genus *Acarina* are also vectors of some viruses. Although they are not insects, mites are included here because of their feeding habits and close relationship with insects.

Aphid transmission has been characterized in the past as either nonpersistent or persistent. Nonpersistent means that the insect can acquire the virus and transmit it within a matter of seconds or minutes, but soon loses the ability to transmit it, unless the virus is reacquired. Persistent means that there is a latent period between acquisition and ability to transmit, and the insect vector maintains the ability to transmit the virus for many days following the removal of the insect from any source of acquisition. However, there are a number of viruses that are neither nonpersistent nor persistent. Therefore, with our present knowledge, it is best to adopt the terms "stylet-borne" and "circulative".

Stylet-borne Viruses

The term stylet-borne refers to viruses that adhere to an insect's stylet after feeding on an infected plant.

Circulative Viruses

The term circulative refers to viruses that are acquired through the insect's mouthparts, swallowed, passed through the gut wall into the blood, and back to the salivary glands, before the insect is infective. Circulative viruses may, or may not be propagative. The term circulative now includes all viruses called persistent under the earlier terminology.

The question is often raised as to whether or not viruses that infect and multiply in both plant hosts and insect vectors are harmful to their insect vectors, since they surely are in host plants. There appears to be little if any noticeable adverse effect on most of the efficient insect vectors, but there is evidence that some viruses are harmful to certain inefficient insect vectors.

Insect With Sucking Mouthparts

Aphids, leafhoppers, treehoppers, whiteflies, and mealybugs all belong to the order Homoptera and feed by means of sucking mouthparts. These insects are well adapted for the piercing of plant tissues and for the removal of the sap or plant juices. Their mouthparts consist of a pair of thread-like mandibles flanking a pair of thread-like interlocking maxillae. These mouthparts form the long flexible stylet bundle. The maxillae slide up and down between the mandibles, and the stylet bundle slides up and down in a grooved labium. Two grooves on the paired inner faces of the maxillae form two channels; saliva is injected down one, and the other is used to draw up a mixture of plant sap and saliva. Gelling saliva is secreted on the surface of the plant around the point where the stylet is injected prior to feeding. As the stylet is inserted deep in the plant in search of food, gelling saliva is intermittently ejected and forms a sheath around the stylet whose function is unknown. Gelling saliva begins to gell when it is secreted and fills intercellular spaces and sometimes whole cells. The gelled saliva on the plant surface around a point of stylet insertion is called a flange. The feeding locations identified by the flanges are often called feeding tracks. For additional information, the reader is referred to the review by Miles, 1968.

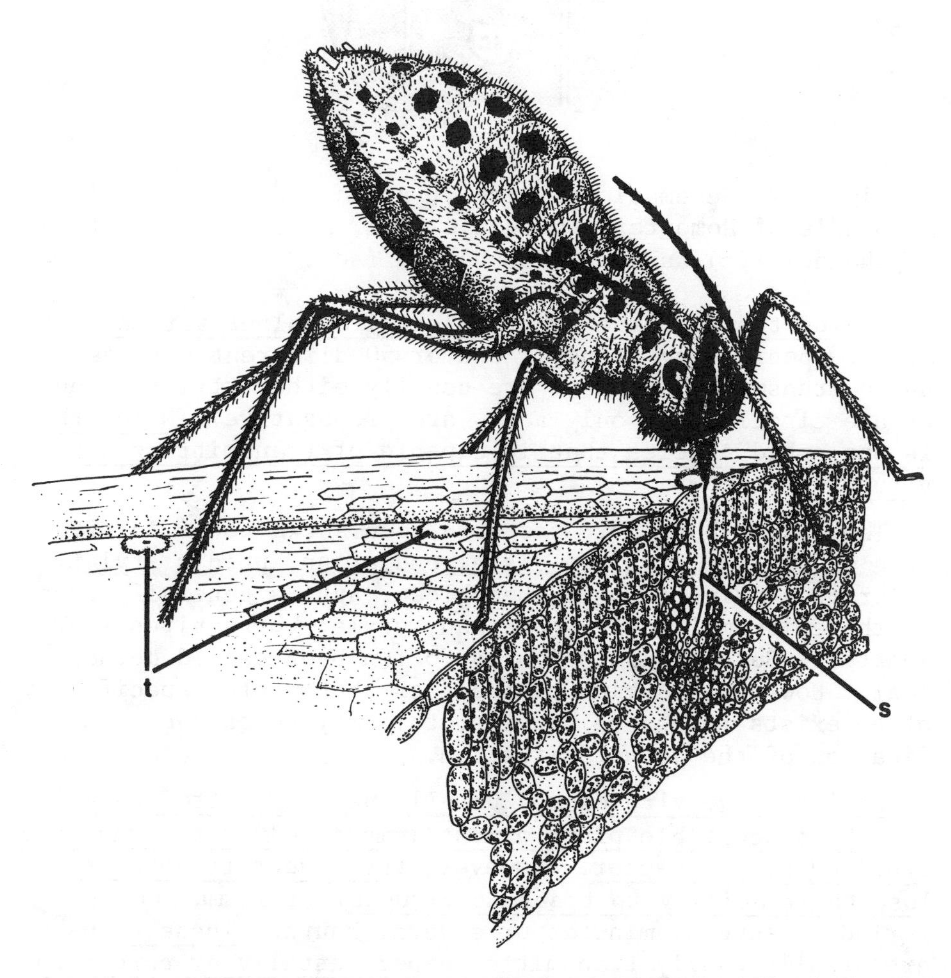

Fig. 5.6 Semidiagramatic drawing of an aphid feeding deep in the phloem of a host plant. The stylet (s) is inserted intercellularly until it reaches the phloem; it is surrounded by a gelled saliva sheath. Flanges of gelled saliva on the leaf surface surround two earlier feeding locations, and are called feeding tracts (t).

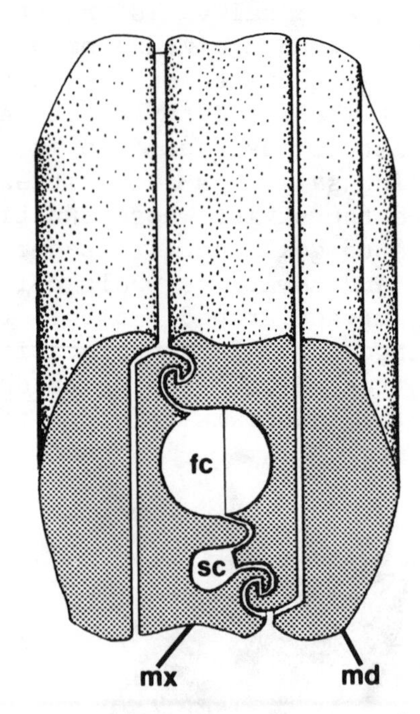

Fig. 5.7 Schematic drawing of a section of the stylet bundle of Homoptera; mandible (md), maxillae (mx), food channel (fc), and saliva channel (sc).

Aphids transmit the great majority of plant viruses. A single species may transmit one or 50 different viruses. Aphid transmitted viruses are usually either stylet-borne or are circulative; only a few are propagative. Often there is one aphid species that is capable of transmitting a given virus.

Vector Specificity Is Common

The stylet-borne viruses have been shown to be carried on a 15 μ segment of the distal end of an aphid's stylet. Prior to feeding deep in the phloem, aphids generally probe in the epidermal layer where many stylet-borne viruses are located. Such transmission has often been called mechanical. However, in view of the fact that vector specificity often exists for these viruses, this may be an over-simplification of the actual situation.

Characteristic Features Of Stylet-borne Viruses And Their Vectors

Stylet-borne viruses may usually be transmitted to a healthy susceptible plant almost immediately after they are acquired by the vector. However, these vectors quickly lose their ability to transmit viruses. This may be a period of several minutes or several hours. These viruses are usually easily transmitted experimentally by pure mechanical means.

The circulative viruses transmitted by aphids are less common, but are more intimate in their association with the aphid vector than are the stylet-borne viruses. There is usually a latent period after feeding before aphids are able to transmit circulative viruses. These viruses are usually located in the phloem and are often difficult to transmit by mechanical means. Vectors of circulative viruses usually retain their ability of transmission over a longer period of time than the stylet-borne viruses.

Characteristic Features Of Circulative Viruses And Their Vectors

Leafhoppers transmit the second largest number of plant viruses. In these insects, the biological association of virus and vector varies from that approaching stylet-borne transmission to circulative or propagative. Propagative viruses are transmitted from an infective female parent to her offspring. When a leafhopper acquires a propagative virus, there is a period of several days before the vector is able to transmit it. Several workers have shown that a virus may be carried through many generations even though the insects do not have access to infected plants and when the females are bred only with virus-free males. Leafhoppers are alternate hosts for propagative viruses.

Insects With Chewing Mouthparts

Grasshoppers, beetles, and earwigs have chewing mouthparts. These insects transmit viruses mainly by the particles adhering to their mouthparts. For a few other viruses, the method of transmission is more complicated. These insects have no salivary glands, and they regurgitate part of the foregut contents during feeding to aid in the digestion process. It is this regurgitation process that appears to bring infective virus particles, ingested during feeding on disease plants, into contact with fresh feeding wounds on healthy plants.

5-7.2 NEMATODES

A few plant viruses have long been known to exist and persist in the soil. One of these viruses causes the disease fanleaf of grape and this was the first virus disease shown to be transmitted by a nematode, *Xiphinema index*. Several other "soil-borne" viruses have since been shown to be transmitted by nematodes.

To date, some thirteen species of nematodes from three genera, all of which belong to the subfamily Tylencholaiminae of the family Dorylaimidae, are known to transmit plant viruses. These nematodes migrate through the soil feeding on the roots of host plants by injecting their stylets into young cortical tissue. Feeding on infected plants is the only known way that nematodes acquire plant disease viruses. The limited information now available indicates that nematodes lose their infectivity slowly. There is no proof yet that the infectivity of any virus is permanently retained.

5-7.3 FUNGI

Primitive Chytrid Fungi

The soil-borne virus that has long been known to cause big vein of lettuce, is now known to be transmitted by spores of the primitive Chytrid fungus *Olpidium brassicae*. The fungal resting spores are capable of surviving long periods of time in dry soil and appear to harbor the virus, permitting the latter to survive. By itself, this virus quickly loses infectivity in desicated soils. The fungus is also a pathogen, and when the resting spores are able to resume their activity in wet or moist soils, motile spores are formed that carry the virus to the host roots. The spores come to rest, germinate, and penetrate the root carrying the virus into the host plant. It is not known how the virus leaves the fungus to infect the host plant.

Zoospores of *O. brassicae* are also vectors for tobacco necrosis virus which infects the roots of many plant species. Two other Chytrid fungi, *Polymyxa graminis* and *Spongospora subterranea*, have also been shown to be vectors of plant viruses. *P. graminis* transmits wheat mosaic virus and *S. subterranea* transmits potato mop-top virus (Rao and Brakke, 1969, and Jones and Harrison, 1969).

Longevity Of Virus

Specific Relationships

Two characteristic features of fungi as vectors are; (1) the ability to harbor viruses for a number of months in air-dry soil, and (2) the highly specific relationship between the fungal vectors and the viruses. For example, there is evidence to indicate that certain strains of *O. brassicae* consistently transmit tobacco necrosis virus, while some strains do not.

5-7.4 INFECTED SEED

Seed transmission of viruses appears to be more common than it was once thought to be. Fulton, 1964, lists sixty-four virus and host plant combinations in which seed-borne transmission occurs. The frequency with which virus appears in the seed varies with both the virus and plant species. As little as 3 percent and as high as 100 percent of seed has been found to carry viruses from their diseased plants. A few viruses have been shown to be transmitted through pollen from infected male plants to flowers of healthy female plants, giving rise to infected seed.

Nematode-borne Viruses Are Usually Seed-borne

Seed transmission appears to be important in the ecology of nematode-borne viruses. Lister and Murant, 1967, found that nematode-borne viruses are characteristically transmitted through seed of crop and weed plants. The weed plants are often symptomless carriers. Infected weed and crop seed germinate and produce diseased plants from which the virus is transmitted by nematodes to healthy crop plants. The virus may be disseminated and overwinters, or oversummers as the case may be, in the seed. Some nematodes lose their ability to transmit viruses in a period of 2 to 3 months in

Fig. 5.8 Late infection phase of bud blight of soybean, incited by tobacco ringspot virus. Infection occurred after the pods formed, causing purple blotches on the pods. Early infection may cause a stunted plant that fails to produce seed. Later infection will usually produce poorly filled pods that may drop prematurely. This virus is seed-borne. *(Courtesy of R. W. Samson)*

fallow soil. For such viruses storage within the seed undoubtedly plays an important ecological role. Viruses such as grapevine fanleaf is retained by its nematode for 8 months or more, and in this situation, the storage of viruses in seed is probably of less importance.

5-7.5 INFECTED VEGETATIVE PLANT PARTS

All Viruses May Be Transmitted By Grafting

All viruses are graft transmissible. In light of this fact, graft transmission has been employed experimentally especially with viruses that have been difficult to transmit by mechanical means. The limitation with this procedure is that generally a plant may only be successfully grafted to a plant of the same genus. Transmission of viruses through parasitic seed plants, such as dodder, is essentially a variation of grafting. The dodder forms an organic bridge from the infected plant to the healthy plant. This technique is used experimentally, but probably is of little importance in nature.

Rule -- Transmission Of Viruses Via Vegetative Propagation

A number of food and ornamental plants are propagated from vegetative plant parts. For example, potatoes are grown from tubers; strawberry plants are multiplied by daughter plants produced on runners; tulips and gladiolus are grown from bulbs; and numerous ornamentals are grown from cuttings. Also, many fruit trees such as apple and citrus, are produced by grafting to various root stocks. When the "mother" plants are virus infected, it follows that all daughter plants, or cuttings taken from them will also be infected. Virus infected root stocks to which young fruit tree stems are grafted also result in virus infected trees. In fact, unless "mother" plants are especially processed to free them of virus, daughter plants are usually virus infected because "vegetatively-propagated plants are generally infected by one or more viruses," (Campbell, 1973).

5-7.6 MECHANICAL MEANS

A few viruses seem entirely dependent upon mechanical means of transmission for their spread. Tobacco mosaic virus is one of the few. It is easily transmitted from infected plants to healthy plants by pruning tools or other implements, by workers' handling the plants, and by the plants' rubbing against each other.

The experimental means of mechanical transmission is essentially the removal of plant sap from a diseased plant and its transfer to a healthy plant. The transfer is accomplished by gently rubbing the sap on the surface of a leaf. This causes minute wounding of the leaf surface and places virus particles in these wounds, thus permitting the establishment of the virus. Not all viruses lend themselves to transmission by this method.

5-8 CLASSIFICATION AND IDENTIFICATION

The present system of naming a plant disease virus and the disease caused by the virus uses the description of the major symptom coupled with the name of the host plant upon which it is first described. This is a situation which confounds the student who is new to the study of plant virology. Probably no confusion would result from this system of naming if each virus had only one host, that being the plant from which it was named. However, many viruses possess a multiple host range, and some have a wide host range. For example, tobacco mosaic virus (TMV) is found in tomato plants and many other hosts, as well as in tobacco. Regardless of the host in which it is found, the pathogen is still TMV although the disease may be called tomato mosaic.

Viruses Have No "Handles" By Which They Are Easily Classified

Most living organisms, animals, green plants, fungi and nematodes, possess characteristics that allow us to precisely catalog them according to the Latin binomial system of classification. The convenient "handles" by which we grasp these biological entities are lacking in the viruses. Bacteria and most fungi can be grown in cultures, and can be readily observed with a compound microscope. Because viruses are ultramicroscopic, their detection in plant tissue or sap is only possible with an electron microscope which is often not readily available. Viruses are handled differently from organisms. They are dealt with more like chemicals. Methods used in protein chemistry are utilized in the assay of viruses and in their purification prior to assaying. This generally involves sophisticated techniques including ultracentrifugation.

Local Lesion Assay

The symptoms of plant virus diseases are useful, within limits, to diagnose the inciting agent. However, symptoms are not precise or always characteristic. The precise identification of viruses requires their extraction from living plants and their purification. The purified virus is examined for infectivity. Often, this is done by means of the local lesion assay, which is a count of the number of lesions produced on a local lesion host by a particular dilution of the purified virus. The infective concentration is roughly proportional to the number of local lesions. Viruses that are stable in purified form may also be examined with the electron microscope and their size and shape determined. Not all viruses readily lend themselves to purification, since they tend to be unstable (lose their infectivity).

Serology

Serological tests are extremely useful in determining both related and unrelated viruses. When purified viruses are injected into a vein of the warm-blooded test animal, such as a rabbit, the serum of the animal forms antibodies. An antibody is a specific protein that is produced in the tissues or fluids and that opposes the action of a foreign protein, which is called an antigen. An antibody combines chemically to the antigen, inactivating the latter. This is an important defense mechanism in vertebrates and some parasites. Blood is withdrawn from the test animal and allowed to clot under refrigeration. The serum is decanted and is ready for use. When this serum is mixed with a partially purified, or purified extract of the same virus that was injected into the test animal, the antibodies of the serum combine with the viral protein forming a visible precipitate. So sensitive is this test to similar proteins that a group of viruses that give similar precipitin reactions with one serum are regarded as strains of the same virus. A virus will not cause a precipitin reaction with the serum formed by an unrelated virus.

Strains Of A Virus Have Similar Precipitin Reactions

Selected References

Bawden, F. C. 1960. The multiplication of viruses. pp. 71-116, In J. G. Horsfall, and A. E. Dimond, (ed.), Plant pathology. Vol. II. Academic Press, New York.

__________. 1964. Plant viruses and virus diseases. Ed. 4. The Ronald Press Co., New York.

Cadman, C. H. 1963. Biology of soil-borne viruses. Annual Review of Phytopathology. 1:143-172.

Carter, W. 1962. Insects in relation to plant disease. Interscience Publishers, New York, pp. 269-692.

Corbett, M. K., and H. D. Sisler, (ed.), 1964. Plant virology. Univ. of Florida Press, Gainesville.

Diener, T. O. 1963. Physiology of virus-infected plants. Annual Review of Phytopathology. 1:197-218.

Goodman, R. N., Z. Kiraly, and M. Zaitlin. 1967. The biochemistry and physiology of infectious plant disease. D. Van Nostrand Co., Inc., Princeton, N. J. pp. 25-41.

Grogan, R. G., and R. N. Campbell. 1966. Fungi as vectors and hosts of viruses. Annual Review of Phytopathology. 4:29-52.

Haselkorn, R. 1966. Physical and chemical properties of plant viruses. Annual Review of Plant Physiology. 17:137-154.

McWhorter, F. P. 1965. Plant virus inclusions. Annual Review of Phytopathology. 3:287-312.

Maramorosch, K. 1963. Arthropod transmission of plant viruses. Annual Review of Entomology. 8:369-414.

Mundry, K. W. 1963. Plant virus-host cell relations. Annual Review of Phytopathology. 1:173-196.

Ossiannilsson, F. 1966. Insects in the epidemiology of plant viruses. Annual Review of Entomology. 11:213-232.

Raski, D. J., and W. B. Hewitt. 1963. Plant-parasitic nematodes as vectors of plant viruses. Phytopathology. 53:39-47.

Siegel, A., and M. Zaitlin. 1964. Infection process in plant virus diseases. Annual Review of Phytopathology. 2:179-202.

Smith, K. M. 1968. Plant viruses. Ed. 4. Methuen and Co. Ltd., London.

Wetter, C. 1965. Serology in virus-disease diagnosis. Annual Review of Phytopathology. 3:19-42.

CHAPTER 6

NEMATODES AS CAUSAL AGENTS OF PLANT DISEASE

6-1 THE WORLD OF NEMATODES

Thread-like

Nematodes may be defined as unsegmented roundworms. The segmentation refers to internal separation of organs and not to the superficial segmentation of the cuticle. The word nematode means "thread-like", and this aptly describes the usual form of these animals. They are also known by several common names such as threadworm, roundworm, eelworm and nema. Nematodes are not closely related to other common worms like earthworms, wireworms, and flatworms; they are in their own separate class and have no close relatives. However, the class Nematoda is a clearly differentiated group of invertebrates that contains several thousand species, which makes this a large class, indeed.

Fig. 6.1 Nematodes

As may be expected in such a large group, there is considerable variation to be found in size, form, life-cycle, feeding requirements and habitat. Most species of soil nematodes feed on bacteria, fungi, and other nematodes. The type of food a nematode uses can usually be determined by examining their stomal (mouth) structures. For instance, a bacterial feeder has a simple unarmed mouth cavity; a predacious nematode has teeth in the mouth cavity; and a fungus feeder has no cavity at all, but does have a primitive type of stylet similar to the piercing and sucking mouth parts of a mosquito or aphid.

Nematodes Everywhere

The most commonly known nematodes are probably the parasites of animals and man. Among the major parasites of man are roundworm, hookworm and pinworm. There are about 32 species of nematodes which attack man, and many others that infest domesticated and farm animals. Nematodes attack all of our wild game animals, as well as fish and birds, and many lower animals including a number of insects.

Nematodes are found at the bottom of fresh water lakes and streams, in salt water, and in polar seas and hot springs. They were thawed out alive from Antarctic ice by members of the Shakleton Expedition. Immense numbers of plant parasitic and free-living nematodes are found in all kinds of soils. Even though soil nematodes are not microscopic, they are seldom seen. The majority are just a little too small to be easily seen with the naked eye, even when separated from the soil. The length of adult plant parasitic nematodes varies from one sixty-fourth of an inch to a few forms which are one-eighth inch.

6-2 THE PATHOGEN

Complex Animals

Plant parasitic nematodes are typically worm-shaped. Some forms change during their life cycles to become nearly round or kidney-shaped. Although they are primitive, tiny forms, they are complex organisms, nonetheless. Their minuscule bodies have a muscular system, a digestive system, an excretory system, and a well-developed nervous system composed of many sensitive fibers and a "nerve center" that appears to serve as a brain since nematodes have no such organ. They have no eyes or nose; however, there are sensory organs that appear to detect certain chemical substances, both attractive and repelling. Nematodes do not have respiratory or circulatory systems, but these functions are carried out in other ways. They do have complete reproductive systems, and both males and females occur in most species; however, it is not unusual for reproduction to occur without males. A number of species are known in which males do not exist.

Life Cycle--

Egg Stage

Four Larval Stages

Adult Stage

The typical life cycle of plant parasitic nematodes is as follows:

a. Females usually lay eggs in the soil or in plants where they feed.

b. Larvae are usually differentiated and hatch from eggs.

c. If their host plants are available, the larvae feed immediately and develop through four larval stages. Each stage is separated by a molt.

d. After the last molt, adult males and females are differentiated. They are now sexually mature and able to reproduce.

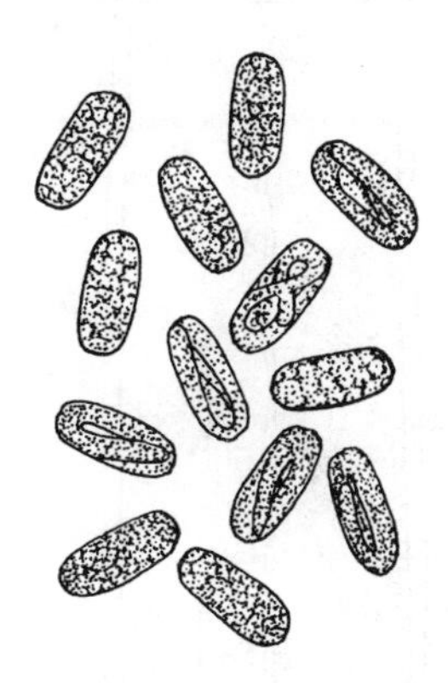

Fig. 6.2 Nematode eggs.

The minimum period of time for a complete life cycle varies from species to species. The minimum life cycle for the root knot nematodes is 18 to 21 days. The ability to survive periods of unfavorable environment also varies from species to species. Since the larvae do not grow and molt unless they feed upon a host, they subsist meanwhile on a reserve food supply that orginated in the egg. This food reserve lasts longer if the soil is cold or cool, but is soon depleted in warm moist soils. For a discussion of nematode survival, see Chapter 12.

Rule --
Parasitism
By Nematodes

Most nematodes that cause plant disease are obligate parasites. They neither feed nor reproduce except on their living hosts. The survival of such a nematode then is dependent upon its reaching a host before the reserve food supply of the egg is exhausted. This means that the nematode larva must seek out a suitable host unless the egg from which it is hatched is deposited on or in a host plant. In some species of cyst nematodes, the hatching of eggs within the cyst is dependent on stimulation received from the root exudates of a host plant. This mechanism ensures the newly emerged larvae of an immediate food source, which greatly enhances its chances for survival.

Nematodes
Swim

Soil nematodes move by swimming in a film of moisture clinging to the soil particles. It appears that most of the nematode movement is random; however, they may be attracted by exudates of roots of some host plants. At any rate, most nematodes probably never move more than a foot or two from the location where they hatch.

Rule --
Plant Damage
By Nematodes

Population dynamics of plant parasitic nematodes are not fully understood. With a few exceptions, plant damage by parasitic nematodes is dependent upon high soil populations. In all temperate climates, the soil population is diminished, but not abolished by freezing of the soil. Also, the cool weather of the spring and fall retards the metabolism of the nematode and lengthens the reproductive cycle. In climates where the soil remains warm the year around, nematode populations may reach fantastic proportions. A single female root knot nematode may produce more than 500 eggs, and if only a limited number of them survive to reproduce again and again, it is easy to see how a great increase in a field population can occur during one growing season.

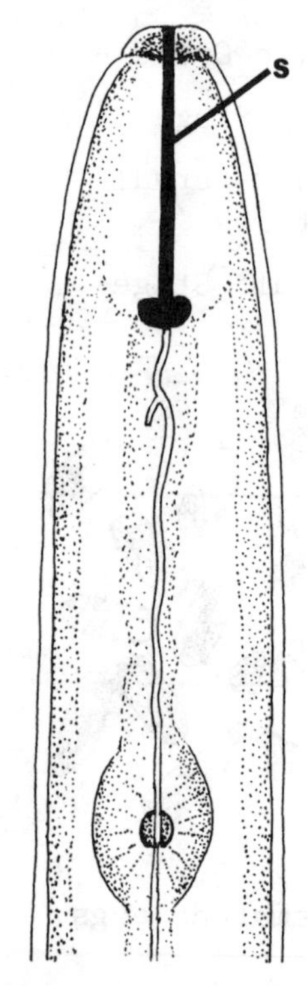

Fig. 6.3
Stylet (s) in
head of nematode.

Some species of plant parasitic nematodes are geographically limited to warm climates; some forms are very specific as to the hosts they will attack, while others have a wide host range. Occasionally a given species of nematode will feed on a plant, but not appear to be able to reproduce. In such a situation, the nematode population does not increase and does little if any damage. Such plants are termed immune to the particular nematode. Other plants often show various degrees of inhibition of reproduction by the nematode and are called resistant. Susceptible plants are those that allow a normal reproduction of the nematode.

The digestive system of nematodes is of considerable interest. All plant parasitic nematodes have a hollow, needle-like spear called a stylet. Behind the stylet is a muscular esophagus that is also associated with several digestive glands. When a nematode feeds, it inserts its stylet into a plant cell and with a sucking action of the

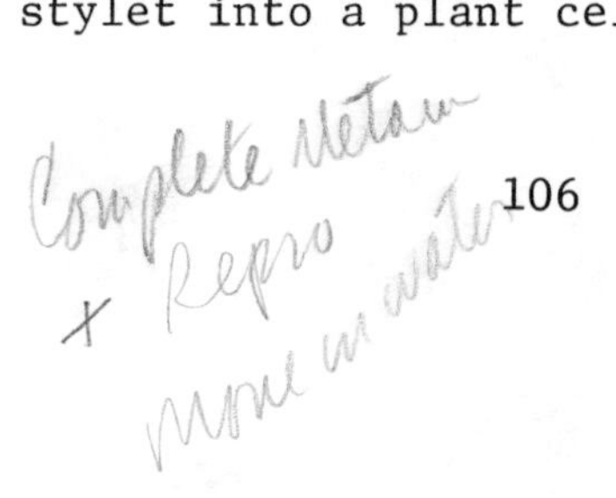

esophagus withdraws plant juices. Generally, a fluid from its digestive glands is secreted through the stylet into the plant cell causing predigestion and liquefaction before ingestion. When the plant juices have been ingested by the nematode, they pass into the intestine where absorption of nutrients occurs.

6-3 PARASITIC HABIT

Ectoparasites And Endoparasites

Plant parasitic nematodes may be conveniently grouped according to their parasitic feeding habits into (1) ectoparasites and (2) endoparasites. Ectoparasitic nematodes feed on the root surface and normally do not enter the root tissue. Endoparasitic nematodes invade, feed, and may or may not become established in the root tissue or permanently attach themselves to it. Some nematodes bury only their heads, or up to the forward third of their bodies, in host roots. These forms are placed in the endoparasites, although some authors consider them as ectoparasites, or place them in a separate group called semi-endoparasites.

Common names have been given to the genera of most plant parasitic nematodes, or in some cases, specific species. However, some genera are without common names.

6-3.1 ECTOPARASITES

Vectors Of Plant Viruses

This ecological group of nematodes has been implicated as agents of plant diseases only since 1951. The common names and genera of some nematodes within this group are: sting nematodes, *Belonolaimus*; pin nematodes, *Paratylenchus*; stubby root nematodes, *Trichodorus*; needle nematodes, *Longidorus*; and dagger nematodes, *Xiphinema*. The last three genera contain the nematodes that have been shown to be vectors of several plant disease viruses.

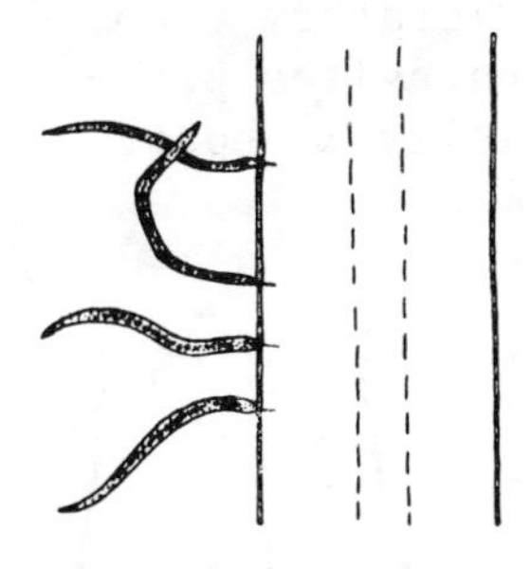

Fig. 6.4 Ectoparasites

These nematodes attack and severely damage many of our cultivated crops by feeding on root hairs and at the outer tips of the roots. They are often found around roots in enormous numbers. Their feeding causes the root tips to stop growing and often promotes the growth of lateral branches giving a witches'-broom effect.

6-3.2 ENDOPARASITES

This group is sub-divided into migratory endoparasites and sedentary endoparasites. Migratory endoparasites migrate within the host and/or between the host and soil. Sedentary endoparasites invade the root or attach themselves to it, but in either case remain sedentary.

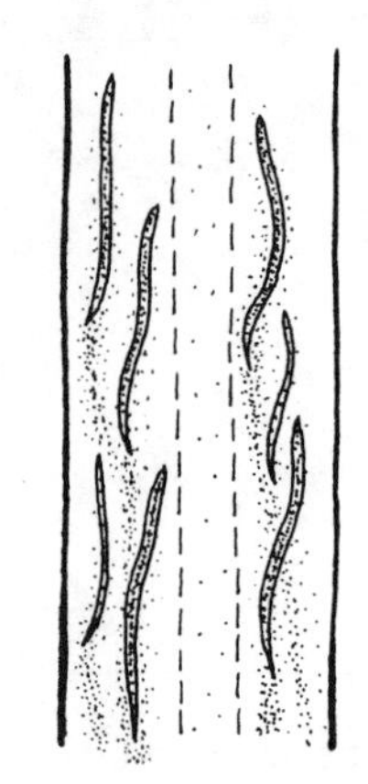

Fig. 6.5
Migratory endoparasites within host.

Migratory endoparasites - two categories of migratory endoparasites may be easily recognized; those that burrow entirely into the host, and those that may burrow partially into the host.

Examples of nematodes that burrow entirely into the host plant are: lesion nematodes, *Pratylenchus*; burrowing nematode, *Radopholus similis*; stem and bulb nematode, *Ditylenchus dipsaci*; and foliar nematodes, *Aphelenchoides*.

The lesion nematodes are common plant parasites. They feed in the cortex of roots and kill the cells on which they feed. Secondary invaders, consisting of both bacteria and fungi, invade the wounded cells and cause a rot and decay that in later stages may girdle the root and sever it. The most widely known disease caused by the burrowing nematode is spreading decline of citrus in Florida. This nematode causes root lesions similar to those infected by the lesion nematode. The stem and bulb nematodes cause a localized type of deformation with considerable twisting and thickening of stems and leaves. Bulbs of affected plants are invaded by bacteria and fungi and become soft and rotted. The foliar nematode, *Aphelenchoides ritzemabosi*, swims up the moist stem of its host, the chrysanthemum plant, and enters the plant through a stomatal opening. It becomes established in the intercellular spaces of leaves, flowers and buds. Under conditions of high moisture, this nematode may also feed as an ectoparasite. Other species of foliar nematodes may feed in host leaf buds and cause small deformed leaves to develop.

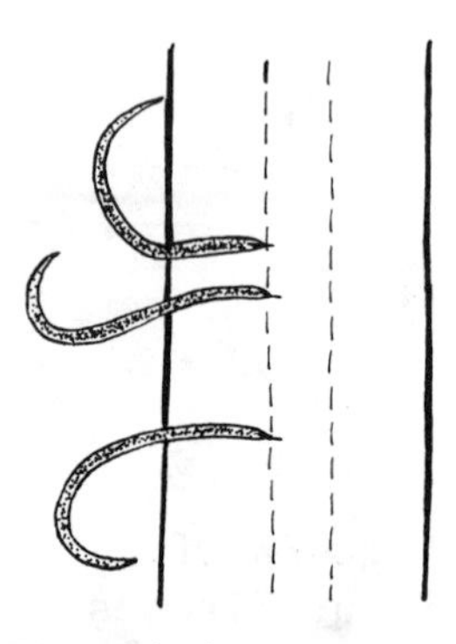

Fig. 6.6
Migratory endoparasites partially in host.

Examples of nematodes that may burrow partially into the host include: lance nematodes, *Hoplolaimus*, and spiral nematodes of the genera *Helicotylenchus* and *Rotylenchus*. They have a variable parasitic habit. Sometimes they are observed feeding as true endoparasites, completely buried in host roots, and at other times, with only the front part of their bodies buried in host roots. They are placed here because of their reported preference to feed on host phloem, which requires at least a partial entry into the host.

Sedentary endoparasites - may be divided into two categories; those that burrow entirely into the host, and those that burrow partially into the host. The most prominent members of sedentary endoparasites that burrow entirely into the host are root knot nematodes, *Meloidogyne*, and cyst nematodes, *Heterodera*. These nematodes invade the root, and typically the adult females swell and become pear- or lemon-shaped, and usually break through the roots to which they remain attached.

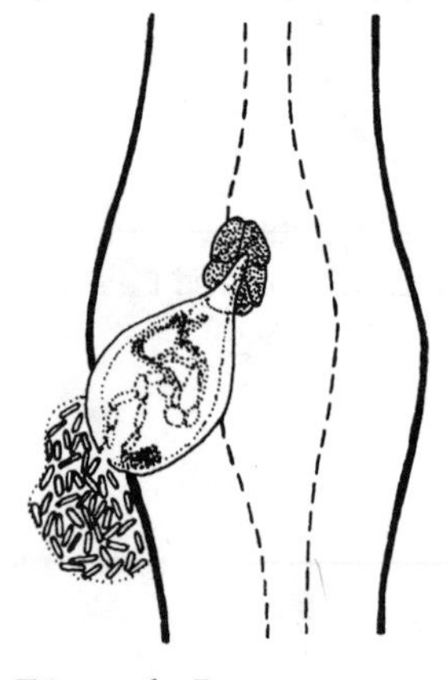

Fig. 6.7
Sedentary endoparasites within host.

Root knot nematodes are common plant parasites and attack over 1700 host plants. They generally invade the host root near its region of cell elongation. These nematodes migrate to the stele and align themselves parallel to it with their heads pointing away from the root tips. The nematodes then become sedentary. They appear to feed on neighboring vascular parenchyma cells, a few of which undergo repeated mitosis without cell division and enlarge to become multinucleate nurse cells. These serve as sources of food for the maturing females. Surrounding parenchyma cells undergo hyperplasia and hypertrophy. The swollen body of the root knot female often ruptures the outer root tissue, and she lays her eggs in a gelatinous mass that clings to the outer side of the root. The adult male is thread-like and may be found in the brownish egg masses. Several nematodes may enter the root in the same area, and their combined effect produces a host response in the form of a root knot or gall. The root knots vary in size according to the size and age of the root on which they are formed. Compound galls of one inch or more in diameter form on larger roots. Generally, various degrees of rotting are associated with the galls. There is a general disruption and blocking of translocated nutrients within the host, and root hairs and small lateral roots quickly die.

Feed On Nurse Cells

Eggs Deposited In Gelatinous Mass

The cyst nematodes include a number of species that have captured many "headlines". The golden nematode, *H. rostochiensis*, is one of these. It is a parasite of potatoes that is difficult to control and was introduced into the United States, and more recently into Canada, from Europe, (see Chapter 20).

Some species of cyst nematodes attack a wide variety of hosts, and other have a narrow host range. The parasitism of the cyst nematodes is similar to the root knot nematodes. However, all or most of the eggs are retained within the enlarged body of the dying female. The cuticle forms a leathery, tough, protective cover that encases the eggs and is termed a cyst. Viable eggs have been hatched from cysts of the golden nematode after remaining in field soil for eight years. Host root diffusates stimulate the hatching of the eggs and act as an attractant for the resulting larvae.

Examples of sedentary endoparasitic nematodes that bury the front part of their bodies into the host include the reniform nematode, *Rotylenchulus reniformis*, and the citrus nematode, *Tylenchulus semipenetrans*. The reniform nematode is so named because the female becomes swollen to a reniform shape, while the male remains thread-like. The posterior portion of the female citrus nematode that lies outside the host also becomes swollen.

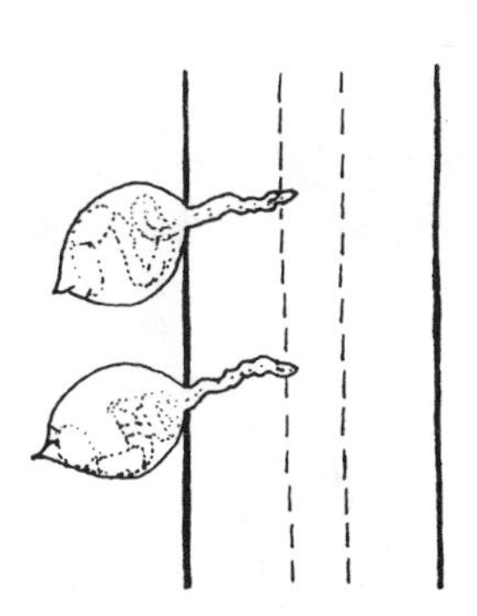

Fig. 6.8 Sedentary endoparasites partially in host.

The citrus nematode causes a disease called "slow decline" in citrus plantings. The terminal buds of host trees die, and the leaves turn chlorotic and are easily wilted when the tree is placed under moisture stress. Twig die-back is common. Both yield and fruit quality is severely affected.

6-4 HOST INJURY AND RESPONSE

Rule -- Nematode Pathogenesis

When nematodes feed upon plants, they inject their stylets into host cells and appear to secrete saliva containing enzymes into the cells. The salivary secretions are highly important in nematode pathogenesis. A nematode feeding upon a host cell generally produces one or more of several types of reaction. The major reactions appear to be: (1) hypertrophy and hyperplasia, (2) dissolution of cell walls, (3) dissolution of middle lamellae, (4) inhibition of growth, (5) necrosis and hypersensitivity, (6) repeated mitoses without cell division.

6-4.1 HYPERTROPHY AND HYPERPLASIA

Hypertrophy Is Common

Hypertrophy is one of the most common host effects. The extent of this reaction is variable, depending upon both nematode and host. A few or many cells may be affected, and they may be slightly or considerable enlarged. Hyperplasia is less common than hypertrophy, although both are probably associated in the formation of most nematode induced galls. These two reactions are clearly evident in nurse cell formation by root knot nematodes.

Host Defense Mechanism

Sometimes cortical cells around root lesions become meristematic and form a suberized periderm around the wound. It may be that this host defense mechanism is a reaction to wounding and is not chemically stimulated by the nematode.

6-4.2 DISSOLUTION OF CELL WALLS

Disintegration of cell walls may take place immediately surrounding the feeding location or extend for a considerable distance into the neighboring tissue. The response is variable, depending upon both nematode and host. This response probably results in the death of the affected host cells. The burrowing nematode may be cited to illustrate disintegration of cell walls. Large areas of host cells around its feeding sites are affected. The cells are destroyed as their walls disintegrate.

Fig. 6.9 Galls on carrot caused by root knot nematodes. *(Courtesy of M. W. Gardner)*

6-4.3 DISSOLUTION OF MIDDLE LAMELLAE

Salivary Pectinase

There is a disagreement as to whether or not the dissolution of middle lamellae actually takes place or not. However, there is some evidence that dissolution of middle lamellae of host cells by a salivary pectinase may be characteristic of some *Ditylenchus* spp. These nematodes penetrate into the intercellular spaces of parenchymatous and cortical tissues of leaves and stems. The middle lamellae of cell walls appear to be dissolved and the cells become separated and assume a spherical shape. Affected cells soon lose their cytoplasmic contents and collapse. Nematodes do not appear able to reproduce in resistant plants that do not undergo cell separation.

Resistant Plants

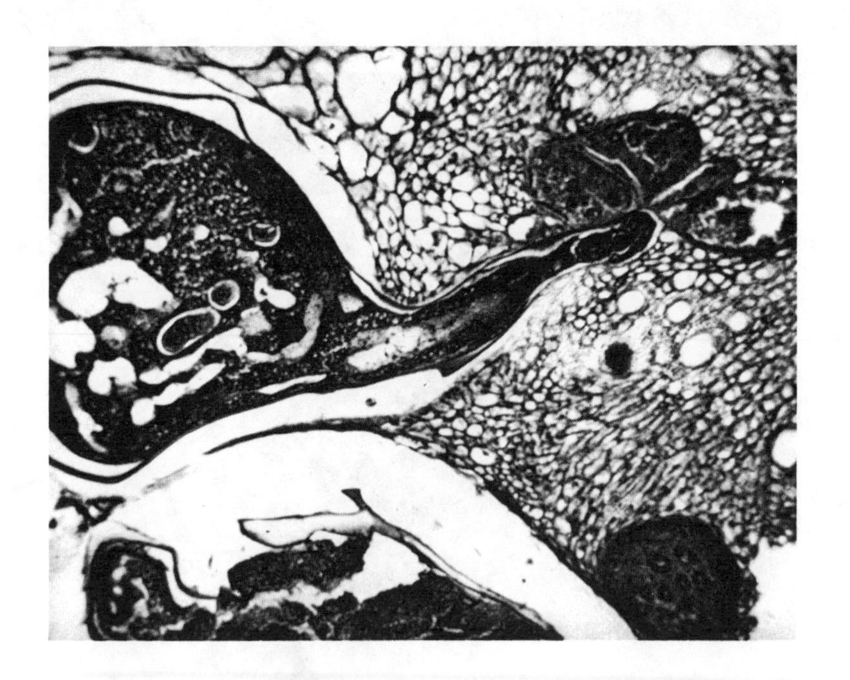

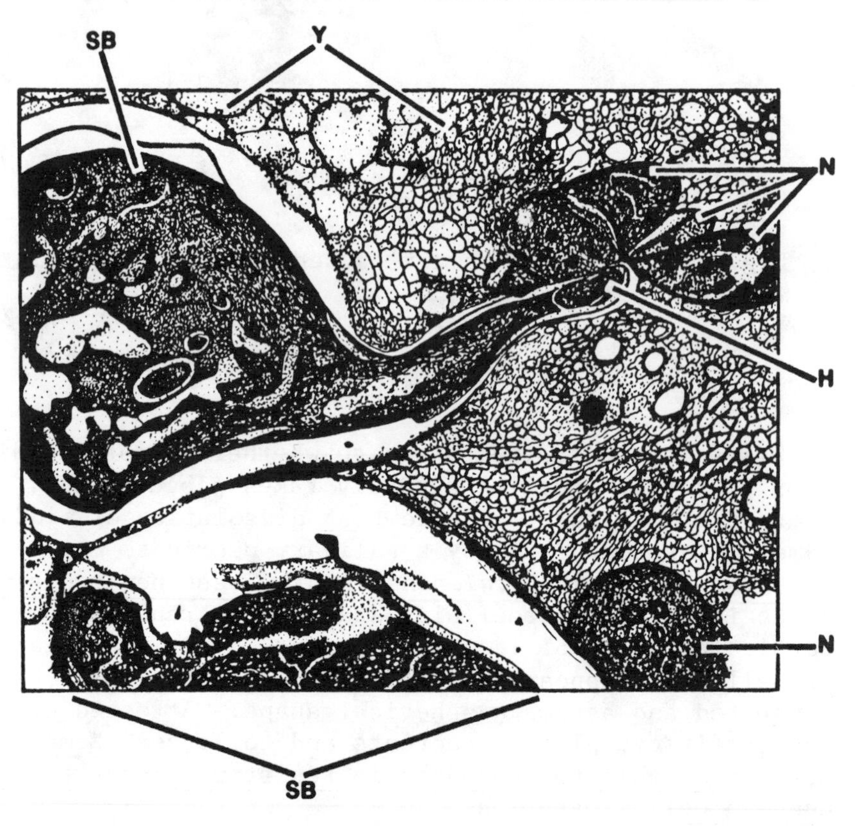

Fig. 6.10 Photomicrograph and drawing of a cross-section through a young tomato root containing adult female root knot nematodes; multinucleate nurse cells (N), nematode head (H), swollen bodies of the nematodes (SB), hypertrophic and hyperplastic plant tissue (Y). *(Photo courtesy of R. J. Green)*

6-4.4 INHIBITION OF GROWTH

The ectoparasitic *Trichodorus* spp, may feed on root tips, causing the apical meristem to stop dividing. The cessation of growth of a single root tip is not serious, but when large populations of nematodes are present around the root system they may cause the cessation of growth of many root tips and quickly stunt the plant. Other nematodes, such as the root knot nematodes that invade root tips may also cause them to stop growing.

6-4.5 NECROSIS AND HYPERSENSITIVITY

Many lesions caused by nematodes become necrotic. Often necrosis is enhanced by the presence of other organisms. However, the feeding of some nematodes alone brings a direct response by the host. This results in a necrosis of host tissue and varying degrees of toxicity to the nematode.

The lesion nematode, *Pratylenchus* spp., migrates along the cortical tissue of roots causing necrotic lesions that may result in a sloughing of the cortex. There is evidence to indicate that brown discoloration in many such necrotic lesions is caused by an accumulation of phenolic compounds. In some cases, polyphenoloxidase is released which then oxidizes polyphenolic compounds into polymerized substances, that appear to be the toxic materials.

Some varieties of plants may support large numbers of lesion nematodes without forming brown necrotic areas. These plants are little affected by the nematodes and are termed tolerant.

Tolerant Plants

Tissue discoloration has also been attributed to hydrogen cyanide released by enzymatic hydrolysis of the glycoside amygdalin. The enzymes involved in this reaction probably originate from both the nematode and the injured host cells.

Hypersensitivity of host cells has been observed in a number of hosts. Apparently this is also related to the ability of the host plant to produce phenolic compounds following nematode infection. Essentially this reaction causes a rapid death of cells near the invading nematodes.

When it is associated with a host response, necrotic browning, whether it is hypersensitive or not, may probably be considered as a mechanism of resistance to nematodes.

Resistant Mechanisms

6-4.6 REPEATED MITOSIS WITHOUT CELL DIVISION

This reaction occurs in the formation of the enlarged nurse cells characteristic of infection by root knot nematodes. Huang and Maggenti, 1969, found that nurse cells are formed by repeated mitosis and enlargement without cell division of several parenchyma cells near the head of second-stage larvae. The initial changes in the parenchyma cells appear about 1 week after nematode entry into the root. These workers discount the widely held earlier belief that nurse cells are formed by the dissolution of adjoining cell walls of numerous cells and the subsequent fusion of cell contents.

6-5 DISEASE SYMPTOMS

Symptoms Seldom Specific

Symptoms of nematode damage are seldom specific, and the general decline of attacked plants often goes unrecognized, especially in the early stages of infection. In the temperate zone this decline may progress over a span of several growing seasons before it is recognized. Once the decline is realized, the diagnosis of the causal agent as to the specific nematode is generally accomplished only in the laboratory. However, general disease symptoms do serve as a preliminary guide in establishing the whereabouts of the pathogen on or in the plant, and often as to what general type of pathogen is involved. Except for the foliar, seed and leaf gall, and stem and bulb nematodes, superficial examination of the foliage is seldom sufficient to determine the causal agent; this is because the root system and underground stems are the organs attacked. Therefore, it is best to divide the discussion into above ground symptoms and below ground symptoms.

General Decline

6-5.1 ABOVE GROUND SYMPTOMS

Rule -- Stunting Caused By Nematodes

A stunting of the top growth is the most common above ground symptom of nematode damage. This causes a reduction in yield and often quality. In the field, the first evidence of nematode attack is a slight stunting in one or more irregular areas of the field. The stunting becomes more obvious as the season progresses. If susceptible crops are replanted in successive years the damage becomes increasingly severe. Since the nematodes are seldom evenly distributed and the decline is most often associated with large numbers of the pathogen, the affected plants are generally in spotty patches throughout the field.

Poor growth may be caused by nematodes in which no above ground symptoms are produced. Invariably, reduced growth means reduced yield and often reduced quality. These losses may be economically important even when the growth reduction is barely noticeable.

Reduced Yield And Quality

Abnormal coloration of the foliage characteristic of a mineral deficiency, often results from nematode infection. The foliage may be a deeper or lighter shade of green, or various shades of red, purple or bronze. A die-back may be present in plantings of citrus, apple, peach, etc., and is often in association with foliar discoloration. The nutritional deficiency is a direct result of root injury.

Color Change

Wilting of plants that have an injured root system often occurs under conditions of a moisture stress. At first the wilting is temporary, generally occurring during the heat-of-the-day and recovering in the cool-of-the-evening. If the root injury is severe, the infected plants may be killed.

Wilting

Abnormal growth of foliage is mainly due to the feeding of the stem and bulb, foliar, and seed and leaf gall nematodes. The distortion may be in the form of twisting, swelling, gall formation, or irregular shaped leaves. Galls may also be formed in the floral parts of some grass species.

Abnormal Foliage Growth

6-5.2 BELOW GROUND SYMPTOMS

A reduced and necrotic root system is the most common below ground symptom of attack by nematodes that feed in or on the root system. The amount of root reduction is variable and the lesions may be in all stages of decay. It is not uncommon to find a lack of fine lateral roots and almost total lack of root hairs. Extensive rotting is the result of other pathogens that have invaded the roots and contributed to the overall root decay.

Rule -- Root Destruction By Nematodes

Other Agents Involved In Decay

A reduced root system without necrosis may occur from the feeding injury by some ectoparasitic nematodes such as *Trichodorus* spp. Sometimes these nematodes appear to inhibit root growth without causing any other symptoms.

Abnormal growth is sometimes found on nematode infected plants that have few obvious lesions and no general root decay. This is often a root proliferation that may appear like a witches'-broom. It is also common to find plants that normally produce a tap-root system to have a sprangled root system. This occurs when the meristematic tip of the tap root is killed or inhibited and lateral roots are formed. Such vegetable crops as table beets or carrots may have a fairly normal appearing foliar system, yet have a sprangled root system which makes then totally unsalable.

Abnormal Root Growth

Galls

Galls may be formed on the roots of most crop plants. The galls may be few or numerous, large or small. They may be rotted or intact. The most common gall is produced by root knot nematodes, *Meloidogyne* spp.

Fig. 6.11 Galls on roots of an older tomato plant caused by root knot nematodes. Notice the lack of secondary roots especially on lower portion of root system. *(Courtesy of R. W. Samson)*

6-6 DISEASE COMPLEXES INVOLVING NEMATODES

Definition

A disease complex is a disease that is caused by the interaction of two or more species or types of pathogens. Nematodes have been implicated in a number of disease complexes with various soil-borne bacteria, fungi, viruses and other nematodes. This discussion will center upon disease complexes involving nematodes plus bacteria, and nematodes plus fungi since they comprise the bulk of diseases with complicated interactions.

Plant parasitic nematodes are principally soil-borne pests, most of which attack plant parts in the soil. The soil is a vast, diverse, media containing enormous numbers of microscopic organisms. On this basis, it should not be surprising that plant infections, especially those occurring in the soil, are often complexes involving two, three or more organisms. In fact, Wallace (1971) states: "Disease in a plant is the result of numerous interacting antecedents (or causes and determinants) and although the influence of the nematode may sometimes overshadow the other factors contributing to disease, it is unlikely in my opinion ever to be the sole cause." The major theme may be restated as: nematodes involved in plant disease are unlikely ever to be the sole cause.

Rule -- Nematodes Unlikely To Be Sole Causal Agent Of Disease

6-6.1 NEMATODES PLUS BACTERIA

Hunt, et al., (1971) reported on the effects of root knot nematodes on bacterial wilt of alfalfa. They found a close relationship between the incidence of bacterial wilt and the presence of the root knot nematode. They indicated the root knot nematode may hasten the colonization of alfalfa plants by the bacterial wilt organism *Corynebacterium insidiosum*. The effect of the root knot nematode plus the bacterium appeared to be additive, not synergistic (greater than additive effect). Both bacterial wilt of alfalfa and root knot are major diseases of alfalfa with similar geographic distribution, so the complex caused by both organisms may be a common phenomena although no data is available on the occurrence of this disease complex.

Wheat spike blight is a disease complex caused by the seed and leaf gall nematode, *Anguina tritici*, and the bacterium *Corynebacterium tritici*. In this disease, the nematode acts as a vector. Bacterial cells cling to the body of the nematode and are inadvertently carried into the host by the invading nematode.

A complicated relationship exists between the foliar nematode *Aphelenchoides ritzemabosi* and the bacterium *Corynebacterium fascians*, in "cauliflower" disease of strawberry. The nematode acts as a vector, but the disease symptoms are different than when either pathogen is present alone. Alone, the nematode causes a rosette type of plant growth and numerous leaf scars as a result of its feeding within the leaf bud. Alone, the bacterium causes gall formation and stunted growth. Together, the pathogens cause the development of numerous axillary buds, extreme stunting and short and swollen petioles (Pitcher, 1963).

One of these examples is a disease complex whose interaction between the nematode and bacterium is rather simple in that the symptoms appear additive. Another example is that of a complicated interaction where the final symptoms are different and more severe than a simple additive effect of the pathogens acting alone.

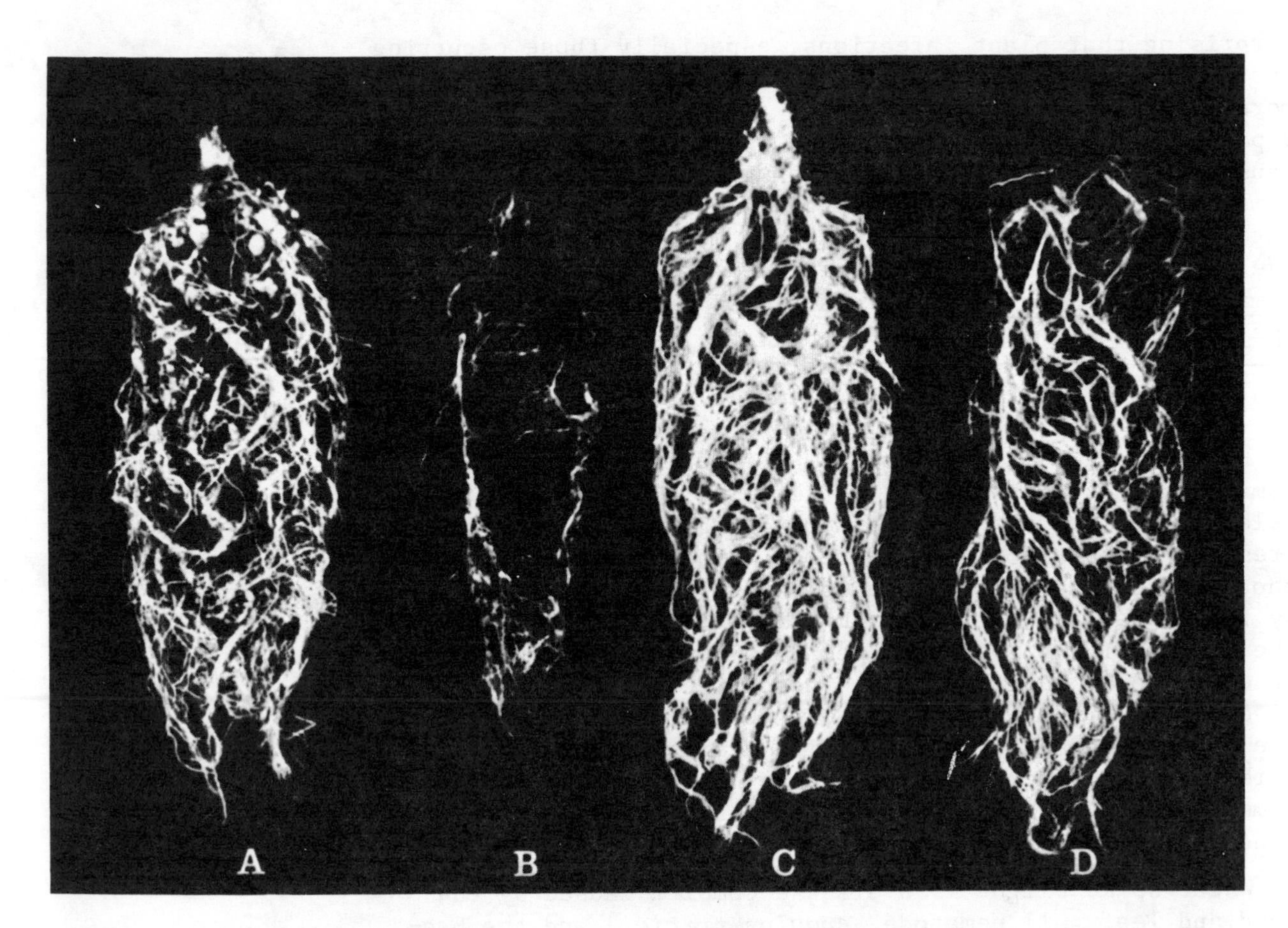

Fig. 6.12 Twelve-week-old tomato roots showing: A. plant grown in sterile soil (240 ml.) plus 6,000 larvae of *Meloidogyne incognita*, that are free of microorganisms; B. plant grown in non-sterile soil plus 6,000 root knot larvae; C. plant grown in sterile soil without root knot nematodes; D. plant grown in non-sterile soil without root knot nematodes.
(Photo courtesy of P. S. Mayol and G. B. Bergeson)

Mayol and Bergeson (1970) working with tomato plants, found great root deterioration caused mainly by a complex whose dominant members were root knot nematodes and various bacteria; less numerous were the fungi *Trichoderma* sp., *Fusarium* sp. and *Rhizoctonia solani*. The number of root knot larvae, 6,000 per plant, were considered to be a moderate inoculum level. The root destruction was obviously synergistic (see Fig. 6.12). This disease complex is caused by a known pathogenic nematode and a mixture of soil-borne bacteria, most of which are only weak parasites at best, and many of which are probably saprophytes living on the necrotic tissue. The role of the four fungi in this complex is not clear. Since they were not dominant members, they may not have been important.

6-6.2 NEMATODES PLUS VASCULAR WILT FUNGI

A great amount of documentation is present in the literature on root knot - *Fusarium* complexes involving numerous hosts. It is well reviewed by Powell, 1971. These complexes involve various species of the nematode and the fungus. Hosts involved include crops such as cotton, peas, tomato and tobacco. Interactions between the two types of pathogens accelerate the disease process, resulting in increased wilt severity. Furthermore, host plants resistant to *Fusarium* are predisposed by the action of the root knot nematodes and are then readily invaded by the fusarium wilt fungus. In several studies thus far reported, host plants are maximally predisposed to the fusarium wilt fungi only after the root knot nematodes are in contact with the host for 3 to 4 weeks before attack by the *Fusarium* (Porter and Powell, 1967).

Root Knot Nematodes Break Resistance To *Fusarium*

The root knot nematode is not the only sedentary endoparasite shown to be a major part of disease complexes. The soybean cyst nematode, *Heterodera glycines*, together with the fusarium wilt fungus cause a severe wilting of soybean. The cyst nematodes also caused the soybeans to become predisposed to *Fusarium*.

A wilt disease complex of tomato and eggplant has been reported to be caused by the migratory ectoparasitic lesion nematode, *Pratylenchus*, and the soil-borne wilt fungus, *Verticillium dahliae*. Other species of the lesion nematode and verticillium wilt fungi have been reported to cause wilt disease complexes on peppermint, cotton, eggplant, pepper, tomato and potato. In most cases, the disease complex was more damaging than the additive effects of the component pathogens.

A few other migratory ectoparasitic nematodes have been implicated in wilt disease complexes, but most of the literature to date points the accusing finger at the sedentary endoparasites such as the root knot and cyst nematodes.

The interactions between nematodes and fungi on wilt diseases are highly involved and much current research is pointed towards explaining them. The study by Faulkner et al., (1970) is pertinant here. These authors worked with the disease complex involving *Verticillium dahliae* and the lesion nematode, *Pratylenchus minyus* on peppermint. Plant roots were separated so that one-half of the roots were placed in a pot inoculated with *V. dahliae* or *P. minyus*, or both. The lesion nematodes feeding on the roots in one pot predisposed the roots in the other pot to attack by *V. dahliae*. Furthermore, the severity of the wilt was also increased.

These studies indicate that disease complexes between nematodes and systemic wilt fungi result in severe plant damage whereby the nematodes cause a predisposition of the host to the fungi pathogen.

Rule -- Nematodes Alter Outcome Of Other Diseases

The discussion above validates a verbal statement made to the author by N. T. Powell: "Nematodes often alter the epidemiology of other plant diseases."

6-6.3 NEMATODES PLUS ROOT ROT AND SEEDLING DISEASE FUNGI

Both sedentary endoparasitic nematodes and migratory ectoparasites have been cited as being involved as members of disease complexes with root rots and seedling disease fungi. Fungi involved here include the tobacco black shank fungus, *Phytophthora parasitica* var. *nicotianae*, and fungi such as *Aspergillus flavus*, which attacks peanuts in the soil; *Rhizoctonia solani*, which is well known to attack roots and lower stems of many plants; and *Pythium ultimum*, a fungus of wet soils widely recognized as causing damping-off of many kinds of seedlings.

One of the most fascinating studies is that of Powell, et al., 1971, who worked with disease complexes in tobacco involving the root knot nematode *Meloidogyne incognita* and several soil-borne fungi. They reported a severe root necrosis on root knot susceptible tobacco when the tobacco was inoculated with the nematodes several weeks before introducing the fungi. Furthermore, none of the fungi induced disease unless *M. incognita* was present. Also of interest was the fact that four of the fungi used in this study were not regarded as pathogens on tobacco. The activity of the root knot nematodes prior to the introduction of the fungi had predisposed the tobacco plants to fungi not normally considered as pathogenic.

6-6.4 DISEASE COMPLEXES -- EFFECTS OF COMPONENTS UPON EACH OTHER

Reduction Of Sedentary Endoparasites

The fungal component of a disease complex often has a considerable effect on the nematode population. A pronounced effect is brought upon the sedentary endoparasitic nematodes; namely, the root knot (*Meloidogyne* spp.) and the cyst-formers (*Heterodera* spp.). This may well be expected since these nematodes are stationary in their feeding and the early death of surrounding host cells result in their death also. Furthermore, the fungal component of the complex often quickly invades the nurse or giant cells upon which these nematodes are dependent. When the nurse cells die, the nematodes do likewise. Therefore, the population of sedentary endoparasites in disease complexes often becomes depressed. In such circumstances, the nematodes which were the dominant pathogens in the initial stages of the disease complex give way to the fungus which becomes dominant in the latter stages of the disease.

When migratory ectoparasites are a part of the disease complex, their population is often increased as a result of interactions with the fungal components. The reasons for the population increase of migratory ectoparasites are unknown. Perhaps the invading fungi alter the roots to make them more attractive and/or susceptible to the nematodes.

Increase Of Migratory Ectoparasites

A word of caution is necessary at this point; there are some notable exceptions to the generalities just presented. Also, there are cases in which the nematode populations do not appear affected by their fungal partners in the disease complex.

Nematodes aid other pathogens in entering the host by providing wounds through which they may enter. Sometimes, nematodes act as vectors and carry or deposit other pathogens (especially certain bacteria and viruses) in the host.

Nematodes, such as the sedentary root knot, often cause a long split-like wound where the swollen body of the mature female ruptures the cortex. This wound is many times larger than the wound caused by a second stage larva as it penetrates the root. Such a large wound is an open avenue allowing the easy entry by both pathogenic and non-pathogenic organisms alike.

We do not understand the nature of the predisposition caused by nematodes, but it is a profound physiological change. Furthermore, it is unlikely that the predisposition plus the wounds on the roots account for the total interaction between the components of many disease complexes. Suffice it to say: plant parasitic nematodes often provide soil-borne fungi with greater number of suitable hosts than they otherwise would have and therefore often have a positive effect upon populations of fungal components of disease complexes.

Root Knot Caused Predisposition Is Profound Physiological Change

Powell (1971), cautions against the tendency to classify the fungal components of these disease complexes as "secondary invaders". This is apt to happen because nematode infection appears to be required prior to invasion by the fungal component. He states a more appropriate term would be "secondary pathogen", which recognizes the fungus follows a primary pathogen yet also acknowledges the pathogenicity of the fungus.

Secondary Pathogen

Selected References

Christie, J. R. 1959. Plant nematodes their bionomics and control. H. & H. B. Drew Co., Jacksonville.

__________, and V. G. Perry. 1959. Mechanism of nematode injury to plants. pp. 419-426, In C. S. Holton, et al., (ed.), Plant pathology problems and progress. 1908-1958. The University of Wisconsin Press, Madison.

Fielding, M. J. 1959. Nematodes in plant disease. Annual Review of Microbiology. 13:239-254.

Jenkins. W. R., and D. P. Taylor. 1967. Plant nematology. Reinhold Publishing Corporation, New York.

Jones, F. G. W. 1959. Ecological relationships of nematodes. pp. 395-411, In C. S. Holton, et al., (ed.), Plant pathology problems and progress. 1908-1958. The University of Wisconsin Press, Madison.

__________, and M. G. Jones. 1964. Pests of field crops. Edward Arnold (Publishers) Ltd., London. pp. 220-273.

Krusberg, L. R. 1963. Host response to nematode infection. Annual Review of Phytopathology. 1:219-240.

Pitcher, R. S. 1963. Role of plant-parasitic nematodes in bacterial diseases. Phytopathology. 53:35-39.

Powell, N. T. 1963. The role of plant-parasitic nematodes in fungus diseases. Phytopathology. 53:28-34.

__________ 1971. Interactions between nematodes and fungi in disease complexes. Annual Review of Phytopathology. 9:253-274.

Sasser, J. N., and W. R. Jenkins, (ed.). 1960. Nematology fundamentals and recent advances with emphasis on plant parasitic and soil forms. The University of North Carolina Press, Chapel Hill.

Spears, J. F. 1968. The golden nematode handbook. U. S. Dept. Agr., Agr. Handbook No 353.

Steinhorst, J. W. 1961. Plant-nematode inter-relationships. Annual Review of Microbiology. 15:177-196.

Taylor, A. L. 1953. The tiny but destructive nematodes. pp. 78-82, In Plant diseases. U. S. Dept. Agr. Yearbook.

Thorne, G. 1961. Principles of nematology. McGraw-Hill Book Co., New York.

Van Gundy, S. D. 1965. Factors in survival of nematodes. Annual Review of Phytopathology. 3:43-68.

Wallace, H. R. 1963. The biology of plant parasitic nematodes. E. Arnold, London.

CHAPTER 7

PARASITIC SEED PLANTS AS CAUSAL AGENTS OF PLANT DISEASE

7-1 INTRODUCTION

Parasitism Variable

We usually think of parasites of plants as being viruses or microorganisms and not higher plant forms. Actually, there are relatively few plant diseases caused by parasitic seed plants, and the degree of parasitism is variable. It is thought that parasitism by seed plants is the result of an evolutionary degeneration of plant forms that were once free-living. Three general levels of such parasitism may be recognized:

(1) The epiphytes, as represented for instance by Spanish-moss, are a group of plants which are physiologically independent, requiring only support and protection from the host. Since these plants cause little damage, they will not be discussed further.

(2) The hemi-parasites, or water-parasites, are plants which are dependent upon their hosts for moisture and minerals; but since they contain chlorophyll in their leaves, they are capable of manufacturing their own food from carbon dioxide and water.

(3) The true parasites have no chlorophyll and therefore are wholly dependent upon their hosts for nourishment. These plants are never green.

7-2 HEMI-PARASITES

7-2.1 WITCHWEED

Pest Introduced

Witchweed, *Striga asiatica*, is a parasitic seed plant that was discovered in the Carolinas in 1956. It was probably introduced into this country from Africa, Australia, or Asia where it has been a serious pest for many years. Witchweed is capable of inflicting severe damage on corn, sorghum, sugarcane, many grasses and sedges, and numerous broadleaf plants. Thus far, severe damage from this parasite has been limited to corn in the Carolinas where many fields have been complete failures.

The witchweed roots become attached to the host roots and drastically reduce the ability of the host to take up water and minerals for its own use. The host plants become stunted, may wilt or turn yellowish, and sometimes die.

The witchweed plant is 6 to 12 inches high and has bright green leaves on a branched stem. Many small red to yellowish flowers are formed; these produce large numbers of seed. The seed is disseminated by wind and water, and falls to the ground where it may remain viable for up to 20 years before germinating. The seed will normally germinate only in the presence of exudates secreted by roots of host plants. Once the seed germinates, its young root grows towards and contacts the host root. It penetrates into and becomes established in the host root where it absorbs nourishment. The young witchweed plant sends up aerial shoots about four weeks after germination.

Seed Needs Host Root Exudate To Germinate

After emergence, the plant turns green and manufactures its own food, but continues to depend partially on its host for water and minerals. Thus, witchweed is transformed from a parasite to a hemi-parasite.

From Parasite To Hemi-parasite

The best menas of control is to plant a crop such as sudangrass which stimulates the witchweed seed to germinate and grow; the crop is them plowed under before the witchweed plants can produce seed. A strict crop rotation program should be followed, and the production of susceptible crops should be avoided.

Control

7-2.2 TRUE MISTLETOE

The true mistletoes found in North America belong to the genus *Phoradendron*. These pests attack broadleaved trees and some conifers and stimulate the growth of abnormal and excessive branching, swellings, and occasionally "witches'-brooms". The true mistletoes generally have leathery yellow to dark green leaves, but occasionally are without leaves. The plants produce berries that contain a sticky mucilaginous material. The berries are disseminated by birds and wind to host trees where they adhere to the limbs and twigs with which they come in contact. The seed germinates on limbs of susceptible host trees, then forms an attachment disc on the bark, and sends a tiny haustorial strand through a lenticel or axillary bud. The haustorium enlarges, and branches develop that extend into the host phloem. These branches are called sinkers. Generally, just the young host branches or twigs are invaded. The sinkers invade the rays and advance radially with the cambium each year. The aerial shoots of the mistletoe that form, do so from the attachment disc, and the stems are jointed and slow growing. Control of this pest is achieved by pruning well below the point of infection or by sacrificing the tree.

Seed Sticky

Young Host Limbs Invaded

Control

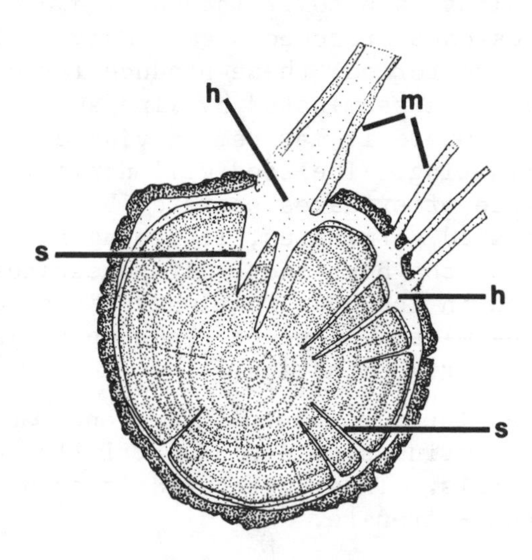

Fig. 7.1 Schematic drawing of a cross-section through a branch infected with mistletoe. The haustorium (h) grows through the cork cambium. Sinkers (s) grow in radial rays, and aerial stems (m) of the mistletoe grow out from the haustoria.

7-3 TRUE PARASITES

7-3.1 DWARF MISTLETOE

Damage Western Conifers

In North America, the dwarf mistletoes are found on many native evergreen trees, with the notable exception of junipers. They cause extensive damage to western coniferous forests by reducing the vigor and growth rate of infected trees and lowering the quality of the timber. Susceptible trees may be infected at any age, with extensive damage and death occurring in young trees. Infected trees develop fusiform swellings on branches and trunks and extensive witches'-brooms.

Aerial Shoots Develop Slowly

The dwarf mistletoes belong to the genus *Arcueuthobium*. They are small plants ranging from one-half inch to occasionally twelve inches. New infections result from seed germinating on limbs of host plants. An absorbing root penetrates the tender bark and sends out a system of sinkers that extend into the phloem of the host. Some of the strands penetrate into the cambium layer of the host and thereby retain the connection with the new phloem as it is laid down year after year. When the system of sinkers is well established, the plant sends out aerial shoots. The shoots may develop the year after infection occurs, or several years later. Flowers are borne on the shoots, and seeds are borne in berries that have a viscus, sticky pulp. Mature berries open and forcibly eject the sticky seeds a short distance. The seeds are carried by air currents to branches or twigs of the same tree, or neighboring trees, where they adhere to the bark. In forest stands, the spread from tree to tree is predominantly downward. Birds also feed upon the berries, and together with other small animals, they probably help in disseminating the seed.

Control

The control of this pest is entirely by removal of infected trees. Since points of infection are not detectable until the shoots are formed, at least two separate cuttings are necessary for eradication.

7-3.2 DODDER

Six Damaging Species

About 44 species of dodder or love vine, belonging to the genus *Cuscuta*, are found in the United States. Of these, only six species cause the most damage. All species of dodder look very much alike and usually must be distinguished in the laboratory.

The major crop loss by dodder is in alfalfa, lespedeza, and clovers, with minor damage to flax, sugar beets, onion and other vegetables, and ornamentals. Some dodders attack only a few different hosts, while others have a wide host range.

Food Absorbed Through Haustoria

A dodder seed germinates in the soil and forms a thin and fragile root, and a 4-5 inch long aerial stem that waves around and around until it contacts a host plant. The stem then attaches to the host plant by coiling around it. Haustoria are sent into the host plant and penetrate the food-conducting tissues of the host. The dodder draws elaborated food through these haustoria for its own growth. The haustoria secrete diastase, which dissolves starchy substances in the host. After the dodder unites with the host, the base section of the dodder stem shrivels, dries, and breaks away from its contact with the ground. The dodder sends out branches that in turn wave about until they contact and attach themselves to a neighboring plant. This cycle is repeated over and over. Tiny flowers are produced in clusters and their seeds fall to the ground. These seeds have been known to remain dormant in the soil for up to five years before germinating.

Lower Yield, Loss of Quality

The economic loss from dodder is due to a reduction of yield, loss of quality, cost of cleaning seed, and interference with machine harvesting. Dodder seed is often harvested with, and contaminates the crop seed. It is costly, often difficult, and sometimes prohibitively expensive to separate dodder seed from crop seed. Because of these factors, dodder has been the target of more legislation than any other weed. The percentage of dodder seed as a contaminant in crop seed is strictly controlled; in some states, all crop seed sold must be completely free of dodder seed.

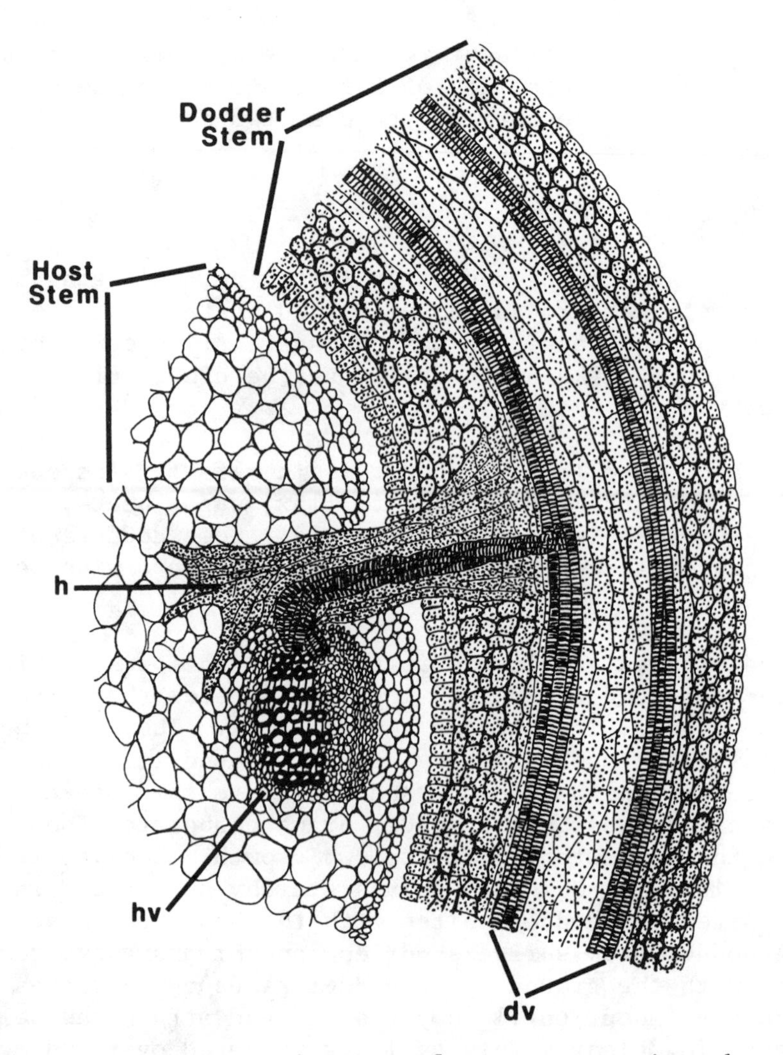

Fig. 7.2 Schematic drawing of a cross-section through a plant stem and a longitudinal-section through a dodder stem. A haustorium (h) is embedded in the host, and the vascular system of the dodder (dv) is connected to the host vascular bundle (hv).

Avoid Introduction Of Dodder

The best control is to avoid introducing dodder in the first place. This is best accomplished by sowing only certified seed, or seed that is known to be free of this pest. Machinery should be cleaned thoroughly after use in infested fields. All hay containing dodder seed should be burned and not used for feed. Manure suspected of containing dodder seed should not be spread on dodder-free land. Dodder growing along fences, ditches, and roadways, should be destroyed by burning or with the use of herbicides. Scattered patches of dodder in fields should be destroyed by cutting the host below the point of dodder attachment, by use of herbicides, or by burning. It is important to kill the dodder plants before they go to seed. Dodder seed that remains on the soil surface after harvest should be destroyed by burning.

Destroy Patches Of Dodder

7-3.3 BROOMRAPES

The broomrapes are all true parasites. They may appear as clumps of whitish, yellowish, brownish or purplish stems rising annually from the soil at the base of their host plants. The stems are 6 to 18 inches high and have bract-like leaves with showy flowers resembling those of a snapdragon. Numerous minute seeds are produced that are carried about by the wind and rain. The seeds germinate in the soil if host roots are present; otherwise, they may remain dormant for up to 13 years. After germination, a root grows to a host root tissue. New roots and a stem of the parasite grow to and fuse with other host roots, forming a nodule which sends out a stem and new roots.

Flowers Resemble Those Of Snapdragon

The damage caused by broomrapes in the United States is not great, and the only crop that suffers considerable injury from them is hemp. However, the crop loss in Europe by attack from broomrapes is extensive and indicative of the potential of these parasites.

Selected References

Anonymous. 1957. Watch out for witchweed. PA-331. U. S. Dept. of Agr.

Boyce, J. S. 1961. Forest pathology. Ed. 3. McGraw-Hill Book Co., New York. pp. 320-343.

Dowson, J. H., et al. 1965. Controlling dodder in alfalfa. Farmers' Bulletin No. 2211. U. S. Dept. of Agr.

Gill, L. S. 1953. Broomrapes, dodders, and mistletoes. pp. 73-77, In Plant diseases. U. S. Dept. of Agr. Yearbook.

__________ , and F. G. Hawksworthy. 1961. The mistletoes, a literature review. Technical Bulletin No. 1242. U. S. Dept. of Agr.

Hawksworth, F. G., and D. Wiens. 1970. Biology and taxonomy of the dwarf mistletoes. Annual Review of Phytopathology. 8:187-208.

__________, and __________. 1972. Biology and classification of dwarf mistletoes (Arceuthobium). Agriculture Handbook No. 401, Forest Service, U. S. Dept. Agr., Washington, D. C.

Heald, F. D. 1933. Manual of plant diseases. Ed. 2. McGraw-Hill Book Co., New York. pp. 858-879.

Lee, W. O., and F. L. Timmons. 1958. Dodder and its control. Farmer's Bulletin No. 2117. U. S. Dept. of Agr.

Walker, J. C. 1969. Plant pathology. Ed. 3 McGraw-Hill Book Co., New York. pp. 525-532.

Wellman, F. L. 1964. Parasitism among neotropical phanerogams. Annual Review of Phytopathology. 2:43-68.

CHAPTER 8

PART I: NONINFECTIOUS PLANT DISEASE

8-1 INTRODUCTION

Adverse weather, lack of nutrients, fertilizer imbalance, air pollution, and chemical residues all take their annual toll of economic plants. Their importance in the total disease picture is all too often overlooked.

The disease classification presented in Chapter 2, cited ten headings under which causal agents of noninfectious plant diseases are grouped. This section is concerned with diseases caused by the first nine of these causal agents:

Causal Factors

1. Low temperatures (plant injury)
2. High temperatures
3. Unfavorable oxygen relations
4. Unfavorable soil-moisture relations
5. Lightning, hail, and wind (plant injury)
6. Mineral excesses
7. Mineral deficiencies
8. Naturally occurring toxic chemicals
9. Pesticide chemicals

8-2 LOW TEMPERATURES

Disorders by freezing are usually thought of as "injury" instead of disease. Injury to perennial plants by low temperatures may occur the year around; however, most cultivated crops are only subject to such damage during the growing season.

Freezing injury is common on many fruit and vegetable crops. The injury due to freezing is caused by ice crystals that form either intracellularly, intercellularly or both. On trees, injury occurs on the side of the trunk exposed to the sun, where the tissues may be killed to the sapwood by freezing. This is the result of rapid and extreme temperature fluctuation from day to nighttime. The injury is most common in late winter when the radiation of the sun causes an increase in temperature of the bark exposed towards it. Then subsequent and sudden freezing temperatures incurred at nightfall may kill the inner tissues. This injury later appears typically, as long cracks in the trunk.

Winter injury to ornamental evergreen trees and shrubs can occur when the soil freezes producing a frost layer down through part or most of the root zone. This prohibits roots from absorbing sufficient moisture to meet the needs of the foliage that is still undergoing tanspiration.

One of the most common forms of low temperature injury is that caused by frost. Fruit crops suffer damage by frost when they are in flower or in fruit. Severe frosts will kill the flowers or fruits. Frosts that occur when the fruit is setting, or soon after, may cause glaring russeting on the apple and pear fruits. Often, the russeting occurs in bands. Sometimes, the entire surface is affected, and cracks may develop as the fruit expands. Occasionally, young leaves may be killed by frost or may be distorted with crinkling and blistering.

Fig.8.1A Russeting of apple skin caused by frost.

Fig.8.1B Internal necrosis of apple fruit caused by frost.

(Courtesy of the Department of Botany and Plant Pathology, Purdue University)

It has long been known that many plants can withstand cooler temperatures than others, and that many varieties of a single species of plant also show this characteristic. Most crucifers grow well in cool climates; corn, cucumbers, and melons demand a warmer environment.

Frost Resistance

Many plants exhibit frost resistance, or hardiness. This occurs when plants repeatedly exposed to temperatures at or just above the freezing point, become conditioned so as to be able to withstand a lower temperature than they originally could. This process appears to be that of a gradual dehydration of the protoplasm which results in a higher concentration of solutes, thereby lowering the freezing point. Crucifers, for example, exhibit this hardiness while tomatoes do not.

8-3 HIGH TEMPERATURES

8-3.1 SUNSCALD AND SCORCH

Many plants are damaged by high temperature and bright sunlight. Young leaves of cultivated annual plants may be permanently wilted and die. They generally turn a light tan to golden brown in color. Such injury is called sunscald. Young transplanted hardwood trees, such as the hard maples, are often affected in this manner. Since these trees are relatively slow growing, their root system generally takes one to two years to become well established; therefore, they are more than normally subject to such disorders. This type of disease of hardwood trees is termed scorch. Damage by high temperature and bright sunlight is always increased manyfold by insufficient moisture in the soil.

Scorch, Common To Hardwood Transplants

8-3.2 WATERCORE OF APPLE

Watercore of apple is likely to occur in fruit that is fully exposed to bright sun and high temperatures. Damage is increased as the apples become overmature. The disease is recognized by the clear, glassy, or water-soaked appearance of affected apple tissue. It may be confined either to the core region or show through to the skin; in both cases, it is always in association with the vascular elements. Watercore is thought to be due to excess sap concentration. The best way to prevent this disease is to pick the fruit at the proper time and not allow it to become over-ripe.

8-4 UNFAVORABLE OXYGEN RELATIONS

Blackheart of potato is the classic disease showing the result of suboxidation or insufficient oxygen during storage. It is characterized by an internal necrosis in the tubers. At first the internal color is pinkish, then tan, brown, dark brown, and finally black. Long periods of storage in poorly ventilated buildings will produce blackheart since the respiration of the potatoes gradually uses up the available supply of oxygen causing internal cell degeneration and discoloration. It is recommended that potatoes be stored in well ventilated buildings at 36 to 40° F. The low temperature reduces the respiration rate of the tubers and also prohibits the growth of bacteria

Storage Disease Caused By Suboxidation

and fungi. Apples, citrus fruits, and most crucifers are also subject to disorders induced by suboxidation.

8-5 UNFAVORABLE SOIL-MOISTURE RELATIONS

There are three general types of soil-moisture problems; they are (1) insufficient moisture, (2) excessive moisture, and (3) irregular supply of moisture.

8-5.1 INSUFFICIENT MOISTURE

Rule -- Interrelated Drought Factors

The first symptom of insufficient moisture is, of course, wilting. In dry periods, many plants commonly wilt during the midday and recover at night. If the period of dryness is extensive, the wilting may become permanent and result in death of the plant. Lack of moisture, high temperatures, dry winds, and bright sunlight are interrelating factors of drought.

Moisture shortage may lessen growth and lower yield. Tuber and root crops may not "size up"; cereals may produce shriveled kernels; fruits may be deformed, small, of poor flavor, or they may shrivel and fall prematurely.

8-5.2 EXCESSIVE MOISTURE

The flooding of soils in some areas is common after heavy rains, and damage to crops occurs if there is inadequate drainage to remove the water promptly. Flooding alters the soil microflora. It encourages the growth of anaerobic bacteria, which deprives crop roots of much needed oxygen. The result is usually a general decay of crop roots and storage organs, ending in a complete crop loss.

8-5.3 IRREGULAR SUPPLY OF MOISTURE

Fluctuations in water supply cause blossom-end rot of tomatoes. The first appearance of injury is a water soaked spot near the blossom-end of the fruit. This spot soon becomes brown and often enlarges until it covers up to one-half the surface of the fruit. The affected area becomes sunken as the disease progresses, and the lesion also darkens in color.

Growth Cracks

Growth cracks in tomatoes appear to be common disorders that occur as the result of extremely rapid growth brought about by periods of abundant rain and high temperatures. Also, if tomatoes have grown under a reduced moisture condition and have started to ripen, they will often crack if a heavy rainfall then occurs. The cracks develop in circles or semi-circles about the stem end, or extend radially down the fruit starting at the stem. The cracks vary in depth and in speed of rupture. If they develop slowly, the surface tissue heals over fairly well. However, if the tissue splits suddenly, large open wounds occur which provide ports of entry for microorganisms, and rotting often follows.

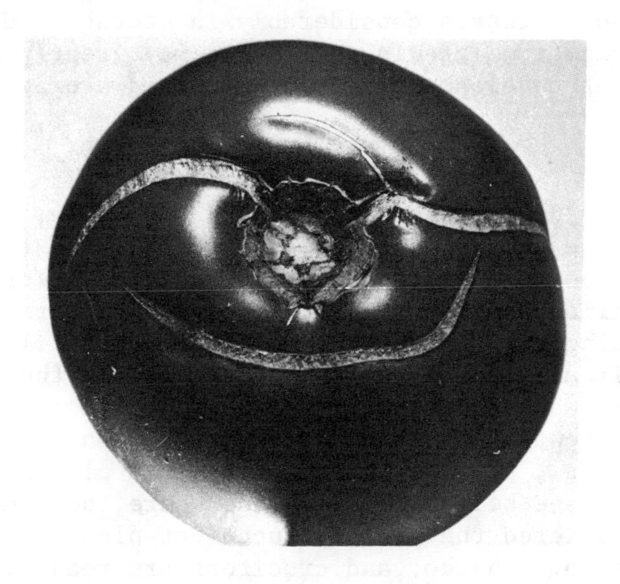

Fig. 8.2A Growth cracks in a tomato. These cracks developed slowly and are healed over. *(Courtesy of R. W. Samson)*

Fig.8.2B Alternaria rot of a tomato caused by the fungus *Alternaria tenuis*. This pathogen has invaded the fruit through open growth cracks at the stem end. *(Courtesy of R. W. Samson)*

Bitter-pit Of Apple

Another disease related to irregularities of moisture and heavy applications of nitrogenous fertilizer is <u>bitter-pit of apple</u>. This disease is characterized by purplish-red spots in the flesh underneath. The spots are one-eighth to three-eighths of an inch in diameter and are irregularly circular and slightly sunken. The disease is so-named because of a bitter off-flavor of the fruit. This disease also increases considerably in storage. Heavy rainfall or irrigation late in the season may greatly increase the amount of development of bitter-pit in storage.

8-6 LIGHTNING, HAIL AND WIND

Lightning

The injury due to lightning is, on the whole, relatively small. The damage to trees is prominent and familiar to almost everyone, while the damage to cultivated crops is not so familiar and is often confused with other causes. When lightning strikes the ground it commonly kills plants within a circular area. Of some interest are the plants at the periphery and a number of feet away which show various degrees of growth reduction and injury. Upon examination of these plants, it will be found that the pith is often killed above and below the soil line where the electrical charge has entered the plant. Succulent plants such as tomato, potato, tobacco, and crucifers are readily damaged.

Hail

Hail causes plant injury in varying degrees, depending upon the crop, stage of plant development, size of hailstones, and length of time the plants are exposed to the hail, and the weather following the hail. Damage can occur ranging from only small holes in a few leaves to complete defoliation and destruction of plants. Large hailstones can split the tender bark of young branches of shrubs and trees leaving many open wounds for entrance of other organisms. Although damage by hail is generally localized, it is not uncommon for entire fields of vegetable crops, field crops, and fruit crops to be destroyed.

Wind

Wind and heavy rain are capable of causing injury of several different forms. Leaves of crops may be ripped and torn or water-soaked, which predisposes plants to invasion by bacterial and fungal pathogens. The splashing rain and wind also may carry pathogens into the stomatal openings of the leaves. Heavy winds often blow crops down, flattening them so they are in contact with soil and are thus subject to rot; mechanical harvesting is made difficult or impossible.

8-7 MINERAL EXCESSES

Plants vary in respect to their demands for soil acidity or alkalinity. Highly acid soils are toxic to some plants and just what is required for others. Crop plants are known as acid-loving, acid-tolerant, or acid-sensitive. Most agricultural plants grow well in soils with a pH ranging from 4 to 8. Acid soils are often injurious because iron and other elements are available in toxic concentrations. Also, it is not uncommon for over-fertilization to occur from time to time which results in toxic excesses.

Soil pH

Saline soils present a serious problem in many arid regions and areas bordering oceans. Sodium chloride, sodium sulfate, and sodium carbonate are the three most common salts, although others may be responsible for damage too. Most crop plants are sensitive to salts; their seeds may not germinate, but if they do, their subsequent growth may be retarded. Plants may show symptoms from stunting to outright death; and they may exhibit chlorosis, wilting, leaf burn, and leaf thickening.

Saline Soils

8-8 MINERAL DEFICIENCIES

It is perhaps fortunate that symptoms of nutrient deficiency tend to be similar in different plants. Otherwise, it would be a difficult task indeed to determine the remedy. At least a dozen mineral elements are required by plants, and each must be in a form available to the plant and in proper balance with the other mineral elements. A deficiency of any one or more of these may cause disease. However, since the mineral requirements of plants are variable, the mineral "balance" that incites a deficiency disease in one plant may not do so in another.

Rule -- Nutrient Deficiency

Mineral Requirements Variable

The minerals in the soil that are required by plants are:

Major Elements	Minor Elements
Nitrogen	Iron
Phosphorus	Zinc
Potassium	Manganese
Calcium	Copper
Sulfur	Boron
Magnesium	Molybdenum

The major elements are those required by plants in relatively large quantities. Iron is required in much smaller concentrations. Zinc, manganese, copper, boron, and molybdenum, are required in still lower concentrations.

Common Soil Deficiency

Nitrogen deficiency is commonplace throughout most of the world, and is often a limiting factor of growth in many fields in the United States. The symptoms are: chlorotic leaves, with yellowish, reddish, or purplish tints developing; failure to size up; leaves falling prematurely; and poor yield in seed or fruit. Severe stunting is common if the deficiency is extreme.

Phosphorus deficiency is second in importance to nitrogen in most areas where soil fertility has not been maintained. However, the symptoms are not always characteristic. Roots may be poorly developed, leaves small and often narrow and necrotic, fruiting is reduced, and overall color is a dark dull green.

Metabolism Of N and P Is Interrelated

The metabolism of nitrogen and phosphorus is interrelated in many ways. If there is a deficiency of phosphorus, inorganic nitrogen is readily absorbed and accumulates in plant tissues. On the other hand, when phosphorus is readily available, the absorption of inorganic nitrogen is depressed. Thus, the application of phosphate fertilizers directly affects the nitrogen balance of the plant. Illustrative of this is the early maturity of plants that often results when the phosphorus level is high, and the late maturity, as expressed by a delay in flowering, when the phosphorus level is low.

Potassium, or potash, deficiency produces rather characteristic symptoms on many plants. The leaf tips and margins become chlorotic, usually starting with the older and lower leaves. Necrotic spots often develop on the leaf blades and margins. On cereals, the chlorosis and necrosis are in the form of interveinal streaking.

Severely affected plants may exhibit dieback of the growing shoots. Yield of seed and fruit is generally poor.

Calcium deficiency is generally expressed by a poor growth and development of all growing points of the plant. Chlorosis and death of the leaf tips and margins may occur, followed by an inward rolling of the leaves. Death may result in extreme cases. Symptoms may be exhibited from the seedling stage to the mature plant, where the formation of fruiting is reduced.

Sulfur deficiency causes plants to become typically off-color. The leaves are generally a light green to yellowish. Sulfur deficiency in field soils is rare.

Magnesium deficiency causes chlorosis. Since this element is an essential component of chlorophyll, this is to be expected. Usually the chlorosis is first evident on the older leaves, and small necrotic spots may develop.

Affected leaves may later die and fall from the plant.

Iron deficiency causes chlorosis of leaves that may become whitish. A mottling of the leaves is also common, along with a twig and branch dieback. An alkaline soil is often responsible for making the iron unavailable, and in such cases this may be corrected by acidifying the soil. Another corrective procedure is to spray the plant with an iron chelate compound.

Iron Is Unavailable In Soils Of High pH

Fig.8.3 Chlorosis of pin oak caused by an iron deficiency. *(Courtesy of R. W. Samson)*

Zinc deficiency is common in many soils in the United States, and especially in the western states. Zinc deficiency is often characterized by a mottled chlorotic appearance of the leaves. Later, the leaves may become extensively necrotic. The leaves are thickened and the plant has shortened internodes. A defoliation and dieback may occur on fruit trees. A zinc deficiency in peach and citrus trees may be overcome by spraying the foliage with zinc sulfate.

Manganese deficiency is not common, but occurs occasionally in acid soils. Two of the better known diseases caused by a lack of this element are "grey speck" of oats and "speckled yellows" of sugar beets. The general symptoms are chlorosis and scattered necrotic spots on the young leaves, and often a dwarfing.

Copper deficiency causes a dieback of leaf tips, wilting, and stunting. This deficiency has occurred in muck soils and also in soils that have been subject to many years of croping.

Boron deficiency causes a dry rot of sugar beets that is called "heart rot". Other crops commonly affected are potato, turnip, cauliflower, tobacco and celery. Damage to the apical meristems is common, and plants generally have poor growth, and death of the terminal shoots is frequent. Since most plants are highly sensitive to excessive amounts of boron, extreme care must be used in corrective applications to the soil.

Molybdenum deficiency is infrequent, but has occurred in some soils on Long Island and on the West Coast.

8-9 NATURALLY OCCURRING TOXIC CHEMICALS

Juglone

There are a number of chemicals produced in plants that are toxic to other plants. One of the best known is juglone (5-hydroxy-1,4-naphthoquinone), produced by black walnut. A related form hydrojuglone, which is oxidized to juglone upon exposure to air, is also formed in the plant. Juglone is toxic to tomatoes, potatoes, alfalfa, apple, and many other plants.

Replant Toxicity

When a crop such as peaches is grown in the same soil for a number of years and is removed for one reason or another, the subsequent plantings often fail to grow well. Scientific reasons to explain these replant problems are just beginning to shed light on the real causes. Such a problem is exemplified by the peach tree replant toxicity. This appears to be caused by the breakdown of the nontoxic amygdalin of peach roots into the highly toxic hydrogen cyanide and benzaldehyde.

The aerobic and anerobic decomposition of various crop residues also cause the formation of a number of chemicals that are toxic in varying degrees to many plants. Although these products have been isolated from decomposing residues in laboratory experiments, little is known about this area, except that plant damage has repeatedly been demonstrated.

A disease of rice has been shown to be caused by toxic concentrations of hydrogen sulfide formed in the presence of ferrous iron under anaerobic conditions. This may occur when the fields are flooded. A number of other substances produced under anaerobic conditions are suspected of causing damage to rice.

8-10 PESTICIDE CHEMICALS

Pesticide chemicals include the following major groups: herbicides, fungicides, insecticides, miticides, soil fumigants (including nematicides and broad-spectrum toxicants), defoliants and desiccants, and growth regulators. Annually, the greatest plant loss is caused by herbicides, although severe plant damage can occur from most pesticide chemicals.

Misapplication

Vapor Drift

Residue

Plant injuries in this category are generally due to: (1) misapplication; (2) phytotoxic vapor drift; and (3) phytotoxic residue. Misapplication often can be attributed to the use of the wrong chemical pesticide, wrong dosage rate, or improper application.

The most widely publicized plant toxicity from pesticides has come from 2,4-D (2,4-dichlorophenoxyacetic acid), or related compounds. This is an effective herbicide for controlling many broadleaved weeds in corn, small grain, grass pastures, fence rows, and on highway, powerline, and railroad rights-of-way. 2,4-D is a hormone type of herbicide, and extremely minute amounts may cause grossly distorted plant growth. Perhaps the main reason for the damage caused by this particular material is that plants may be damaged by phytotoxic vapor drift that results from application to near-by fields, etc. Tomatoes and grapes are especially susceptible to injury by vapor drift. On tomatoes, the symptoms are often epinasty, followed by a bending of the growing points, a twisting of the stem, severe curvatures of the leaves, and general distortion of the new growth. The formation of numerous small roots along the basal portion of the stalk is common, also. Moderately affected plants may not set fruit on exposed flower clusters. However, unless there is a repeated exposure, these plants will quickly outgrow the effects and return to a normal growth habit.

Grape vines show severe foliar distortion on new leaves. The leaves are small and twisted, with prominent yellow veins, and are fan-shaped. A severe exposure may kill the vines. A moderate exposure will often result in this distortion occurring the year after exposure.

In almost every case, injury from pesticides may be avoided if the label directions are followed.

Fig.8.4 Distortion of young grape leaves caused by exposure to 2,4-D. *(Courtesy of R. W. Samson)*

PART II: AIR POLLUTION TOXICITY

8-11 INTRODUCTION

Air pollution toxicity to plants has been known for over a century; however, it has only come into prominence and received the attention of the scientific community since 1945. Throughout the world, air pollution damage to crop and ornamental plants is increasing along with a corresponding increase of population, industrial expansion and energy usage. In the United States, the incidents are more frequent, although less severe, than those locally associated with the smelting of ores earlier in this century. With intensified concern over world food supplies, additional consideration will have to be given to all aspects of air pollution toxicity to plants.

Air Pollutant Defined

An air pollutant has been defined by Wood, 1970, as: "any factor mediated by the atmosphere that causes an unwanted effect." This is a broad definition that includes particles, gases, dusts, odors, radiation and noise. Airborne pollen, fungal spores, bacteria, viruses and insects may cause unwanted effects, but are seldom considered as air pollutants. The air pollutants that cause plant disease are mainly particles, dusts, and gases, with the latter being the most important.

Air Pollutants Originate From Natural And Man-related Sources

Contrary to popular opinion, the sources of air pollutants are not all from man-related activities. Many of the same types of air pollutants that emanate from man-related activities are also formed in nature. For example, lightning discharges produce ozone; while sulfur dioxide, hydrogen sulfide and fluorides are emitted from various natural sources. Forest fires of natural origin, volcanic eruptions and vegetation emissions are major sources of naturally occurring air pollutants and will continue to be major sources in the future. These naturally occurring air pollutants undoubtedly are important in their direct effects on plants, but probably are of greater importance in their reaction with pollutants emitted from man-related sources. Man-related sources of air pollutants are principally from all types of energy release activities, such as ore-smelting and refining, petroleum refining, industry, heating, generation of electricity, transportation, etc.

Primary Air Pollutants

Air pollutants are generally termed as primary or secondary. Primary air pollutants originate at their source in a form that is toxic to plants. Examples of primary air pollutants are sulfur dioxide (SO_2), nitrogen oxides (NO_x), particulates, hydrogen chloride (HCL), aldehydes, chlorine (Cl_2) and hydrogen

Secondary Air Pollutants

fluoride (HF). Secondary air pollutants are formed in the atmosphere, usually photochemically, by reactions among pollutants. Peroxyacetyl nitrate (PAN) and ozone (O_3) are examples of secondary air pollutants. Another term useful in discussing air pollutants is the term point source. A point source refers to a fixed emission source, such as might come from a stationary manufacturing plant; pollutants such as sulfur dioxide, particulates and hydrogen fluoride usually arise from point sources. In contrast to the point source is the moving source. Various vehicles of transportation, such as autos, buses, trucks, and airplanes are major moving sources of air pollutants.

Point Source

Moving Source

Chronic Plant Damage

Most of the plant damage suffered by plants because of air pollutants is a growth suppression without clear-cut symptoms. This is a chronic form of plant damage; that is, a disease lasting a long time. This type of plant disorder is highly insidious in nature, because the lack of good growth is seldom recognized as a disease. When plants are grown in carbon-filtered air versus unfiltered air, and placed in localities subject to air pollutants, they provide the best evidence of retarded growth.

Acute Plant Damage

Air pollutants may also cause acute damage. Acute damage is attended by symptoms of some severity and comes speedily to a crisis. To cause acute disease, air pollutants are present in higher concentrations than those that cause chronic disorders. Acute symptoms are sometimes easily recognized and suggest the causal agent, because when acting alone, each phytotoxic pollutant tends to produce its own acute symptom pattern.

Plants vary in their sensitivity to air pollutants. Furthermore, considerable variation often exists between plant varieties within a given species. Use has been made of this variation of sensitivity. Plants that are highly sensitive (susceptible) to air pollutants are often grown in various urban and rural locations as indicators to detect the presence or absence of certain air pollutants. Another use of this plant variation is the growing of resistant or tolerant (less sensitive) varieties in areas where air pollution toxicity is probable.

Air Pollutants Damage Plant Leaves

In respect to acute plant damage, air pollutants have their primary effects within and upon leaves, and to a much lesser extent upon flowers. This should be expected since gaseous pollutants in the air close to the leaf enter the leaf through open stomata as a result of normal gaseous exchange. Acute plant damage by an air pollutant is a function of concentration times exposure time, (concentration x exposure).

Concentration x Exposure

The higher the dosage, the lower the required exposure period. Therefore, the threshold concentration at which an air pollutant causes plant damage is presented as a concentration of the pollutant at a sustained exposure time. Pollutant concentrations are usually given as ppm (parts per million--in surrounding air), pphm (parts per hundred million) or ppb (parts per billion).

Source
Transport
Effect

An air pollutant originates from a source, is transported in the atmosphere and causes an unwanted effect(s) in plants. Therefore, in order to diagnose air pollution damage in plants and establish controls and remedies, a thorough understanding of the following topics is necessary: (1) common air pollutants, their sources and effects; (2) interactions of air pollutants; (3) transport and dispersion of air pollutants; (4) diagnosis of plant damage; and (5) control of damage by air pollutants.

8-12 AIR POLLUTANTS

The common air pollutants in the United States that are phytotoxic to plants are:

1. Particulates (soot, fly ash, dusts)
2. Chlorine (Cl_2)
3. Hydrogen chloride (HCl)
4. Ethylene
5. Fluorides
6. Nitrogen oxides (NO_x)
7. Sulfur dioxide (SO_2)
8. Ozone (O_3)
9. Peroxyacyl nitrates (PANs)

The most important air pollutants, those that cause the greatest plant damage, are fluorides (mainly hydrogen fluoride), sulfur dioxide, ozone and the peroxyacyl nitrates.

8-12.1 PARTICULATES

Particulates have long been known to cause plant damage. Particulates such as fly ash, acid soot, and lime and cement dust have been recognized as causing plant disease for many years. However, plant damage by particulates is usually localized around and predominantly down-wind of sources such as:

Combustion of coal and fuel oil
Cement Production
Lime kiln operation
Incineration
Agricultural burning

Electrostatic precipitators have been developed to control the particulate emissions from a variety of industrial processes, and this can be accomplished with efficiency rates up to 99 percent.

Particulate Plant Damage

Plant damage by particulates usually consists of necrotic leafspots where the particles contact the leaf. Leaves coated with a layer of particulate matter often turn necrotic and fall to the ground prematurely.

8-12.2 CHLORINE AND HYDROGEN CHLORIDE

Chlorine (Cl_2) damage occurs infrequently by accidental spillage of storage tanks, or railway or truck tank-cars. Chlorine is sometimes found in the atmosphere around swimming pools and sewage disposal systems where it is commonly used as a disinfectant. Plant damage by chlorine is found close to the pollutant source.

Chlorine Damages Plants Nearby

Hydrogen chloride (HCl) is liberated by various industrial processes, and in the vapor state it is similar in action to that of chlorine vapor. However, occasionally a hydrochloric acid (HCl) mist is formed which is unlike HCl vapor. Hydrochloric acid mist burns leaves upon contact, leaving necrotic (often bleached) spots.

Chlorine and hydrogen chloride originate from the following sources:

- Refineries
- Glass-making
- Incineration and scrap-burning
- Accidental spillage

Combustion of polyvinyl-chloride (which is used in large quantities to manufacture packaging materials and wire insulation), results in the emission of HCl and many other compounds. The burning of large quantities of polyvinylchloride in open dumps still frequently occurs. However, in controlled incineration, techniques are available to effectively scrub HCl from flue gasses. Such scrubbing techniques are also available to remove HCl from industrial flue gasses.

Symptoms Of Chlorine And HCl Damage

Chlorine and hydrogen chloride cause necrotic spots or areas between leaf veins. Affected tissue may appear bleached. Marginal leaf burn is frequent. In affected areas, both upper and lower epidermis and the mesophyll between them become necrotic. Injured leaves frequently fall prematurely. Middle-aged (mature) leaves are usually injured first, followed by the oldest, then the youngest leaves. This damage is similar to that caused by sulfur dioxide. The acute injury threshold is about 0.1 ppm for 2 hours.

8-12.3 ETHYLENE

Ethylene is an unusual air pollutant. It is one of the few hydrocarbons that may cause direct plant injury without undergoing photochemical conversion with nitrogen oxides. Also, ethylene is a plant hormone.

Ethylene An Urban Problem

Ethylene is found principally in the atmosphere of large cities and urban centers, because a primary source of ethylene is motor vehicle exhaust. Other sources include the combustion of natural gas, fuel oil and coal.

Symptoms Of Ethylene Damage

Typical symptoms of ethylene injury to plants include stunting, leaf abnormalities, premature senescence, reduced flowering and fruit production. Ethylene exposure to budding and flowering plants often causes sepal withering and necrosis, flower dropping and failure of flower buds to open properly. All leaf tissues are damaged by ethylene. Abeles and Heggestad, 1973, reported concentrations of 0.7 ppm of ethylene in the air at the center of Washington, D.C., and 0.039 ppm in areas outside the circumferential beltway. Using controlled environment chambers, they reported plants grown in these ethylene concentrations exhibited typical ethylene injury. Since ethylene is not removed by common carbon filtration, these authors also used $KMnO_4$ to remove the ethylene by oxidation in their control chambers, to compare plant reaction with and without ethylene. The acute injury threshold is about 0.05 ppm for 6 hours.

8-12.4 FLUORIDES

Flurodies - Absorbed, Translocated And Accumulated

Fluoride compounds in either gaseous or particulate form can accumulate outside or inside leaves of plants and cause considerable damage. The fluorides are highly unusual in respect to their action as cumulative poisons. They are absorbed and translocated to leaf tips and margins, but are not translocated from leaves to other parts of the plant. Although high concentrations of fluoride have accumulated in leaf tips, all other parts of the plant contain little fluoride. Prolonged air exposure to levels of 0.5 ppb fluoride may result in injury to several species of plants.

Gaseous fluoride such as hydrogen fluroide (HF), causes most of the fluoride damage suffered by plants. The deposition of particulate fluoride compounds on leaf surfaces is not harmful while it remains in the solid state. However, when the solid fluoride is dissolved by dew or a light rain, the soluble fluoride is readily absorbed, translocated, accumulated, and causes toxicity.

Another unusual aspect is that some plant species accumulate high concentrations of fluoride without apparent harm. For example, cotton is such a plant; it can tolerate as much as 4,000 ppm fluoride without harm.

Symptoms Of Fluoride Damage

The fluorides cause a leaf margin necrosis on most broadleaved plants. Necrotic spots or blotches often occur within the leaves of certain plants. There is usually a sharp line of demarcation between the necrotic and healthy tissue. Early defoliation is common in some plants. On narrowleaved plants, leaf necrosis occurs first at the tips and progresses downward. Citrus trees exposed for several months to fluoride, exhibited leaf chlorosis, produced smaller than normal leaves, lower fruit yields and had reduced vigor.

On conifers, fluorides typically cause a tip chlorosis that may progress becoming a tip necrosis. Since these same symptoms are often caused by other agents, care should be used in making diagnosis as to the cause.

The gladiolus variety Snow Princess is highly sensitive to fluorides and has been used as an indicator plant to detect fluoride pollution. If the leaf tips become necrotic, samples of the necrotic leaf tissue are analyzed for fluoride proving that fluorides were present in the atmosphere.

Fluoride compounds originate from the sources listed here:

- Aluminum reduction processes
- Brick and ceramic plants
- Fiber-glass production
- Iron smelting
- Phosphate fertilizer production
- Refineries
- Rocket fuel combustion
- Steel manufacturing plants

Fluorides - More Toxic Than Most Pollutants

Generally, fluorides originate from the molten cryolite bath in the manufacture of aluminum and from impurities in the raw materials used in other industries. Fluorides are toxic at much lower concentrations than most other pollutants; and although the annual tonnages produced do not compare with some other pollutants, they still represent a major problem.

8-12.5 NITROGEN OXIDES

Nitrogen oxides, especially nitrogen dioxide (NO_2), have been long recognized as air pollutants. For a number of years the nitrogen oxides have been known to undergo a photochemical reaction with gaseous hydrocarbons to form the phytotoxic ozone and peroxyacyl nitrates (PANs). However, within the last few years, the nitrogen oxides have been recognized as directly causing plant damage.

Symptoms Of NO_2 Damage

Acute symptoms of plant damage by nitrogen dioxide appear as sharply defined white or brown, irregular lesions that form along leaf margins and/or between the veins (intervenial). Middle-aged leaves appear to be more sensitive than younger or older leaves. Both the leaf palisade and spongy mesophyll cells are affected and collapse. These symptoms are similar to those caused by sulfur dioxide. Concentrations of 2.5 ppm of nitrogen dioxide for 4 hours have been known to cause acute symptoms on young bean, tobacco and tomato plants. Plant growth may be retarded (chronic damage) by continuous exposure to 0.5 ppm nitrogen dioxide. This level has been recorded in the atmosphere of Los Angeles, and is therefore, a realistic figure.

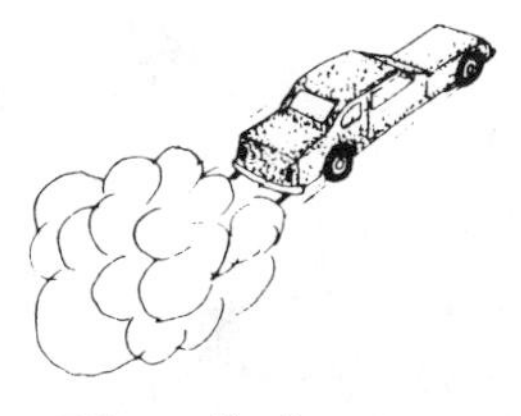

Fig. 8.5 Auto exhaust, a major source of nitrogen oxides

Nitrogen oxides are generated by the high temperature combustion of gasoline, coal, fuel oil and natural gas. Probably the most important source of nitrogen oxide is the automobile engine.

8-12.6 SULFUR DIOXIDE

SO_2 - A Prominent Pollutant

Sulfur dioxide (SO_2) is one of the most prominent phytotoxic air pollutants. For well over a century, sulfur dioxide has been known to be toxic to plants. It commonly causes chronic as well as acute plant damage. At subacute, or chronic concentrations, sulfur dioxide causes chlorotic, reddish-brown or white-bleached blotches or areas on leaf blades between the veins. Acute symptoms consist of intervenial bleaching of leaf tissues, which often appear white or tan then later turn brown. Characteristically, the veins remain green.

Symptoms Of SO_2 Damage

On conifers, sulfur dioxide typically causes a leaf-tip necrosis beneath which is a rather distinct chlorotic band. Older, damaged needles frequently defoliate prematurely.

The major sources of sulfur dioxide are given in the following list:

Combustion of coal
Production and combustion of petroleum and natural gas

Manufacturing and utilization of sulfuric acid and sulfur
Smelting and refining of ores

The combustion of coal represents the major source of sulfur dioxide, and the amount emitted depends upon, among other things, the sulfur content of the coal. The sulfur content of coal in the United States ranges from less than 1% to 6%. Low sulfur coal is considered to contain no more than 2% sulfur.

Coal-burning power plants represent the most important single source of sulfur dioxide. Numerous instances of vegetation damage have been associated with this source. Copper, lead, nickel and zinc ores also have been commonly associated with emissions of sulfur dioxide.

The threshold of phytotoxic concentrations of sulfur dioxide is variable. Plants are more susceptible to damage under these conditions:

Conditions That Promote High Plant Susceptibility

- Intense light
- High relative humidity
- Adequate soil moisture
- Moderate temperature

These conditions usually occur during midday hours (about 10 a.m. to 2 p.m.), which are most often met during late spring and early summer. A partial explanation of this phenomenon is that under these conditions, stomata are open wide, allowing the maximum entry of sulfur dioxide.

Plants are less susceptible (more resistant) to sulfur dioxide during periods of darkness and periods of slight wilting. The probable (or at least partial) explanation for this resistance, is that both darkness and wilting cause stomata to close and thereby reduce the entry of sulfur dioxide.

Older plants are more susceptible to sulfur dioxide damage than are young plants. Also, middle-aged leaves are more sensitive than are either young or older leaves.

Fig. 8.6 Industrial emissions of SO_2

Mesophyll cells of the leaves are affected first; the chloroplasts become bleached and the cells become plasmolyzed, also. The threshold for plant damage by sulfur dioxide is about 0.3 ppm for 8 hours. However, sulfur dioxide is frequently present with other phytotoxic air pollutants which in combination interact in respect to toxicity towards plants. In such combinations, much lower amounts of sulfur dioxide are known to cause plant damage.

Besides damage by gaseous sulfur dioxide, plant injury may also be caused by sulfuric acid or a combination of both. During the combustion of fossil fuel, one part of sulfur trioxide is formed for about 30 parts of sulfur dioxide and continues to be formed during mixing of the combustion products in the atmosphere. The sulfur trioxide then may unite with atmospheric moisture to form misty droplets of sulfuric acid. When the acid mist contacts plant leaves, it causes sharply defined brown to tan necrotic spots or blotches. Initial injury is to the upper epidermis, followed by injury to the mesophyll and then the lower epidermis. Such acid mist is often formed in certain industrial locations during periods of weather inversions.

8-12.7 OZONE

Ozone In Urban Smog

Ozone (O_3) is a major constituent of "urban smog", and has been recognized as a phytotoxic agent for over 100 years. It is a naturally occurring pollutant. Also, ozone is formed photochemically as a result of man-related activities, and ozone thus formed has the dubious distinction of causing the most plant damage. In the United States, the inefficient gasoline engines of automobiles daily emit tons of waste hydrocarbons and nitrogen dioxide into the atmosphere. This gaseous mixture is energized by ultraviolet light, during the daylight hours, with the resultant formation of ozone. Even though the unstable ozone soon decomposes, it is often formed faster than it is lost. With the coming of night, ozone ceases to be formed and its concentration steadily decreases.

Ozone - A Secondary Pollutant

Ozone formed from man-related activities is a secondary pollutant. It is formed in a gaseous cloud of hydrocarbons and nitrogen dioxide in and around a metropolitan network and often drifts a number of miles away. Therefore, plant damage from ozone may also occur a number of miles from the source of the primary emissions.

Ozone is present in the outer layers of the stratosphere, but only rarely, if at all, gets to earth. Possibly, ozone is carried down to earth behind polar cold fronts, by jet streams, or settles to earth during periods of anticyclonic conditions.

Electrical discharges from lightning also cause the formation of ozone. However, it is not known if enough ozone can be produced during thunderstorms to cause plant injury. Ozone injury to crops and trees far from population centers has been well documented, but the actual source of the damaging ozone is unknown. Perhaps the ozone accumulated from violent thunderstorms.

Volatile hydrocarbons emitted from foliage of green plants and radiated by ultraviolet light is another source of ozone. The emission of volatile sbstances through the aerial portion of green plants is well-known. Rasmussen, 1972, states: "The annual contribution of forest hydrocarbon emissions to the air pollution on a global basis is reflected in the 175×10^6 tons of hydrocarbons from tree foliage sources as compared to the 27×10^6 tons from man's activities." To put it simply, emissions from forests are 6.5 times greater than that from man-related activities. However, the hydrocarbon emissions from man-related sources are concentrated around population centers, which in turn leads to locally higher ozone concentrations than probably would occur from forest emissions.

Forest hydrocarbon emissions consist of monoterpenes, pinenes, limonene and isoprene; however, the exact fate of these gaseous olefins in the atmosphere is unknown. The blue haze over forests such as is common in the Smoky Mountains, probably consists to a large extent of ozone and related compounds formed photochemically from vegetation emissions.

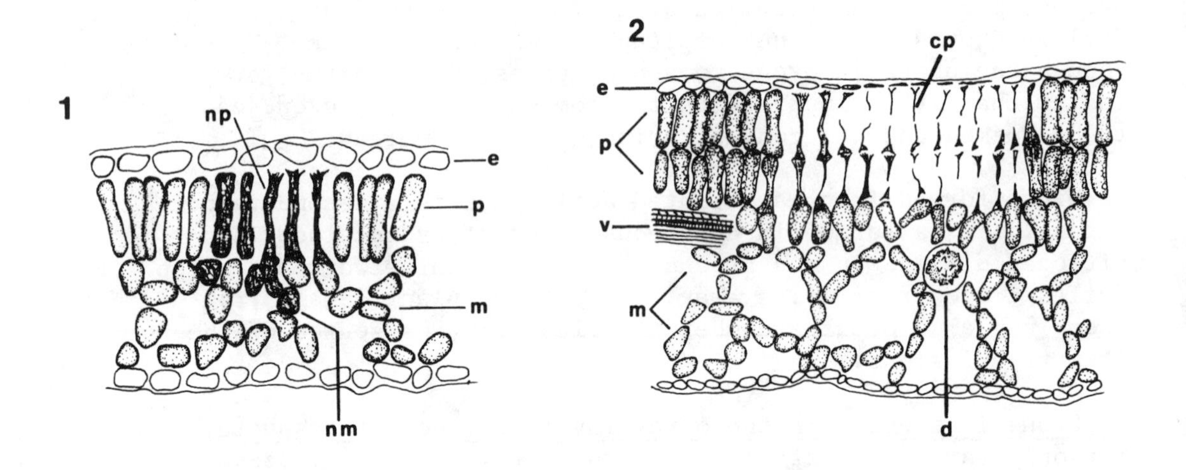

Fig. 8.7 Drawings of sections of leaves damaged by exposure to ozone. 1. Bean leaf has necrotic palisade tissue (np) containing pigments. In underlying spongy mesophyll (m), a few cells are necrotic (nm), the epidermis (e) remained normal. 2. Spinach leaf showing collapsed palisade tissue (cp) overlain by collapsed epidermis. Normal epidermis (e), pallisade layers (p), spongy mesophyll (m), section of vascular bundle (v) and normal druse (d), which is a globular compound crystal with many projecting component crystals. *Redrawn from A. C. Hill, M. R. Pack, M. Treshow, R. J. Downs, and L. G. Transtrum, 1961, Phytopathology 51:356-363)*

Symptoms of plant damage by ozone varies between and within species to a certain extent. Not all symptoms appear in any one plant. One of the initial symptoms is the "watersoaked" leafspots. Such watersoaking is common in tobacco. These areas of water congestion are visible on both sides of the leaf and may or may not become necrotic. If the damage is not severe, the injured plant cells recover, which appears to occur in resistant tobacco varieties.

Initial O_3 Injury On Upper Leaf Surface

The next symptom stage is a brown to black coloration of tiny spots or flecks along veinlets (as in tobacco), which later bleach-out and become lighter. In some plants, such as petunia, wide interveinal bleached-out bands occur. This injury is initially confined to the upper, or palisade parenchyma cells. However, if the leaf has undifferentiated mesophyll (as in grasses), the damage may be seen on either surface. Another symptom of acute ozone damage is a chlorosis of older leaves, often accompanied by their premature defoliation.

Initially, chloroplasts of the ozone affected palisade cells gather near the middle or ends of the cells. The cell walls then usually collapse. Sometimes, the chloroplasts become granulated and form a jelly-like consistency. Damaged cells frequently develop a reddish-brown pigment which later bleaches-out to a lighter color.

On conifers, ozone causes a leaf-tip (needle-tip) chlorosis that progresses to a tip necrosis with a chlorotic banding and mottling below the necrotic area.

Ozone-type symptoms may also be caused by exposure to mixtures of nitrogen dioxide and sulfur dioxide (see 8-13).

The tobacco variety Bel W3 is so sensitive to ozone damage, that it has been used as an indicator plant to detect ozone in the atmosphere. This tobacco variety has been injured by as little as 0.035 ppm of ozone after a 4-hour exposure.

Older plant leaves are most susceptible to ozone damage, while young leaves are less so. Ozone enters through the stomata, most of which are on the lower leaf surface. What remains to be explained, is how and why the upper layer of the mesophyll, the palisade layer, is injured initially, while the lower layer, the spongy mesophyll, is not.

Damage by ozone, like the damage by sulfur dioxide, is correlated with conditions that cause the stomata to open fully. At least partial resistance is imparted by conditions that close stomata.

8-12.8 PEROXYACYL NITRATES

In the early 1940's, a new plant disease called "silver leaf" that affected the lower leaf surfaces of many plants, was noted in the region of Los Angeles, California. Subsequently, the disease was attributed to the local air pollution, called "smog". Later research confirmed that the constituents of smog that caused silver leaf were secondary pollutants called the peroxyacyl nitrates (PANs).

PANs Are Secondary Pollutants

The peroxyacyl nitrates represent a homologous series of compounds the first of which is peroxyacetyl nitrate (PAN), the second is peroxypropionyl nitrate (PPN) and the third is peroxybutyryl nitrate (PBN). PAN appears to be more prevalent than either PPN or PBN. The PANs are produced by photochemical reactions between ozone and waste hydrocarbons, and from the reaction of unsaturated hydrocarbons and oxides of nitrogen in the presence of light. The volatile hydrocarbons emitted from foliage of green plants have also been shown to react photochemically with oxides of nitrogen to form PANs.

Both the PANs and ozone are common constituents of smog. They are photochemical oxidants that have a common ancestry in their primary chemicals, gaseous hydrocarbons and nitrogen oxides, both of which are principally emitted from automobile exhaust. The combustion of fossil fuels may also release the primary chemicals. Ozone and PAN have been shown to cause plant damage throughout the world but are of particular importance in North America. PANs appear to be more important as constituents of smog in California than ozone, while ozone is of greater consequence in the eastern United States. Why this is so has not yet been explained.

PANs Cause Silvering Of Lower Leaf Surface

The PANs cause similar acute disease symptoms on exposed plants. These oxidants are extremely toxic to many plant species, especially to young leaves. Acute symptoms of injury by the PANs include bronzing, silvering and glazing on lower leaf surfaces, and is called "silver leaf".

The plant leaf tissues affected by this damage are the spongy mesophyll cells, the lower epidermis cells, or both. Affected cells become plasmolyzed. Chloroplasts of the spongy mesophyll lose their integrity and the cytoplasm often becomes darkly pigmented. The silvering of the lower leaf surface has been attributed to dehydration and shrinking of the spongy mesophyll and filling of the attendant space with air (Bobrov, 1952).

Concentrations of PAN as low as 0.01 ppm for a six-hour exposure have caused plant injury. However, it would appear that the usual injury comes from a higher concentration and a lower exposure time.

8-13 INTERACTIONS OF AIR POLLUTANTS

Ozone And Sulfur Dioxide Mixtures

Several phytotoxic air pollutants have been shown to interact when present in mixtures and produce an enhanced, or synergistic action which lowers the injury threshold of green plant leaf tissue. Perhaps one pollutant predisposes the plant to injury by the second. Ozone and sulfur dioxide together produce such an enhanced effect. Menser and Heggestad, 1966, reported that ozone and sulfur dioxide together at concentrations of 0.03 and 0.3 ppm, respectively, after a 2- or 4-hour exposure, caused damage to leaves of Bel W3 tobacco plants. Alone at these concentrations and time exposures, neither pollutant caused injury. The damage to the tobacco plants was similar to that produced by ozone. Other mixtures of ozone and sulfur dioxide have shown the same enhanced or synergistic activity upon such plants as white pine (Dochinger, et al., 1970), and peanuts (applegate and Durrant, 1969).

Nitrogen Dioxide And Sulfur Dioxide Mixtures

Tingey, et al., 1971, worked with interacting mixtures of nitrogen dioxide and sulfur dioxide and found enhanced activity upon exposure of tobacco, tomato, bean, soybean, radish and oats. Some of the symptoms which these authors observed were similar to those caused by ozone; pigmented lesions, upper-surface bleaching, necrotic lesions, chlorosis, and chlorotic stippling. The similarities between the plant damage caused by mixtures of nitrogen dioxide and sulfur dioxide with that caused by ozone, probably explain why ozone-type damage has been reported in several instances when high concentrations of ozone were not present. In addition, some of the damage caused by mixtures of nitrogen dioxide and sulfur dioxide were suggestive of sulfur dioxide injury. Therefore, great caution must be used in making field diagnosis of plant damage due to specific pollutants.

8-14 TRANSPORT AND DISPERSION OF AIR POLLUTANTS

Atmospheric conditions often affect both the quantities and kinds of air pollutants emitted within a given region. For example, on a cold day, more fuel is consumed for space heating than on a warm day. As a direct result of this, increased sulfur dioxide emissions occur during the cold day and fewer emissions occur on the warm day. Atmospheric conditions also determine the behavior of air pollutants after they are emitted and until they reach receptor plants. It is with how atmospheric conditions affect air pollutants that the balance of this discussion is concerned.

When an air pollutant leaves its source (such as a stack or chimney), it normally rises. The more it rises, the lower the concentration of the pollutant at ground level. Both meterological and non-meterological factors govern the amount of rise. Two major meterological factors are wind speed and air temperature. A strong wind reduces the effective height to which effluent gasses rise. The greater the temperature differential between the effluent gas (usually hot) and the outside air (usually cool), the greater the tendency of the effluent to rise. Major non-meterological factors are: (1) the quantity of the pollutant emitted; and (2) the height of the source. An additional meterological factor is the rate of temperature lapse (which is the decrease of air temperature with height).

Temperature Lapse

Air is relatively transparent to short-wave radiation of sunlight. Therefore, as sunlight penetrates the atmosphere it does not substantially heat the air. However, sunlight is reflected from the earth's surface back into the atmosphere as a radiation of longer wave-length that is more readily absorbed by the layer of air near the ground, thereby heating it. The heat later becomes diffused through the lower layers of the atmosphere from below, upwards. Therefore, the air temperature is highest at ground level and decreases with increasing height--the temperature lapse. In this state, convection currents are set up as the warm air rises. Such a condition is termed "unstable". The temperature lapse extends to the top of the troposphere, a layer of the atmosphere which extends 10 km (about 6.2 miles) above the surface of the earth. However, inversion layers may form near the ground level or at higher elevations. Within an inversion layer, the temperature increases with height to the top of the layer, above which the usual lapse rate is again encountered. Inversion layers often form with the approach of night when the sky is cloudless, thus allowing radiation from the ground to escape into space. Loss of heat by radiation cools the ground and in turn cools the air lying nearest to the ground. Therefore, a temperature inversion is set up; over the cold air near the ground lies the air at a high temperature--which extends to the top of the inversion layer. Inversion layers are common near the ground at night over rural areas. Elevated inversion layers occur frequently in some regions and sporadically in others.

Unstable Weather

Inversion Layer

Inversions - Common At Night

Inversion layers are termed "stable" since they resist vertical mixing of air. Also, they resist any kind of mixing or vertical penetration of effluent. Inversion layers thus restrict the height to which an effluent can rise and cause it to spread out horizontally. In unstable air, the effluent rises until a stable layer is reached, often at very high elevations.

Air pollutants are carried by winds and dispersed by

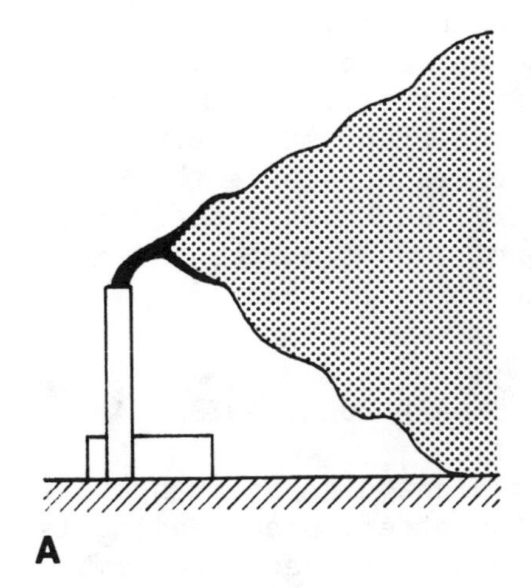

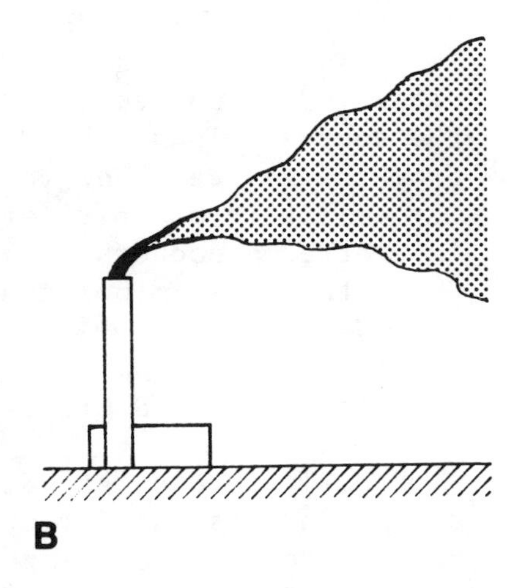

Fig. 8.8 Diagram of the plume dispersion from a tall chimney. A. Dispersion due to mixing caused by strong convection, with low wind shear. B. Dispersion due to mechanical mixing caused by wind shear, with low convection.

Air Turbulence Caused By Convection And Wind Shear

atmospheric "turbulence". The greater the wind, the smaller the concentration of the pollutant. Air turbulence consists of both horizontal and vertical "eddies"--which are formed by heating from below (convection) and by wind shear (mechanical turbulence). Such turbulence causes a mixing of the effluent with air and reduces the concentration of the effluent. Of course, the stronger the turbulence, the more the pollutants are dispersed. Heating produces convection--which occurs whenever the temperature decreases rapidly with height. Wind shear, or mechanical turbulence, occurs when wind changes with height. Mechanical turbulence increases with wind speed and is greater over rough terrain than over smooth terrain.

From this discussion it is easily concluded that greater concentrations of pollutants are possible when the atmospheric conditions are stable--such as under an inversion layer. When atmospheric conditions are unstable, the chances of pollution damage are probably slight.

Emission stack height is important. Pollutants emitted at ground level can quickly build up to toxic concentrations during inversions. However, if the same pollutants are emitted from a stack several hundred feet high, the pollutants reaching the ground even during an inversion will be less concentrated, since considerable dillution of the "plume" will have taken place.

The pattern of vegetation injury may be likewise affected by the elevation of the pollutant source. A ground-level source causes the greatest amount of plant damage close to the source and decreases with the distance from the source. Thus, there is a gradient of plant injury away from the source. However, near an elevated source there may be little or no vegetation injury, but damage occurs in scattered pockets at some distance away from the source where the maximum pollutant concentration is reached.

8-15 DIAGNOSIS OF PLANT DAMAGE BY AIR POLLUTANTS

Usually an accurate diagnosis of plant disease by specific air pollutants is difficult. All too often, plant damage is not symptomatic for a specific pollutant; indeed, the symptoms may overlap those caused by drought, frost, mineral deficiencies, nematode and insect attack, plant viruses and even chemical pesticide toxicity. Also, mixtures of pollutants may cause symptoms that mimic damage caused by specific pollutants. Only occasionally will the symptoms on certain plants clearly indicate the cause to be a particular pollutant. Wood, 1971, suggests the following list of factors be considered when making field diagnosis of air pollution damage:

1. Species affected
2. Area affected
3. Pattern of occurrence of affected individuals
4. Nature of pollutant emissions
5. Pollutant levels and duration in the atmosphere
6. Foliar analyses
7. Special considerations - long standing vs. recent sources

8-15.1 SPECIES AFFECTED

Air Pollutants Affect Numerous Unrelated Species

Normally, when a rural area is exposed to a toxic concentration of an air pollutant, numerous and often unrelated species are affected. Air pollution may be suspected of being the causal agent in such a situation because few agents affect such a variety of plants. The types of damaged species may be compared with the LIST OF PLANT SPECIES SENSITIVE AND TOLERANT TO SPECIFIC AIR POLLUTANTS, included in the back of this book.

If plant damage occurs in a farming community, weed species in the rows and along the edges of the field should be examined for symptoms, also. If only a single crop plant species is damaged and no injury is found on surrounding weeds, the chances are the causal agent is not a pollutant.

8-15.2 AREA AFFECTED

A primary air pollutant that originates from a ground level point source causes plant damage in an area near that point source. The damage is usually most severe next to the source, and lessens as the distance from the source increases. If the region around the source is flat, the affected area is often elliptical and predominantly in the direction of prevailing spring and early summer winds, (Wood, 1971). In the eastern United States, this is usually northeast of the emission source.

If the emission source is elevated, usually there is no plant damage close to the source, but the pattern of damage is characteristically some distance down-wind of the source and often occurs in scattered pockets. The scattered pockets of plant injury occur because of the dispersion of the emission plume.

In hilly country, a pattern of affected plants is seldom uniform. Often plant damage is found in valleys, at particular elevations, or in low pockets.

Areas of damage from photochemical pollutants are predominantly down-wind from metropolitan networks of highways and freeways, and may be a number of miles away.

This discussion then leads to the conclusion that: the area of plant damage from an air pollutant is related to the emission source.

8-15.3 PATTERN OF OCCURRENCE OF AFFECTED INDIVIDUALS

Wood, 1971, states: "Within a given species population, there is a wide range in the sensitivity of plants to specific pollutants." (Note: the underlining is the author's). Therefore, since sensitive non-crop plants typically occur in a random nature within the population, there are no "hot-spots" of plant damage; instead, randomly spaced plants are affected. Of course, such random spacing varies considerably, but is especially typical of wind-pollinated species.

8-15.4 NATURE OF POLLUTANT EMISSIONS

If an air pollutant is suspected of causing plant damage, emission sources up-wind of the affected area should be in-

Correlate Plant Damage With Emission Source

vestigated. If the damage is symptomatic of sulfur dioxide, but no source for sulfur dioxide exists, the damage probably was caused by another agent. Plant damage by primary air pollutants should be correlated with specific emission sources.

8-15.5 POLLUTANT LEVELS AND DURATION IN THE ATMOSPHERE

Even though emission sources are found near areas of damaged plants, the evidence is still circumstantial. Before the emitted pollutant(s) can be indited, it may be necessary to show that the pollutant can accumulate in sufficient concentration to cause plant damage. Sometimes, this may be done by placing various instruments in or near the damaged plants, or by growing "indicator" plants in the same area. The use of instruments is expensive and must be regularly monitored. Plant indicators are inexpensive, but do not provide data on amounts of pollutants, or the length of the exposure period. However, their use is often all that is available. Indicator plants may indicate the presence of a pollutant and this is often sufficient to indite the pollution source.

In many major cities of the United States, information on local atmospheric pollution levels may be obtained through state or federal monitoring agencies.

8-15.6 FOLIAR ANALYSES

Only the air pollutants sulfur dioxide and fluoride lend themselves to foliar analyses, because affected plants accumulate sulfur and fluorine, respectively. Analytical analyses for these compounds are available and are useful in making diagnoses. However, caution must be used in interpreting sulfur analyses of plant tissue, because sulfur is a major mineral element required by green plants. It may occur in soluble forms in the soil and become accumulated in plants in direct relation to its availability. Therefore, in order to interpret a sulfur analysis from plants suspected of being damaged by sulfur dioxide, additional information is needed. Sometimes, background data on the normal sulfur level in plants in the area may be obtained. Usually, it is necessary to make many foliar analyses of plants over a wide area in order to establish that the sulfur content in leaves of affected plants is related to atmospheric sulfur dioxide. Data must indicate a relationship between the sulfur levels of plants in the affected zone and the source.

Fluorine analysis of foliage does not present the dilemma that sulfur does, because fluoride compounds in the soil are usually insoluble and unavailable to plants. Usually only necrotic leaf tips and margins are analyzed, because this is where the fluorine accumulates. Background fluorine in vegetation is usually below 10 ppm (in leaf tissue). If fluorine levels above this amount are found in affected plants, a hydrogen fluoride or other fluorine source should be suspected.

8-15.7 SPECIAL CONSIDERATIONS - LONG STANDING vs. RECENT SOURCES

If a pollutant is new, or is suddenly emitted in greater quantities in a rural or urban area, a wide variety of plant species may be damaged by exposure during fumigation periods. However, if the pollutant has been in a community for a number of years, only isolated and usually newly planted or transplanted plants may show pollution damage. This latter situation is common to most of the large cities in the United States. The reason for this phenomenon is that sensitive plants slowly died out, leaving only tolerant or resistant plants.

8-16 CONTROL OF DAMAGE BY AIR POLLUTANTS

The most obvious means of preventing air pollution plant diseases is by reducing or eliminating the pollutants at their sources. A second means is to grow only tolerant or resistant plants. A third method is to use chemical agents to close stomata or otherwise provide protection.

Undoubtedly, man will never see the day when he has complete control (stoppage) of harmful gaseous emissions from his activities. However, with weather information readily available, it is now possible to forecast weather conditions that lead to dangerous levels of air pollutants. During such periods, certain industries and utilities could be required to reduce their emissions by switching to low sulfur coal and/or by reducing or stopping their activities. What is sorely needed, is the establishment of air quality standards and subsequent legally enforced control of emissions at their sources to meet the standards.

Breeding programs may be established to develop tolerant or resistant varieties. There is a great likelihood of achieving success with such programs because of the wide variation of pollutant susceptibility in plant species. Although most mechanisms of pollution resistance are unknown, the closure of stomata appears to be the basis of the resistance to ozone damage in certain onions and tobacco.

In the near future, chemical protectants may be available to reduce the damage by air pollutants. Experimentation with some chemicals that close, or partially close stomata indicate such substances may well be beneficial. Other chemical agents applied as soil drenches and/or to the foliage have given promising results, but no protective chemical is available for use at this writing.

8-17 VEGETATION AS A SINK FOR AIR POLLUTANTS

Considerable data now exists to indicate that vegetation plays an important role in cleansing certain toxic atmospheric pollutants from the air. Bennett and Hill, 1973, exposed alfalfa in growth chambers to hydrogen fluoride, sulfur dioxide, ozone and nitrogen dioxide. These authors found that hydrogen fluoride, sulfur dioxide, and nitrogen dioxide were removed efficiently by the upper portion of the plant canopy as well as by the immediate subsurface vegetation. The hydrogen fluoride was most effectively absorbed. In an earlier report, Hill, 1971, stated that vegetation cleansed the air of numerous pollutants, claiming the order for the most efficient cleansing to the least efficient; is as follows:

Hydrogen fluoride
Sulfur dioxide
Chlorine
Nitrogen dioxide
Ozone
PAN
Nitric oxide

Rule -- Green Plants, Sinks For Pollutant Removal

These and other studies indicate that vegetation plays an enormous role in acting as a natural "sink" for the removal of pollutants from the atmosphere.

Selected References

Part I

Boyce, J. S. 1961. Forest pathology. Ed. 3. McGraw-Hill Book Co., Inc., New York. pp. 36-71.

Doolittle, S. P., et al. 1961. Tomato diseases and their control. Agriculture handbook No. 203, Agr. Res. Ser., U. S. Dept. Agr.

Heald, F. D. 1933. Manual of plant diseases. Ed. 2. McGraw-Hill Book Co., New York. pp. 58-247.

Hollis, J. P. 1967. Toxicant diseases of rice. Louisiana State University and Agricultural and Mechanical College, Agricultural Experiment Station. Bulletin No. 614.

McMurtrey, J. E. 1953. Environmental, nonparasitic injuries. pp. 94-100, In Plant diseases. U. S. Dept. Agr. Yearbook.

Olin, C. R. 1967. Freezing stresses and survival. Annual Review of Plant Physiology. 18:387-408.

Patrick, Z. A., T. A. Toussoun, and L. W. Koch. 1964. Effect of crop-residue decomposition products on plant roots. Annual Review of Phytopathology. 2:267-292.

Sprague, H. B., et al., (ed.). 1964. Hunger signs in crops. Ed. 3. David McKay Company, New York.

Stakman, E. C., and J. G. Harrar. 1957. Principles of plant pathology. The Ronald Press Co., New York. pp. 50-64.

Walker, J. C. 1969. Plant pathology. Ed. 3. McGraw-Hill Book Co., New York.

Wallace, T. 1961. The diagnosis of mineral deficiencies in plants by visual symptoms. Ed. 2. Chemical Publishing Co., Inc., New York.

Part II

Darley, E. F., and J. T. Middleton. 1966. Problems of air pollution in plant pathology. Annual Review of Phytopathology. 4:103-118.

Heggestad, H. E. 1968. Diseases of crops and ornamental plants incited by air pollutants. Phytopathology. 58:1089-1097.

Hepting, G. H. 1968. Diseases of forest and tree crops caused by air pollutants. Phytopathology. 58:1098-1101.

Hindawi, I. J. 1970. Air pollution injury to vegetation. Publication AP-71. U.S. Government Printing Office, Washington, D. C.

Rich, S. 1964. Ozone damage to plants. Annual Review of Phytopathology. 2:253-266.

Smith, M. E. 1968. The influence of atmospheric dispersion on the exposure of plants to airborne pollutants. Phytopathology. 58:1085-1088.

Treshow, M. 1968. The impact of air pollutants on plant populations. Phytopathology. 58:1108-1113.

Wood, F. A. 1968. Sources of plant-pathogenic air pollutants. Phytopathology 58:1075-1084.

CHAPTER 9

MARKET PATHOLOGY

9-1 INTRODUCTION

Definition

The protection of plant produce from harvest to consumer, or processor, is the specialized area of plant pathology called "market pathology".

Setting, Environment, Physiological State Changed

Harvesting terminates concern for many of the field diseases discussed in previous chapters, continues concern for some of them, and initiates concern for others. Post harvest disorders are not altogether different from those that take place in the field; however, (1) the setting is changed, (2) the environmental conditions are altered, and (3) the harvested plant products are in a different physiological state. The setting is no longer that of the field, but is in storage sheds and warehouses, packaging and processing plants, and shipping vehicles. The environmental conditions are quite unlike those of the field. Stored products are crowded and packed together. For the most part they are protected from the wind, rain and sunlight. Furthermore, almost all fruits and vegetables are stored at controlled humidities and low temperatures.

Fig. 9.1 Green mold on oranges in storage prior to marketing, caused by *Pencillium digitatum*. *(Courtesy U. S. Department of Agriculture)*

The physiological state of plant products after harvest is generally that of either dormancy, semidormancy, living but non-growing, or dead. Examples of dormant and semi-dormant plant products are various crop seeds and cereal grains. Each seed contains a living embryo possessing meristematic regions. Examples of living but non-growing plant products are fruits and vegetables that have virtually ceased all cell division at the time of harvest. Examples of dead plant products are tobacco leaves and cotton lint.

Estimated $500,000,000 Annual Loss

The extent of damage and loss from storage diseases varies considerably with the product, the disease or diseases, and the storage conditions. It is estimated that losses from market diseases in the United States each year total one-half billion dollars. Storage is not normal for living plants, nor is the dense crowding that occurs in storage. Both of these conditions predispose produce to various types of diseases. Storage diseases are divided into disorders of physiological or infectious origin.

9-2 POST HARVEST DISEASES OF PHYSIOLOGICAL ORIGIN

9-2.1 SUBOXIDATION

Suboxidation causes internal breakdown of plant products such as potatoes, apples, crucifers, lettuce, citrus fruits, and others. This group of diseases was discussed in Chapter 8 under the heading Unfavorable Oxygen Relations.

Provide Adequate Ventilation

To control these diseases, affected products should be harvested at the proper degree of maturity, handled carefully, precooled prior to storage, stored at optimum temperature and humidity, and provided with adequate ventilation during storage.

9-2.2 AROMATIC ESTERS

Aromatic Esters Cause Scald

Scald of apples is caused by a toxic accumulation of aromatic esters that are produced as metabolic by-products from normal respiration. A number of fruits are subject to this disease; however, apples are the most susceptible. This disease appears as a superficial browning of the skin that becomes visible in storage or during marketing. Severe scald may be followed by internal breakdown, and decay fungi often invade the affected tissue.

Oiled Papers Control Scald

This disease can be controlled by using oiled papers in packing. The oil absorbs and retains the volatile esters thereby reducing the toxicity. Of course this procedure is not completely preventative. There is no substitute for proper harvesting, handling, and storage procedures.

9-2.3 TEMPERATURE AND HUMIDITY

Unfavorable storage temperatures may incite a number of disorders and predispose plant produce to others.

Chilling Or Freezing

Cold temperatures may cause either chilling or freezing injury. Chilling injury may result even though the storage temperature is well above the freezing point of the particular produce. Commodities vary as to the temperature that will cause chilling. Freezing injury occurs when the temperature drops below the freezing point of the product. Both chilling and freezing injury cause an internal breakdown of the tissues resulting in an off-flavor. More advanced symptoms are scald, and surface and internal decay. Cold injury predisposes the products to rot and decay by pathogenic organisms. Symptoms of cold injury develop rapidly at elevated temperatures which occur when the products are later removed from storage.

Increase Respiration

High temperatures during storage accelerate the respiration rate of stored products. This may promote injury such as that caused by the accumulation of aromatic esters and suboxidation.

Speed Moisture Loss

High temperatures also promote excessive moisture loss through transpiration which often causes shriveled tissues, wrinkling and deep cracks.

Seeds stored under conditions of high humidity and temperature may be attacked by various mold fungi, many of which have a very low moisture requirement.

If a succulent vegetable, such as lettuce, is stored for any length of time at a low humidity, it will lose moisture. If excessive moisture is lost, spoilage will soon occur.

9-3 POST HARVEST DISEASES OF INFECTIOUS ORIGIN

All Stored Products Have Surface Borne Contaminants

Many post harvest diseases are caused by bacteria or fungi that produce blemishes or decay in stored products. Physiological disorders will often predispose stored products to attack by microorganisms. It is particularly important to realize that all harvested plant products are surface contaminated by microorganisms. All vegetables, fruits, grains, tubers, and storage roots contain a high level of inoculum when they leave the field and enter storage. The handling during harvesting, processing, packing and transporting results in redistribution of surface borne organisms. Many of these organisms can do no damage unless their host substrate is wounded during the several handling and shipping operations. For other organisms, all that is needed to promote growth and subsequent spoilage is a suitable warm storage temperature.

Some plant products contain internal pathogens caused by the infections that occurred in the field prior to harvest. If these infected products are not placed in cool storage, the internal microorganisms may continue to grow, resulting in considerable spoilage.

9-3.1 BACTERIAL DISEASES

Bacteria are responsible for a large share of the losses suffered from market diseases. In fact, bacteria cause a higher percentage of the total market disease loss than they do of the total field disease loss.

Bacteria Cause Two Types Of Storage Diseases

In general, two types of post harvest diseases are caused by bacteria: (1) soft rots, caused by various species of the genera *Pseudomonas* and *Erwinia*; (2) blemishes, internal discolorations, and decay, caused by various species of the genera *Corynebacterium*, *Erwinia*, *Pseudomonas*, and *Xanthomonas*. Of these two groups, the soft rots are the most important in terms of product loss. The soft rot group of bacteria attacks almost all vegetables and is capable of causing serious rotting within a few hours. It does not, however, attack the tree fruits.

On fleshy, succulent leaves and stems, the first appearance of soft rot is small water-soaked spots, which later appear slimy or oily. A rapid softening and disintegration of the affected tissues follow. Soft rot spreads rapidly from diseased tissue to healthy tissue. In one day's time, the soft rot may advance to a point of complete tissue disintegration, resulting in a slimy, wet mass.

Soft Rot Bacteria Attack Most Vegetables

Potato tubers often become dark in affected areas and take on a blistered appearance; this is followed by a softening of the underlying tissues. An offensive, foul odor due to secondary bacteria living on the decomposing slimy mass is the final stage of potato soft rot.

On root crops, the first symptom of soft rot is a water-soaked appearance of infected parts. The invaded tissue becomes soft and often sloughs-off from the firm tissue underneath.

Soft rot is often caused by a complex, or combination, of several species of bacteria. The infection by soft rot bacteria in the field may not be evident at the time of storage or during shipment to market. However, precautions must be taken to prevent its development by maintaining low temperatures during storage and shipping. A temperature of 40°F., is recommended for potatoes, and near 32°F., for most leafy vegetables and root crops. Low temperatures only delay rotting; with a rise in temperature, the rotting resumes.

The bacteria that cause blemishes, internal discoloration, and decay of stored plant products are organisms that have infected the growing plants in the field. They often are the same bacteria that cause wilting and spotting of growing plants in the field.

Infection Prior To Harvest

All fruits and vegetables harvested from fields known to contain bacterial spot and wilt diseases should be examined carefully in order to cull out all those that show disease symptoms. Produce should be stored under refrigeration as is recommended for prevention of soft rot.

9-3.2 FUNGAL DISEASES

The diseases caused by fungi in storage are conveniently studied by dividing the stored products into two groups: those having (1) a low moisture content, and (2) a high moisture content. Cereal grains and other seed crops make up the major portion of products possessing a low moisture content. Most fruits, vegetables, tubers and root crops contain a high moisture content.

Stored Seed

The typical field-fungi that invade seeds as they are developing and maturing in the field require a high relative humidity of 90 percent or more in order to grow. In the cereal grains, seed at equilibrium would possess a moisture content of at least 30 percent (on a dry-weight basis). After harvest, grains are stored with moisture contents of 15 to 16 percent (on a dry-weight basis) or lower. The field fungi generally sporulate and slowly die because the moisture content is too low for them to grow. However, a new microflora of fungi, the storage fungi, may attack the grain. These are typically species of *Aspergillus* and *Penicillium*.

Microflora Of Stored Grain Is Different From That On Green Plants

Most of the different species of storage fungi require a narrow and particular range of moisture content within the seed for their growth. Therefore, as the moisture content changes, the attacking species of storage fungi change. The damage they cause may be one or more of the following: (1) reduce germination, (2) discolor the embryos, (3) produce heating, and (4) produce toxic by-products.

Reduce Germination

The reduction of germination appears to be partially caused by invasion of the embryos by the storage fungi. However, the respiration of the seed increases with both moisture and heating, and these adversely affect the viability of the embryo.

Discolor The Embryos

Seed of cereals that have darkened embryos are called "damaged" or "sick", and are heavily discounted in commerce. "Sick" embryos are dead, and are ramified by the storage fungi.

Heating

The heating may be caused by storage fungi and/or insects feeding on the grain. Heating is a direct response to metabolic activity of fungi and insects. The higher the moisture content of the grain, the more likely it is to become injured by heating. Once heating occurs, spoilage is swift and certain. If heating is allowed to go unchecked, the spoilage will spread throughout the grain mass.

Toxic By-products

Some storage fungi produce toxins that are of great concern because of their carcinogenic effect and extreme toxicity. A great deal of publicity was given to the death of 100,000 turkey poults in England in 1960. By the following year, the cause was traced to peanut meal infested with the storage fungus *Aspergillus flavus*, which produced a toxin called "aflatoxin".

Other toxins by other storage fungi are known and are currently the subjects of intense investigations. All of these toxins appear to be destroyed by heat, and are not carried over in processed foods.

Control

In stored grain and seed, disease control generally can be accomplished by careful harvesting and handling; thorough cleaning of the seed; storage in weathertight structures; maintenance of optimum storage temperatures and humidity; and most important, storage of only low moisture grain.

High Moisture Comodities

Crops with a high moisture content may be subject to attack by fungi belonging to such genera as *Alternaria*, *Helminthosporium*, and *Septoria*, etc., which cause spots and various irregular discolorations of the affected parts. Other species of fungi belonging to the genera *Aspergillus*, *Penicillium*, *Botrytis*, and *Rhizopus*, cause molds on stored produce. The molds may be alone, or in association with other fungi, and/or bacteria. Low temperature storage will limit the growth of all fungi and is therefore recommended.

Fig. 9.2 Neck rot of onions, caused by *Botrytis* sp. Infection usually occurs in dormant fleshy scales near the neck, prior to harvest. This disease is most severe when cool moist weather prevails before harvest. The disease progresses rapidly during storage. *(Courtesy of M. W. Gardner)*

Fungal infection may occur in the field, and the pathogens may continue to grow in storage and produce a number of disorders. An example of this is brown rot of stone fruits, caused by *Monilinia fructicola*, which can cause extensive rotting in just 24 hours if the temperature is near 70°F., and the humidity is high. Peaches, plums, cherries, and apricots are most commonly affected by this pathogen. Control is achieved by pre-cooling the fruit below 50°F., immediately after picking, then maintaining storage temperatures at 32°F.

Store Above Chilling Point

Not all commodities can be stored at the preferred temperature of 32°F. For instance, bananas are susceptible to chilling injury if they are stored at a temperature of 55°F. Such commodities must be stored at temperatures above their chilling point.

Apples containing scab lesions, caused by *Venturia inaequalis*, do not store well. The fruit lose moisture through the superficial lesions and become wrinkled and shriveled. Apples infected just prior to picking may not show scab lesions at harvest; however, the lesions may develop in storage, and the fruit may then become shriveled and dehydrated. Again, the control is pre-cooling and maintenance of cool storage temperatures.

Phytophthora infestans, the causal agent of late blight of potato and tomato, can incite a rot of potato tubers and tomato fruit in storage. Infection takes place in the field and develops readily in storage if the temperature is high, and slower if the temperature is low. Control is best achieved through a proper fungicide program in the field so that only clean potato tubers are stored.

Chemical fumigants are used to prevent the growth of storage molds during shipment of commodities such as citrus and grapes. Fungi from the genera *Penicillium* and *Botrytis* are among the most important of these storage molds.

9-3.3 VIRUS DISEASES

Virus Diseases In Storage Are A Result Of Field Infection

Virus disease symptoms, the most common of which is mosaic, often appear on stored vegetables. Often the symptoms are masked at harvest time and appear after a period of storage. The virus symptoms reduce the market value of produce. Virus diseases in storage are always a result of field infection, and there are no means of control of these diseases in storage. Control is achieved in the field through the use of resistant varieties and insect vector control.

9-4 SUMMARY OF GENERAL CONTROL MEASURES

Post harvest diseases may be prevented or controlled by: (1) preventing infection by pathogens in the field; (2) harvesting at the proper stage of maturity; (3) careful handling to prevent mechanical injury; (4) the use of proper sanitary measures to keep the products clean, including the storage warehouse; (5) storage under aerated and closely controlled humidities and refrigerated temperatures; and (6) the use of proper chemical treatments to disinfect the stored products.

Selected References

Eckert, J. W., N. F. Sommer. 1967. Control of diseases of fruits and vegetables by postharvest treatment. Annual Review of Phytopathology. 5:391-432.

Harvey, J. M., and W. T. Pentzer. 1953. The values of fumigants. pp. 844-850, In Plant diseases. U. S. Dept. Agr. Yearbook.

McColloch, L. P. 1953. Postharvest virus diseases. pp. 822-826, In Plant diseases. U. S. Dept. Agr. Yearbook.

________________. 1953. Injuries from chilling and freezing. pp. 826-830, In Plant diseases. U. S. Dept. Agr. Yearbook.

Ramsey, G. B. 1953. Mechanical and chemical injuries. pp. 835-837, In Plant diseases. U. S. Dept. Agr. Yearbook.

____________, and M. A. Smith. 1953. Market diseases caused by fungi. pp. 809-816, In Plant diseases. U. S. Dept. Agr. Yearbook.

Smith, W. L., and B. A. Friedman. 1953. The diseases bacteria cause. pp. 817-821, In Plant diseases. U. S. Dept. Agr. Yearbook.

__________. 1962. Chemical treatments to reduce post-harvest spoilage of fruits and vegetables. Botanical Review. 28:411-445.

Stakman, E. C., and J. G. Harrar. 1957. Principles of plant pathology. The Ronald Press Co., New York. pp. 332-398.

Winston, J. R., et al. 1953. Using chemicals to stop spoilage. pp. 842-844, In Plant diseases. U. S. Dept. Agr. Yearbook.

Wright, T. R. 1953. Physiological disorders. pp. 830-834, In Plant diseases. U. S. Dept. Agr. Yearbook.

____________, and E. Smith. 1953. Cuts, bruises, and spoilage. pp. 837-842, In Plant diseases. U. S. Dept. Agr. Yearbook.

CHAPTER 10

ENVIRONMENT--ITS INFLUENCE ON THE DEVELOPMENT OF INFECTIOUS DISEASES

10-1 INTRODUCTION

Plant Disease Triangle

In order for an infectious disease to occur, there must be a susceptible host available in a vulnerable state; the causal agent must be present and in a condition capable of inciting infection; and the environment must be favorable for the infection and establishment of the pathogen in the host. The environment then is a requisite factor for the development of an infectious disease.

There are two areas of environment, above ground and below ground, or the aerial environment and the soil environment. A plant is exposed to both areas and, therefore, both have a direct influence on the plant. The aerial environmental factors are, temperature, humidity, light rain, hail, snow, wind, etc. The soil environmental factors are the physical and chemical composition of the soil as well as temperature and moisture. The pH of the soil solution and the compaction of the soil are also important.

It would be erroneous to consider the effects of environment only as they pertain to the time of actual infection and development of disease. This is because the environment has a direct bearing upon the survival and well-being of both the pathogen and perennial hosts throughout the entire year. This discussion is separated into: (1) limitations imposed by climate, (2) environment just prior to and during the disease period, (3) environment during the balance of the year and (4) environment during the development of a foliar fungal disease.

10-2 LIMITATIONS IMPOSED BY CLIMATE

Regional Location Of Pathogen Due To Climate

Of great significance is the fact that climate places a limitation on the geographical location within which a pathogen can successfully survive. The climate is often a barrier against the spread of a pathogen from region to region.

In any particular region there is a rainfall and temperature average for each season of the year. This mainly determines the climate of that region. Generally speaking the length of the growing season, together with the average temperature and rainfall during the growing season, determine the crops of that region. However, this does not necessarily mean that pathogens have the same regional distribution as their hosts. The pathogens may have either a wider, more restricted, or the same distribution as their hosts. For example, the potato late blight fungus, *Phytophthora infestans*, is generally restricted to cool moist climates. It cannot survive the hot summer temperatures of the southern United States. However, the fungus is reintroduced on infected seed tubers each year, and since potatoes are grown as a winter crop in this region, the pathogen remains an annual problem.

The cotton root rot fungus, *Phymatotrichum omnivorum*, which can attack so many different host plants, is restricted to the southwestern United States because it cannot tolerate freezing temperatures or acid soils. It is highly destructive in the warm, alkaline, or neutral soils of the Southwest. This fungus is restricted, therefore, by climate and an environmental factor of the soil.

High Moisture And Temperature

Bacterial pathogens are also affected by environmental factors. A high soil moisture and soil temperature favor the severity of infection by the bacterium *Pseudomonas solanacearum*, the causal agent of southern bacterial wilt. This disease affects potato, tomato, tobacco, pepper, eggplant, peanut, and many other hosts. The disease is limited to areas and countries of the world where the soil never freezes.

10-3 BEFORE AND DURING THE DISEASE PERIOD

We should distinguish between the effects of environment upon, (1) the host, (2) the pathogen, and (3) the interaction of host and pathogen.

10-3.1 EFFECT UPON THE HOST

The Predisposed Plant

The situation whereby certain environmental conditions occur prior to infection and make a plant more vulnerable to disease is called predisposition. We tend to think of a predisposed plant as one that has been weakened or is unable to "untrack" its defense mechanism in a normal manner. For example, the plant may fail to form corky tissue to repair the wound caused by the pathogen, or it may fail to grow new roots to compensate for the roots destroyed by the pathogen.

Under conditions ideal for plant growth, the plant is often able to tolerate a certain degree of infection without apparent ill effect. However, under conditions that are less than ideal for plant growth, the plant is generally predisposed to infection and is less able to tolerate

Moisture Stress

infection. Such a situation exists when plants are placed under a moisture stress. The longer the plant is under the moisture stress, the more liable it is to become damaged by pathogens.

A period of wet, cool, cloudy weather will often retard the maturation of the host and keep it in a susceptible state for a longer than normal period of time. Warm, dry sunny weather may hasten the maturation of the host and frequently permit it to escape infection. Many crops are affected in this manner.

Water-soaking

Numerous authorities have cited water-soaking of leaves as predisposing the host to infection by many bacterial pathogens. The water-soaking of the leaves essentially is a replacement of intercellular air spaces and substomatal chambers with water. With moisture also on the outside of the leaf, there is then a continuous "avenue" of water from outside to deep inside the leaf. Both motile and non-motile parasitic bacteria find this watery "avenue" to their liking as they enter into the interior of the host leaf. Wind-driven rain, and shock, caused by a sudden drop in temperature, play a role in causing water-soaking.

Other environmental factors such as wind, rain, hail, and wind-blown sand, often have a direct influence on the well-being of the plant. Hail can cause many large wounds and is often responsible for furnishing avenues of entrance for pathogens such as *Ustilago maydis*, which causes common smut of corn. Wind-blown sand can cause an enormous amount of microscopic wounds on the host foliage, fruit, and stems. This damage is a blanket invitation to pathogens in the vicinity, to "come in".

10-3.2 EFFECT UPON THE PATHOGEN

Rule -- Moisture Requirement For Entry Of Many Pathogens

The first and most obvious environmental factor upon bacterial and fungal pathogens just prior to infection is moisture. Most bacterial and fungal foliar pathogens require free surface moisture in order to enter their hosts. Therefore, it is not surprising that many of these pathogens enter their hosts at night when free surface moisture is present as dew or light rain. Spores of fungi, such as the organisms that cause potato late blight and bean anthracnose, require a film of moisture in which to germinate and penetrate the host. However, some fungi such as the apple scab organism, have spores that may germinate in water and also in high humidity in the absence of free water. A few fungi such as the causal agents of powdery mildew of cucumber and powdery mildew of cereals and grasses, are capable of germination and penetration without free moisture.

Splashing and wind-blown rain is undoubtedly responsible for "driving" bacterial pathogens through open stomata into the substomatal chambers. Once inside, the pathogens are free to multiply, and infection may be quickly established if the plant is susceptible and the temperature favorable. In the majority of cases, bacterial disease infection is favored by warm temperatures of 65-85° F., and moist conditions where the relative humidity is 85% or above.

Warm, Moist Weather

Given the ideal moisture conditions, the limiting factor for spore germination and infection is usually temperature. This has been well worked out for many fungi including the apple scab fungus, *Venturia inaequalis*. The following table shows the time required for ascospore infection at various temperatures by this fungus:

The approximate number of hours of continued wet foliage required for primary ascospore apple scab infection at different air temperature ranges.

Average air temperature range during wet period.	Number of hours of continued wet period required for primary apple scab infection.
33-41° F.	48 hours
42	30
43-45	20
46-50	14
51-53	12
54-59	10
60-75	9
76-78	13

(Mills, W. O., and A. A. Laplante, 1951)

The optimum temperature range for infection is quite a broad one, 58-75° F., but infection can occur down to freezing! Few other fungal pathogens possess the ability to infect their hosts over such a wide latitude of temperature. For *V. inaequalis*, moisture is more often the limiting factor of infection than is temperature.

Temperature also plays a role in determining the type of spore germination such as occurs in some fungi of the Phycomycetes. For instance, the sporangia of the potato late blight fungus, *Phytophthora infestans*, will liberate motile zoospores at 12 to 15° C.; but at temperatures of 21-24° C., the sporangia germinate directly by the formation of a germ tube instead of zoospores.

Rule -- Temperature And Moisture

For both the aerial and the soil environments, temperature and moisture are the main factors influencing the development of most infectious diseases.

Rule -- Limiting Factors

Either moisture or temperature can be limiting in the infection and establishment of disease. If one is favorable, the other is generally limiting. If both are favorable for any length of time, the disease can spread and become severely damaging.

A further point should also be made of the fact that the optimum temperature for spore germination is not always the same as the optimum temperature for growth following germination.

Many plant parasitic nematodes are capable of surviving within a wide range of soil temperatures. In the southern United States, where the soil never freezes, high populations of nematodes may abound the year around. In the northern states the freezing of the soil reduces the population each winter. Therefore, in the spring, the soils of the northern states seldom contain the high level of nematodes that may be found in the southern states.

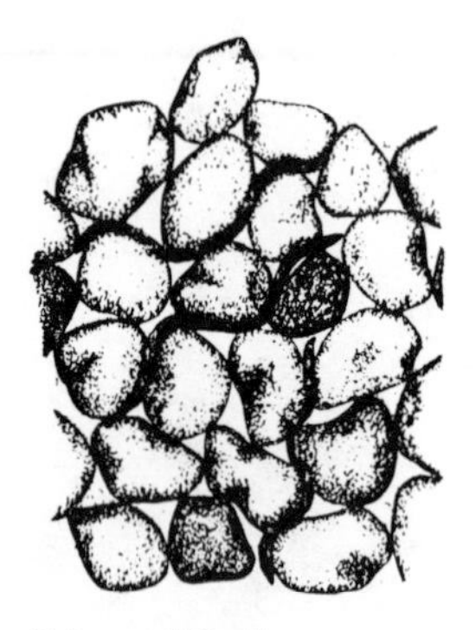

Fig. 10.1 Nematodes swim in moisture of soil pores.

The physical structure of the soil may be a limiting factor in the ultimate nematode population. For example, clay loam and silty clay loam soils cannot maintain as high a nematode population as sandy soils. This is because the nematodes exist in the pores between the soil particles. If the pore size is smaller than the diameter of the bodies of the nematodes, the nematodes are trapped and become prisoners in their own environment. This is what occurs in the clay loam and silty clay loam soils. The pore sizes in sandy soils are very large and they are not limiting. When given the proper hosts, temperature and moisture, nematodes may build up huge populations in sandy soils.

Except for resting forms, nematodes cannot survive long in dry soils or flooded fields. Soil moisture sufficient for good plant growth is about optimum for most soil nematodes.

Temperature Important In Virus Diseases

Temperature is the main factor influencing virus diseases. There is a multiplicity of effects. Viruses may be deactivated at high temperatures, and therefore be limited by regional or seasonal climate. Temperature may affect the concentration of virus particles in the host as well as the expression of disease symptoms. Of course, temperature and other factors that affect the insect vectors will indirectly affect the spread and occurrence of virus diseases.

10-3.3 EFFECT UPON THE INTERACTION OF HOST AND PATHOGEN

Disease results from the interaction of the host and pathogen. Therefore, the optimum temperature for disease development is often different from the optimum temperature for growth of the pathogen in culture. For example, the optimum temperature for the growth of the pathogen may also be optimum for growth of the host. If this temperature were maintained, the host could better tolerate the presence of the pathogen and the maximum disease development would not occur. The optimum temperature for disease development may occur at low temperatures that restrict the growth of the host.

Some Pathogens Need A Non-vigorous Host

Generally, the obligate parasites including the rust fungi prefer a well nourished or vigorous host in which to grow. The free-standing moisture and high humidity required for spore germination and penetration is no longer needed. What is needed is a temperature that is optimum for the best vegetative growth of the host. The pathogen lives with the host until it has ramified certain host tissues and is ready to sporulate. Sometimes a special set of conditions is needed for the maximum sporulation, also.

Rust Fungi Prefer A Vigorous Host

There are three combinations of temperature and humidity that favor a large number of plant diseases. Briefly summarized they are: (1) hot and dry; (2) warm and moist; and (3) cool and wet.

Hot-Dry, Warm-Moist, Cool-Wet

10-4 THE BALANCE OF THE YEAR

For perennial hosts in the temperate zone, an abnormally severe winter may injure them and reduce their vigor for for up to a year or more. Drought, hail, freezing, flooding, and sleet are factors that are capable of injuring the plant and reducing its vigor. This reduction of vigor makes the plant vulnerable to attack by pathogens for as long as the plant remains in this condition. By definition, this is actually predisposition.

Reduction Of Plant Vigor

At the end of the growing season, the pathogen then faces survival until the next growing season. In the southern United States, this may not be a problem for the pathogen since it may have other hosts available the balance of the year. In the temperate zone, however, the pathogen must survive freezing temperatures. Frequently, the amount of inoculum that is available the following spring is dependent upon the severity of the winter weather. A mild winter generally favors the survival of abundant inoculum. There is a greater likelihood that a disease will reach epidemic proportions when there is a large quantity of this primary inoculum, than when there is only a small amount.

Survival Of The Pathogen

The causal agent of bacterial wilt, or Stewart's wilt of corn, *Erwinia stewartii*, is carried overwinter in the bodies of the corn flea beetles. These insects serve as vectors for this pathogen. So it follows that mild winter temperatures that allow the insects to survive in large numbers will favor disease development the following growing season. This same situation may exist in bacterial wilt of cucurbits, caused by *Erwinia tracheiphila*. In this disease the overwintering vectors, two species of cucumber beetles, are thought to harbor the pathogen.

A number of fungal spores will not germinate unless they are exposed to freezing temperatures of alternating wet and dry periods. The fall, winter, and early spring weather of the temperate zone is, therefore, a necessity for some pathogens. For example, mummified fruit such as peaches,

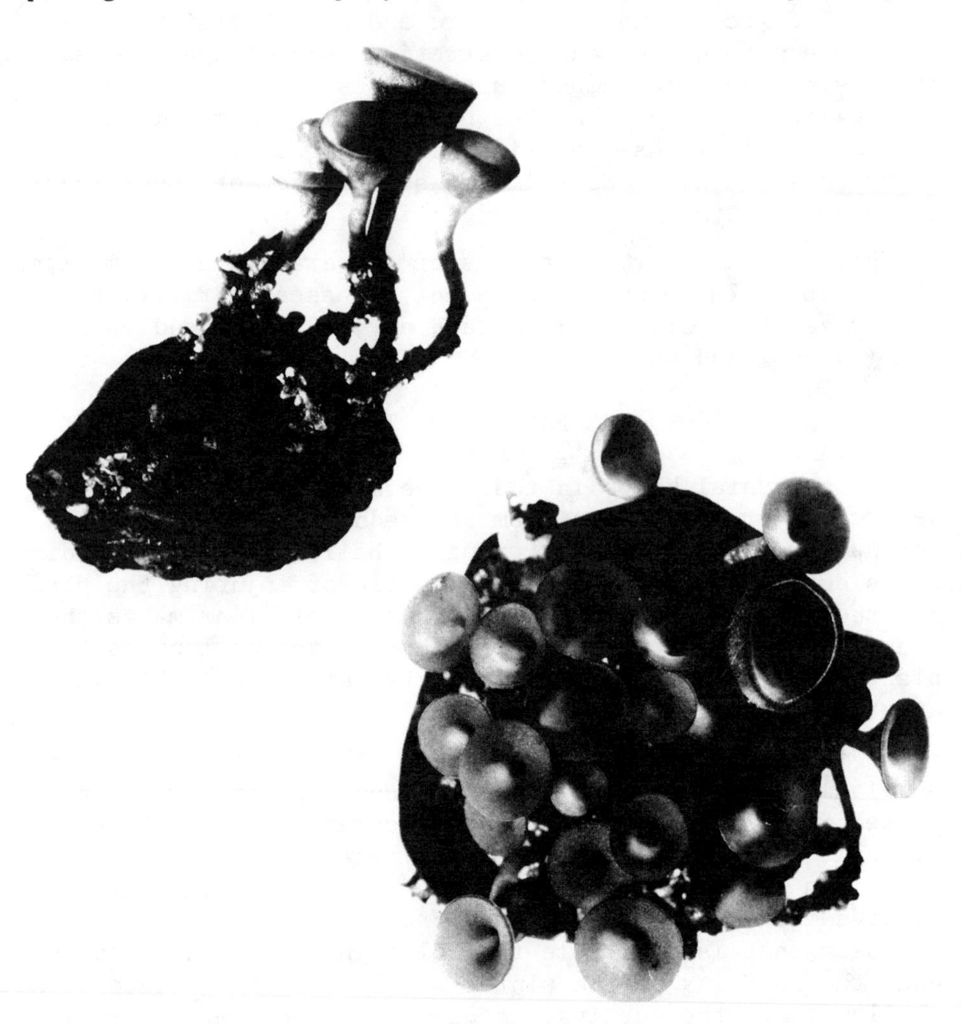

Fig. 10.2 Apothecia of the brown rot fungus, *Monilinia fructicola*, arising from mummies of a cherry-plum hybrid. *(Courtesy of E. G. Sharvelle)*

plums and cherries infected with the brown rot fungus, *Monilinia fructicola*, drop to the ground where they overwinter as pseudosclerotia. In the spring, the pseudosclerotia send up the apothecia in which the sexual ascospores are borne. The ascospores are part of the primary inoculum. An abundant soil moisture is necessary for the formation of the apothecia that develop well at temperatures of 17-20° C., but are soon killed if the temperature reaches 30° C.

It is obvious, therefore, that temperature and moisture are critical at several times during the life cycle of most pathogens. Each pathogen is different as to its requirements, and a condition that may be limiting to one pathogen may not be limiting to another.

10-5 ENVIRONMENT INFLUENCES THE DEVELOPMENT OF FOLIAR FUNGAL DISEASE

Most foliar fungal diseases take place when a proper sequence of environmental (mainly weather) conditions occur. A model of the development of a foliar fungal disease is illustrated in Fig. 10.3, in which the sequence of events during the development of the disease is presented in 5 distinct periods.

Number 1 is the pre-inoculation period. During this time, the host is producing its first susceptible tissue. Usually this tissue consists of the first seasonal leaves or seedling leaves, depending upon the host. Frequently, of importance at this time is the environmental conditioning that causes the predisposition of the host. This is often required for disease development. Environmental conditions that often predispose plants include: low soil temperatures; low air temperatures; high moisture levels; low moisture levels; unbalanced fertility; reduced light, etc. Of course, predisposition is not always required.

During this period, the fungus is also undergoing growth, usually reproductive growth. Overwintering fruiting structures mature and prepare to release or discharge their spores (the primary inoculum). Once the spores are airborne, they are carried by air currents and a few of them alight on a susceptible host (inoculum arrival). Major factors influencing the fungus during this period are temperature, humidity, rain, wind and light. The conditions for spore maturation are often not the same as those for spore release (or discharge), and subsequent spore flight.

The spores that form the primary inoculum contact susceptible host tissue, thereby initiating the usually brief period 2. Sometimes, these spores terminate their flight by

Fig. 10.3 Sequence of development of a model foliar fungal plant disease. *(after R. D. Schein, 1963)*

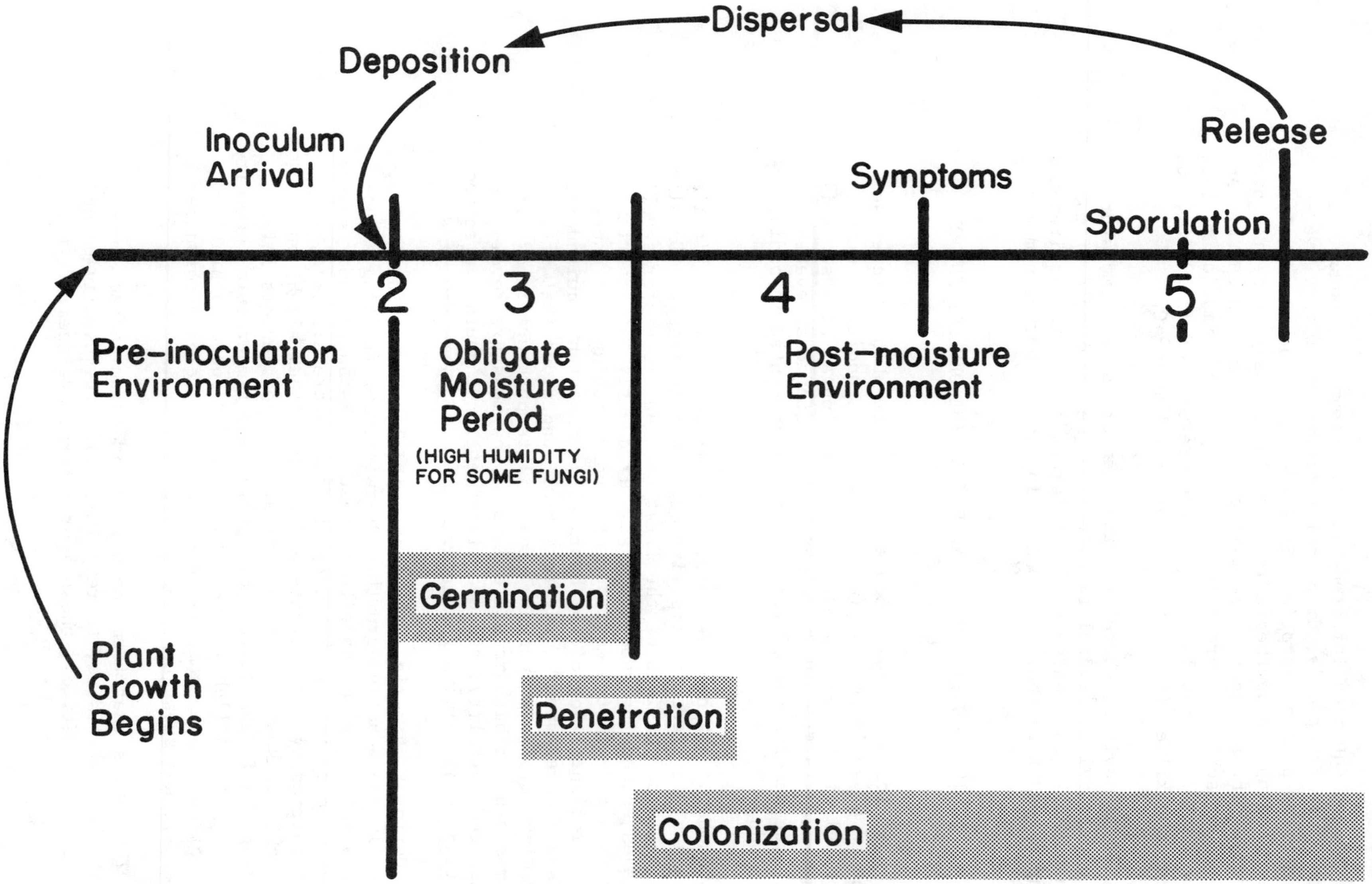

being swept from the air by falling rain. While the spores are on the foliage, they may be subject to wash-off by rain, become desiccated and die because of low humidity, or may be killed by a lethal exposure to ultraviolet light. Sometimes, spores are released from their source under conditions that also permit their germination.

Period 3 is the obligate moisture period, so termed because free moisture is often required for spore germination, germ tube growth and penetration. Many of the lower fungi also require free moisture for the liberation of zoospores and their swimming, germination and penetration. These processess are also temperature dependent. Most fungi possess a rather narrow temperature range within which their spores will germinate. An exception to this is *Venturia inaequalis*, the apple scab fungus (see page 177). A fungal spore will take longer to germinate and penetrate its host near the lower end of its temperature range than at the upper end of the range. Period 3 is seldom longer than 48 hours and usually is much shorter.

The germination period of the fungal spore is the most vulnerable point in the life cycle of a fungus. The germinating spore is extremely sensitive to changes in the environment. The fungi that do not require free-moisture during this period, usually do require a high humidity. Light and darkness are also factors during spore germination and penetration.

The first phase of colonization is in period 4 where both the pathogen and host are affected by the same environmental factors as in period 1. However, the factors that favor optimum disease development in period 4 are not always identical to those that are optimum for period 1. Disease symptoms appear in this period and the fungus soon initiates a reproductive phase of growth (period 5). Temperature and moisture are often paramount environmental factors during the growth of fruiting structures and subsequent spore formation. Often a high humidity is required for abundant sporulation (the secondary inoculum).

Spore release may be an "active" or a "passive" process. Usually, spores are airborne and become disseminated. Of course, during this spore release and transport, the spores are affected by the same factors that influence the spores of the primary inoculum. Many spores, especially thin-walled spores, are subject to desiccation during flight if the humidity is low. Also, light-colored spores are frequently sensitive to ultraviolet radiation of sunlight during flight, while dark-colored spores are usually much less sensitive. Viable spores may alight on susceptible host tissue, where they germinate and cause infection, etc., thereby repeating the disease cycle. A disease cycle may be repeated 6 to 10 times (or more) during the growing season.

The length of a fungal disease cycle, therefore, is dependent upon the sequential progress of changing conditions that influence each period of disease development, and each period is dependent upon the success of preceding ones.

Selected References

Agrios, G. N. 1969. Plant pathology. Academic Press, New York. pp. 161-172.

Hepting, G. H. 1963. Climate and forest diseases. Annual Review of Phytopathology. 1:31-50.

Miller, P. R. 1953. The effect of weather on diseases. pp. 83-93, In Plant diseases. U. S. Dept. Agr. Yearbook.

____________, and M. J. O'Brien. 1957. Prediction of plant disease epidemics. Annual Review of Microbiology. 11:77-110.

Stakman, E. C., and J. G. Harrar. 1957. Principles of plant pathology. The Ronald Press Co., New York. pp. 265-269 and 301-321.

Walker, J. G. 1969. Plant pathology. Ed. 3. McGraw-Hill Book Co., New York. pp. 632-667.

Yarwood, C. E. 1959. Microclimate and infection. pp. 548-556, In C. S. Holten, et al. (ed.), Plant pathology problems and progress 1908-1958. The University of Wisconsin Press, Madison.

____________. 1960. Predisposition. pp. 521-562, In J. G. Horsfall and A. E. Dimond, (ed.), Plant pathology. Vol. I. Academic Press, New York.

CHAPTER 11

THE PATHOGEN--PRODUCTION AND SPREAD OF INOCULUM

11-1 INTRODUCTION

It is fascinating to speculate on the evolution of pathogenic microorganisms. Undoubtedly, it took many centuries for these organisms to adapt themselves to the "parasitic habit". There has been an evolution of pathogens as well as an evolution of green plants. As many green plants have become extinct, in all probability, so have pathogens. The pathogens with us today have evolved characteristics necessary for their successful survival in an age of mass crop cultivation. Knowledge of these characteristics may be utilized in studying the influence of the pathogen on the development of infectious disease.

Successful Evolution Of Plant Pathogens

The very basis of the need for plant pathology is the fact that plant pathogens prosper and persevere year after year. Pathogens are obviously able to survive the cold winters of the temperate regions and the hot summers of lands near the equator. They survive from one growing season to another; and in some instances, persist for many years without the presence of a suitable host.

Immediately prior to their attack on host plants, pathogens exist in some form and quantity. This is the inoculum of the pathogen. Knowledge of this area of phytopathology is important, because infectious diseases cannot occur without inoculum. Also, a great many control measures are aimed directly at reducing the inoculum.

Rule -- Necessity Of Inoculum

What constitutes the inoculum, its source, release, and dissemination are factors that affect the success or failure of pathogens. Therefore, they also have a direct bearing on the development of disease, and a limited, generalized discussion of each factor is warranted.

11-2 INOCULUM PRODUCTION

11-2.1 TYPES OF INOCULUM

The complete assembled virus particles serve as inoculum of infectious viral pathogens and are the units of dissemination. Plant pathogenic bacteria typically form only vegetative cells. An exception to this is the genus *Streptomyces*, which forms highly resistant endospores. Therefore, the vegetative bacterial cells, and occasionally the endospores form the inoculum of bacterial pathogens. L-forms of some plant pathogenic bacteria may possibly serve as inoculum in certain cases. However, as yet, there is no proof that this occurs.

Fungi Produce Several Types Of Inocula

Fungi produce a wide variety of different types of inocula. The most common of these are various sexual and asexual spores and less common are mycelial forms. The latter may be hyphal fragments, dormant mycelium inside seeds or other host parts, or rhizomorphs and sclerotia. These serve the same purpose as asexual spores.

Inoculum of parasitic nematodes may be one or more of several life forms: the adult, the larval stage, or the egg. Of course, the inoculum of mistletoes is in the form of seeds.

Inoculum of some pathogens may be motile and capable of "micromovement" as in the case of some bacteria and zoospores of fungi. Most nematodes are capable of limited movement, and some are able to move a distance of one foot in search of a host. Except for nematodes, most inoculum entities are incapable of locomotion.

11-2.2 SOURCE OF INOCULUM

The term inoculum source refers to the origination of inoculum prior to its release and dissemination. The source is not as variable as one may think. Inoculum comes mainly from infected living hosts; dead plants or plant refuse; infested soil; infested and/or infected seed or other stored plant parts used for propagation; insects; and contaminated crates, shipping containers, storage areas, and equipment. Some nematodes and spores of lower fungi may also be considered to be sources of inocula of some viruses.

Rule -- Major Inoculum Source

Far-and-away the largest inoculum reservoir resides in living plants. These living plants may be infected annual or perennial hosts, alternate hosts, or weed hosts. The examples that could be selected to illustrate the living plant as inoculum sources are seemingly endless. Most viruses are harbored in living plants, and very often these hosts are not the economic crop plants,that are seriously affected. Weed hosts serve as the inoculum source for maize dwarf mosaic virus (MDMV) of corn. This disease appeared in Ohio in 1962 and quickly spread through several midwestern states. It is now found in several eastern, southern, and southwestern states. MDMV is aphid transmitted, and the host range is confined to the grass family. Johnson grass is the most important weed host, because it is a perennial, and the virus is able to overwinter in its rhizomes.

Some vascular disease pathogens, such as the causal agent of verticillium wilt, also colonize many weed hosts. These weeds, such as lambsquarter, pigweed, etc., serve as an important source of inoculum for this pathogen. An example of an economic host that harbors inoculum after becoming infected is the apple or pear tree that has cankers containing the fire blight bacteria. This is only one instance where the inoculum is in the host plant that the pathogen will reinfect under proper conditions.

Rule -- Secondary Inoculum Source

The secondary infection cycle that promotes so many epidemic diseases of annual tissue is highly important, and note should be taken of the principle: the sources of the secondary inoculum for plant diseases that have a secondary infection cycle are diseased, living or dying hosts. Since secondary inoculum comes from living hosts, it is easy to see why host plants form the bulk of the inoculum source, when viewed as a whole.

Examples of diseases for which plant refuse is a source of inoculum are also numerous. The fungal causal agent of apple scab overwinters as immature perithecia in fallen leaves. The bean anthracnose fungus overwinters in dead leaves, stems, and also in infected seed. Other pathogens for which seed is an inoculum source are the fungi causing loose smut of wheat and of barley, and downy mildew of soybean, and the bacteria causing bacterial blight of bean, etc. Viruses are often spread by infected seed or other plant parts that are used for propagation.

Soil is the source of inoculum for soil-borne pathogens. It forms an enormous inoculum reservoir for all groups of infectious pathogens. Such pathogens are usually called soil inhabitants or soil invaders. The soil inhabitants are usually strongly competitive with the other normal microflora of the soil and often survive many years in the complete absence of suitable hosts. Soil invaders, on the other hand, are generally less competitive with soil microflora and are usually shorter-lived. Many species of *Fusarium*, *Phytophthora*, *Pythium*, and *Rhizoctonia* are soil inhabitants. The cotton root rot fungus, *Phymatotrichum omnivorum*, the soft rot bacteria, some viruses and of course nematodes are soil inhabitants also.

Soil Inhabitants

Soil Invaders

Soil invaders include fungi such as: *Diplodia maydis*, the incitant of diplodia disease of corn; the common corn smut fungus, *Ustilago maydis*; and various species of *Colletotrichum*, the causal agents of anthracnose.

Fig. 11.1 Anthracnose of muskmelon, caused by *Colletotrichum lagenarium*. Fruit lesions are characteristically round or oval, sunken, and bear pink spore masses in acervuli. This fungal pathogen is often a soil invader. *(Courtesy of R. W. Samson)*

Resident bacteria: some pathogenic and non-pathogenic bacteria have been shown to persist and multiply on surfaces of leaves and fruit, in leaf axils and on and between bud scales of expanding buds of plants, without causing disease. The fire blight bacterium, *Erwinia amylovora*, has also been shown to multiply and persist in twigs and stems of symptomless apple and pear tissues (Keil and van der Zwet, 1972a). Leben, 1973, used the term "resident" to describe pathogenic bacteria that multiply in association with all parts of the healthy plant. He also used the term "resident phase" for the stage of growth in the life cycle of pathogenic bacteria that takes place in the rhizosphere, within the plant, or on the shoot surface.

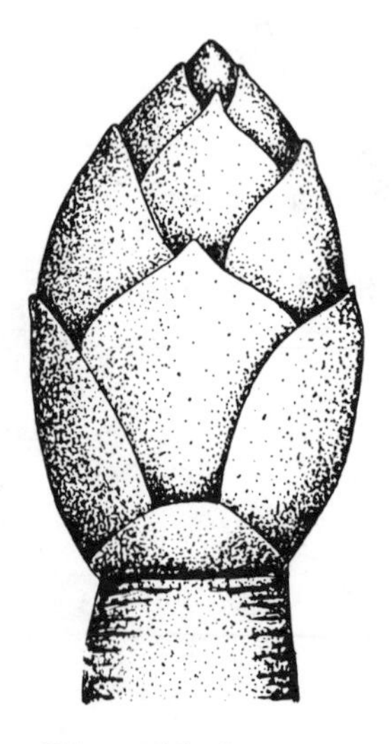

Fig. 11.2 Expanding bud.

The colonization of expanding buds by bacteria may be a common phenomenon of green plants. Inner parts of buds offer a moist, sheltered environment that permits multiplication of bacterial residents. Presumably, there are some nutrients available here. Such a location probably enables a greater multiplication and a more stable microflora than on leaf surfaces that are exposed to drying and ultraviolet rays.

Since pathogenic bacteria are in no way harmful to the plant while in the resident phase, their significance could be easily overlooked. Of prime importance is the fact that populations of these pathogens increase, and this materially increases the inoculum potential. This is to say, all of this inoculum is in the host; or is on the host and close to, or immediately adjacent to potential infection sites. In respect to the particular healthy host upon which the resident phase bacteria are multiplying, the chances of this host's becoming infected are materially increased. This inoculum is multiplying upon the host and does not undergo the fantastic numerical losses attendant with dispersal and deposition of inoculum from distant sources. Even a slight numerical increase of inoculum upon the host surface is equivalent to an enormous numerical increase of inoculum produced at a source distant from the host.

The sites of infection available to pathogens in external residence are injured and uninjured trichomes, stomata, and epidermal wounds so often caused by wind-blown sand. These sites are mentioned so as to emphasize their close proximity to the resident bacterial pathogens.

Although cells of the fire blight organism, *Erwinia amylovora*, are known to persist in living tissues of its hosts, the mechanism which triggers the change from the resident phase to the pathogenic phase is still unknown.

Other bacterial plant pathogens that appear to possess a resident phase include, *Pseudomonas morsprunorum*, the incitant of bacterial canker of plum; *P. syringae*, which causes a foliar blight of numerous hosts; *P. glycinea*, the incitant of bacterial blight of soybean; and *Xanthomonas vesicatoria*, the bacterial spot of tomato organism. Bacterial pathogens in residence are probably of great importance in the epidemiology of a number of diseases.

Resident Phase Important In Epidemiology Of Bacterial Diseases

11-3 INOCULUM SPREAD

11-3.1 RELEASE OR DISCHARGE OF INOCULUM

Many inoculum entities such as fungal spores, are carried from place to place by air currents and winds. They may be carried locally for a few feet, for several hundred feet, or over distances of several hundred miles. The question arises, how are these spores released in the first place? To study release of inoculum, we first need to categorize the inoculum as to passive release or forcible discharge.

Passive release: the teliospores of *Ustilago maydis*, the corn smut fungus, and *U. tritici*, the incitant of loose smut of wheat, are matured as dry powdery masses. The delicate membranes that surround the spore masses break away, and the spores are scattered by plant movements, the slightest breeze, or gusty winds. *Puccinia graminis*, the stem rust fungus, forms urediospores in pustules that push up through the host epidermis; the spores are detached at maturity.

The zoosporangium of the potato late blight fungus is the inoculum unit of dissemination. At maturity, the zoosporangia are readily and simply detached from sporangiophores and are airborne in a slight breeze, or scattered by splashing rain.

Bacteria are often exuded from infected plant parts, lesions or borders of cankers in sticky ooze. This ooze contains countless bacteria embedded in a sticky matrix. The bacteria may be splashed around by rain, or are carried upon the feet or mouthparts of insects attracted to it. Also, the ooze may dry down to a thin film. In this latter condition it is thought the dried film may "flake" off and become airborne and thereby clumps of bacteria are disseminated as units. Keil and van der Zwet, 1972b, have shown that bacteria of *Erwinia amylovora*, the fire blight pathogen, may be extruded from lesions and canker margins in the form of dried aerial strands. The aerial strands appear to be another form of the ooze, since both are composed of about four-fifths matrix and one-fifth bacterial cells. Dried aerial strands are easily broken into fragments by wind and become airborne,

Fig. 11.3 Fire blight of apple, caused by *Erwinia amylovora*. Droplets of bacterial exudate are clearly visible. These droplets contain masses of bacteria that are disseminated by splashing rain. This is an example of passive release of inoculum.
(Courtesy of the Department of Botany and Plant Pathology, Purdue University)

or are disintegrated by fine misty rain. Such strands of bacteria are therefore, disseminated by both wind and rain. The broken strands contain a number of bacterial cells which become disseminated as units. A passive release is characteristic of all bacteria.

Rule -- Release Of Bacteria

A reason for the success of passive release of many forms of inoculum is that they are usually produced above the ground. When such inoculum is dislodged and falls toward the ground, it may be carried aloft by the slightest air current.

Forcible ejection: the dwarf mistletoes and many fungi possess mechanisms of forcible ejection of their inoculum. For example, the sticky seeds of the dwarf mistletoes are often discharged with an explosive force as they mature.

Many fungal spores are borne inside highly specialized fruiting bodies, such as mushrooms, that are often formed at the ground level. The spores of such fruiting bodies must somehow escape, or be forcibly expelled if they are to become useful entities of propagation.

The ascospores of many Ascomycetes are borne in vertical rows in asci. At the top of each ascus is an opening called an ostiole. Ascospores may be forcibly ejected from asci, like water from a toy squirt gun or air escaping from a balloon. They may be propelled from 1/4 inch to 7 inches above the fruiting body. Ascomycete fungi that form the open, cup-shaped apothecia have a simultaneous expulsion of thousands of ascospores occurring at once, the discharge forms a visible spore cloud, and is commonly called "puffing".

Fig. 11.4 Puffing of ascospores

Ascospores that are borne in perithecia are not subject to puffing because the ostiolar openings in perithecia only permit the extrusion of one ascus at a time. In the spring, perithecia of the apple scab fungus, *Venturia inaequalis*, discharge their ascospores during periods of rainfall. Water wets the fallen leaves containing the perithecia and causes a chemical reaction within the asci. Increased internal pressure generated by the reaction elongates the asci. One at a time, the asci protrude out of the ostiole. The ascospores within the extended ascus are ejected into the air and the ascus snaps back, or retracts, and another ascus pushes out through the ostiole. The process is repeated over a several-week period until all spores have been discharged. In some other ascomycete fungi, ascospores are extruded from mature perithecia in gelatinous masses when the fruiting body is exposed to ample mositure. This is similar to the expulsion of gelatinous masses of conidia from mature pycnidia of such fungi as *Septoria* species. Moisture is absorbed by the gelatinous matrix within the pycnidia and triggers the quick discharge of the spores.

Ascospores Ejected Or Extruded From Perithecia

Ascospores Ejected From Cleistothecia

Powdery mildew fungi form the completely closed cleistothecia. This is an overwintering, dormant, or resting structure. Not all powdery mildew fungi discharge their ascospores in exactly the same manner. However, the following example will serve to illustrate this group of fungi. In the spring, the expanding asci rupture the cleistothecial wall and push outward. The elastic wall allows the opening to enlarge, and the protruding ascus is suddenly expelled with force as the jaws of cleistothecial wall suddenly snap shut. The ascus itself breaks apart releasing the ascospores to the vagarious action of air currents.

In the higher Basidiomycetes, fruiting structures reach a unique pinnacle of effectiveness for producing and liberating large and often massive numbers of the sexual basidiospores. This is apparent in the fruiting structures that contain exposed spores on the surface of gills, tubes or spines. The latter are often called gill fungi, pore fungi, and tooth fungi, respectively. They may be shelf fungi or of typical mushroom shape. During the development and growth of these fruiting bodies, a marvelous sensitive mechanism orients them so as to place the gills, tubes or spines in a vertical position. Each basidiospore is borne on a short projection called a sterigma, at the tip of a basidium. Each basidium usually bears four, or sometimes two, basidiospores. As a microscopic basidiospore matures, it is forcibly discharged horizontally a distance several times its length. The basidiospore then falls downward until it is below the lower edge of the fruiting body or cap, and then it may be carried aloft by the capricious air currents.

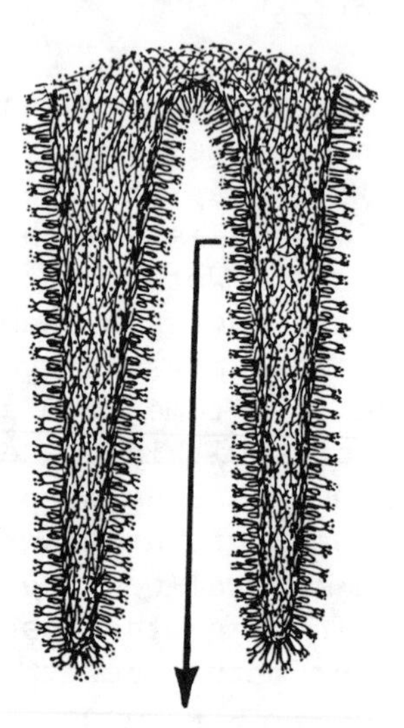

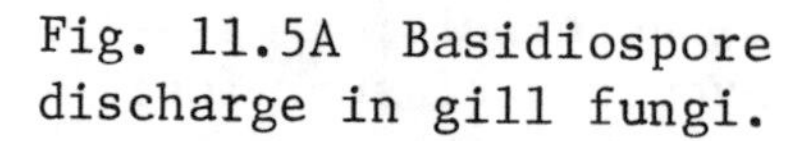

Fig. 11.5A Basidiospore discharge in gill fungi.

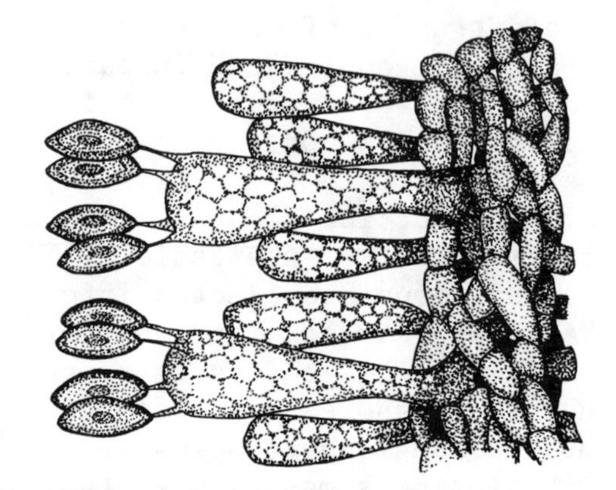

Fig. 11.5B Basidia and basidiospores of gill fungi.

These examples illustrate some of the major mechanisms of spore release and discharge, and are not intended to be all inclusive.

11-3.2 DISPERSAL AND DEPOSITION

Inoculum dispersal may be thought of as either local or distant dissemination. Local dissemination may occur by wind, water, insects, birds, small fur bearing animals, by man, and by combinations of these factors. Distant dissemination may occur by wind, insects, man and probably migratory birds. Wind is a principal means of inoculum dissemination. Fungal spores are so minute and have such a large surface area for their mass, that they fall with amazing slowness in still air. When there is even the slightest breeze, the spores may be swirled about and carried for a surprising distance. The most gentle breeze is all that is needed for the local spread of many fungal spores. Strong, turbulent winds are responsible for carrying some types of spores one to six miles high and for distances of thousands of miles. How far can they be blown and still remain viable? How long can they remain aloft and still remain viable? These and many other similar questions remain unanswered.

Wind Spreads Fungal Spores

The classic example of long distance dissemination via the airborne route is the urediospores of the wheat stem rust fungus. Each spring urediospores are blown northward from northern Mexico and many of them alight on young susceptible wheat plants in Texas and Oklahoma. In repeated infections, the urediospores hedge-hop northward following the emergence of the wheat crop into Kansas, hence to Nebraska, to South Dakota, North Dakota, and lastly to Saskatchewan. Of course, local spread within a field is also by air currents that carry urediospores from infected to healthy plants and also cause repeated infections of individual plants.

Wind dissemination and subsequent deposition of inoculum may be thought of as being "semi-random", meaning that the inoculum will be carried down wind, up and away; to come down, who knows where.

Wind Causes Semi-random Dispersal

Rain and windblown rain are responsible for the local spread of many fungal spores and bacteria. Since the presence of standing moisture is often a general requirement for infection, there is a greater probability of infection occurring when these pathogens are dispersed during rainy periods. The first raindrops of a spring or summer shower carry fungal spores down out of the air, terminating their airborne flight and bringing many of them in contact with susceptible hosts. Again, these spores generally cannot bring about infection without moisture on the host foliage, and it is indeed advantageous for them to be deposited onto a premoistened host.

Rain Spreads Fungal Spores And Bacteria

Raindrops scatter bacteria as they splash into the drops of exuded ooze. In this same manner, rain disperses ascospores and conidia that are extruded from their fruiting

Surface Adherence By Gelatinous Matrix

bodies in a gelatinous mass. The gelatinous matrix in which such inoculum is embedded serves a dual prupose, that of aiding the inoculum to adhere to host surfaces, and helping retain essential moisture.

Surface Adherence By Electrostatic Charge

Another major mechanism to assist the adherence of some fungal spores to the host surfaces with which they come in contact, is the electrostatic charge they possess. Conidia of powdery mildew fungi have a high moisture content and they carry such an electrostatic charge.

Dwarf mistletoe plants that grow high up in their host trees discharge sticky seeds that are subject to being blown by the wind to a lower branch of the same tree or to lower branches of neighboring trees. The sticky substance on the seeds helps them to stick to any branch they contact; this assures their survival. Since this seed is exceptionally heavy it is not carried far except by very strong winds The seed of dodder is also a heavyweight, when compared to most inoculum, and may fall to the ground at maturity or is harvested with crop seed. Dodder is often widely disseminated as a contaminant in crop seed.

Insects Spread Viruses And Some Fungi And Bacteria

Insects are responsible for disseminating a number of bacteria and fungi, and most of the viruses. Of great importance is the fact that insects are also agents of inoculation. When insects feed on the host plant they leave behind a quantity of inoculum in the open wound thus introducing the pathogen inside the host. Non-flying insects move about on the same plant and crawl to adjoining plants. Flying insects are capable of flying to adjoining plants and to plants in neighboring or near-by fields as well. Some flying insects may also be windblown and carried for a number of miles before they can alight on a host plant.

The slimy, sticky ooze characteristic of many bacterial diseases is often oderiferous and attracts insects. Some insects transport the bacteria on their legs and bodies and carry it locally to other plants.

The dissemination and subsequent deposition of inoculum by insect vectors may be thought of as being "methodical", relating to the fact that there is a relatively good probability of this inoculum finding its way into a host plant. This is because insect vectors usually feed on a limited number of specific hosts. Therefore, many insect vectors seek out plants to feed upon that are susceptible hosts for the pathogens they transmit.

Man Spreads Pathogens

Man himself has been responsible for the spread of disease agents; in fact, man is one of the foremost agents of long distance dissemination of foreign pathogens. Before countries established quarantine laws, pathogens were carried back and forth across the oceans, spreading diseases far and wide. The introduction of some foreign pathogens into the United States is discussed in Chapter 20. Within the United States, pathogens are often disseminated from state to state or location to location in infected seed and transplants, infested soil, and contaminated crates, bags, and equipment. Man is also responsible for spreading some fungi and bacteria within fields by walking between the rows when the foliage is wet and brushing the foliage as he goes.

Seed-borne Pathogens

Seed and plant parts used for propagating may carry pathogens internally (infected), or externally (infested). While a number of fungal pathogens are external and a few are internal, the viral and most bacterial pathogens are internal.

Birds And Small Fur-bearing Animals Spread Pathogens

The extent of dissemination conducted by birds and small fur-bearing animals is not known. These animals undoubtedly carry a large number of bacteria and fungi on their bodies, feet, and head parts. Birds are well known for their dissemination of some parasitic seed plants, the seeds of which pass through their intestines and are deposited in the droppings on branches of host trees. Birds are also known to disseminate cysts of the soybean cyst nematode *Heterodora glycines*, which are deposited in a viable state in the droppings.

11-3.3 TIMING OF INOCULUM RELEASE

Spore Release Coincides With Host Susceptibility

Many fungal spores would be of no use in propagating their species if they were not released at a time when their hosts were susceptible, or when the temperature and moisture were suitable for infection. Such critical timing is necessary for some, but not all, fungal plant pathogens. The passive release of teliospores of *Ustilago tritici*, the incitant of loose smut of wheat, coincides with the flowering of normal healthy wheat plants. Infection of the host can only occur in the flower, and the host becomes resistant to infection about a week after pollination. Therefore, only a short period of time exists when infection can occur.

Spore Release Coincides With Favorable Environment

In some plant diseases, the timing of inoculum release and dispersal must coincide with the proper environment that will allow the pathogen to enter the host. If this inoculum were released in an unfavorable environment, it would most likely be lost, since it is usually sensitive to desication and is short-lived. This situation is especially apparent in diseases whose pathogens are dependent upon their own growth processes to enter their hosts. Such timing of inoculum release is seen in the discharge of ascospores from perithecia of *Venturia inaequalis*, the apple scab fungus. The release of ascospores occurs when there is abundant free standing moisture and a corresponding high humidity. The moisture is necessary to allow the spores to germinate and infect the host. Furthermore, the ascospores are thin-walled and highly susceptible to death by desication.

Such timing is not required by pathogenic nematodes, viruses, and bacteria and fungi that are transmitted by vectors; or for most soil-borne fungi.

Selected References

Broadbent, L. 1959. Insect vector behavior and the spread of plant viruses in the field. pp. 539-547, In C. S. Holton, et al., (ed.), Plant pathology problems and progress 1908-1958. The University of Wisconsin Press, Madison.

__________. 1960. Dispersal of inoculum by insects and other animals, including man. pp. 97-135, In J. G. Horsfall and A. E. Dimond, (ed.), Plant pathology. Vol. III. Academic Press, New York.

Burchill, R. T. 1966. Air-dispersal of fungal spores with particular reference to apple scab (Venturia inaequalis (Cooke) Winter). pp. 135-140, In M. F. Madelin, (ed.), The fungus spore. Vol. XVIII of the Colston Papers. Butterworths, London.

Colhoun, J. 1966. The biflagellate zoospore of aquatic Phycomycetes with particular reference to Phytophthora spp. pp. 85-92, In M. F. Madelin, (ed.), The fungus spore, Vol. XVIII of the Colston Papers. Butterworths, London.

Corke, A. T. K. 1966. The role of rainwater in the movement of Gloeosporium spores on apple trees. pp. 143-149, In M. F. Madelin, (ed.), The fungus spore. Vol. XVIII of the Colston Papers. Butterworths, London.

Garrett, S. D. 1966. Spores as propagules of disease. pp. 309-318, In M. F. Madelin, (ed.), The fungus spore. Vol. XVIII of the Colston Papers. Butterworths, London.

Hirst, J. M. 1959. Spore liberation and dispersal. pp. 529-538, In C. S. Holton, et al., (ed.), Plant pathology problems and progress 1908-1958. The University of Wisconsin Press, Madison.

Ingold, C. T. 1960. Dispersal by air and water -- the take-off. pp. 137-168, In J. G. Horsfall and A. E. Dimond, (ed.), Plant pathology, Vol. III. Academic Press, New York.

___________. 1965. Spore liberation. Clarendon Press, Oxford. pp. 1-140.

___________. 1966. Aspects of spore liberation: violent discharge. pp. 113-132, In M. F. Madelin, (ed.), The fungus spore. Vol. XVIII of the Colston Papers. Butterworths, London.

Leben, C., et al. 1968. The colonization of soybean buds by Pseudomonas glycinea and other bacteria. Phytopathology. 58:1677-1681.

Muskett, A. E. 1960. Autonomous dispersal. pp. 57-96, In J. G. Horsfall and A. E. Dimond, (ed.), Plant pathology. Vol. III. Academic Press, New York.

Schrodter, H. 1960. Dispersal by air and water -- the flight and landing. pp. 169-227, In J. G. Horsfall and A. E. Dimond, (ed.), Plant pathology. Vol. III. Academic Press, New York.

Stakman, E. C., and J. G. Harrar. 1957. Principles of plant pathology. The Ronald Press Co., New York.

Wolf, F. A., and F. T. Wolf. 1947. The fungi. Vol II. John Wiley and Sons, Inc., London. pp. 166-265.

CHAPTER 12

THE PATHOGEN--INOCULUM DYNAMICS AND ENTRY INTO PLANTS

12-1 DYNAMICS

12-1.1 AMOUNT OF INOCULUM

When we become aware of the success enjoyed by pathogens at our expense, we have to conclude that all too many of them are highly effective in the art of propagation and survival. In order to be so successful, pathogens must produce enormous quantities of inoculum, and indeed they do.

Reproduction vs. Size Of Organism

A principle of biology is: in respect to all living organisms, the smaller the organism is, the greater is its rate of reproduction. There are exceptions to this principle, of course; however, for the vast majority of living organisms, it holds true. Applying this to plant pathogens, we find that they have a much greater rate of reproduction than do their hosts.

The simplest illustration of logarithmic growth is a bacterium that divides by fission to produce 2, 4, 8, 16, etc., cells. Logarithmic growth is characteristic of unrestricted growth and development of populations of living organisms. However, all living organisms have restrictions upon their continued growth and are usually unable to maintain logarithmic reproduction for an extended period of time.

Reproductive Potential

The reproductive rate of individuals of any group of similar organisms never equals the biotic potential.

Pathogens that possess secondary infection cycles cause diseases that may be called multiple-cycle diseases. We are highly concerned about these diseases, since their pathogens are capable of logarithmic increase of inoculum during the growing season and thereby incite epidemics. These pathogens are also sometimes termed "explosive" to indicate their sudden ability to cause widespread disease. Characteristically, pathogens that lack a repeating stage increase intermittently with time during the growing season, and incite single-cycle diseases.

Corn smut is a single-cycle disease, and most of the inoculum is released near the end of the growing season. However, vast quantities of spores are produced annually by this fungus. In fact, one cubic inch of gall contains about six billion spores. At this rate, a gall of average size would then contain around 25 billion spores. Furthermore, one acre of corn, with galls on 2 percent of the plants would produce an astonishing 10,000,000,000,000 (10 trillion) smut spores. Since over 70 million acres of corn are produced in the United States each year, we get a clearer picture of the colossal amount of corn smut inoculum produced each year. Of course, not all plant parasitic fungi produce such astronomical quantities of spores, nor do their hosts occupy such vast acreages. However, this is illustrative of fungal potential.

Fig. 12.1 Corn smut, a single-cycle disease, is caused by *Ustilago maydis*. Galls occur on any part of the plant where an embryonic tissue is exposed to infection. This includes stem, leaf, axillary bud, tassel, or ear as is shown above. Galls are at first light colored but become dark, owing to formation of spore masses. *(Courtesy of A. J. Ullstrup)*

Multiple-cycle foliar pathogens, include such fungi as those that cause stem rust of wheat, apple scab, anthracnose, powdery mildews, etc. These plant pathogens possess the capability of logarithmic inoculum production. A specific illustration is the work of Lin, (1939), whose technicians laboriously and tediously counted conidia and pycnidia of the celery late blight fungus, *Septoria apii*, (now *S. apiicola*). This author found an average of 3,675 spores per pycnidium and an average of 56 pycnidia per lesion. Therefore, about 200,000 spores are formed in an average lesion that is incited by a single infective spore! When just a small number of these spores incite separate infections, the increase of inoculum is logarithmic.

Nematodes are restricted in population growth because of their limited dispersal and the lack of adequate host roots. It is not surprising to find the relative increase in numbers of nematodes is greatest when the number of nematodes is low; but over a number of years, this increase is not logarithmic. According to Spears, (1968), the USDA assumes about a tenfold increase of cysts of the golden nematode per year in fields in the Long Island area that are under

Golden Nematodes Increase Tenfold Per Year

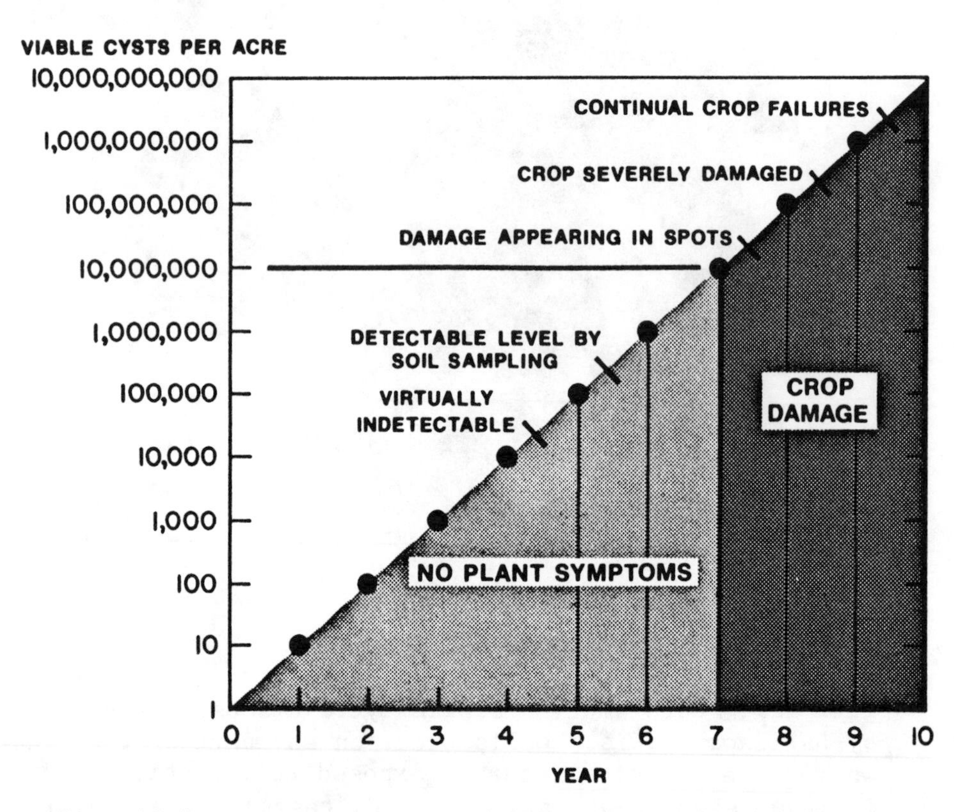

Fig. 12.2 Increase in golden nematode populations in a continuous potato culture, assuming a tenfold increase. *(After Spears, 1968)*

continuous potato production. They further find that spotty plant damage occurs when there are about 10 million viable cysts per acre. The potato crop is severely damaged when there are 100 million to one billion cysts per acre. Continual crop failures occur when there are one billion to 10 billion cysts per acre. Furthermore, it is possible for a potato plant to have 50,000 cysts, and each cyst may contain up to 500 eggs, (see Fig. 12.2).

Viruses produce unbelieveable amounts of infectious entities. However, the bulk of these are never transmitted from their hosts, and are destroyed as the host dies. The major limiting factor of spread of virus inoculum is the population of efficient insect vectors.

Rule -- Limiting Factor Of Virus Spread

12-1.2 SURVIVAL OF PATHOGENS AND THEIR INOCULUM

Within the fungi there is a great variation as to the ability of various spore forms to survive. Thick-walled, and often dark colored spores are often capable of survival over long periods. Sometimes these spores are dormant and cannot germinate before a period of rest. This dormancy may be controlled by internal chemical or physical means, or by external chemical or physical means. Dormancy is an aid to survival.

Dormancy Aids Survival

Fungal spores residing in the soil may be prevented from germinating and growing if essential nutrients are lacking, by antagonistic microflora that may or may not secrete antibiotics, or by fungistatic products of decomposition of various plant residues. On the other hand, susceptible host roots may secrete products that stimulate and/or enable the spores to germinate and grow. Such a situation is to be found in many of the cyst forming nematodes of the genus *Heterodera*. The cysts are exceptionally long lived, and often require an excretion from susceptible host roots in order to stimulate the hatching of eggs and breaking of the cysts.

Some spores of plant parasitic fungi are formed only near the end of the growing season with the advent of dieing of the host, or when weather is unfavorable for further vegetative growth. Many of these spores are capable of surviving periods of inclement weather in which the fungal mycelium is killed. We may rewrite this: many plant pathogenic fungi survive unfavorable growing periods in a dormant stage, which is more resistant than the vegetative stage. Sometimes this is an immature sexual sporocarp, or an asexual sporocarp. It may also be a mycelial modification such as a sclerotium or rhizomorph, both of which are highly resistant structures.

Rule -- Dormant Stage

Many fungal spores are thin-walled, hyaline, fragile, and often short lived. These spores usually require specialized environmental conditions to permit their germination and penetration into the host. Such spores may be more sensitive to environment than the mycelium from which they were produced. These spores often serve as the secondary inoculum.

When we think of survival of plant pathogens, we should consider their means of survival during periods of unfavorable growth. In the temperate zone, this is usually during winter; in warm climates, this is often during the hot dry summer.

Temperature Sensitivity

Most fungi will grow between 40 to 130° F., but the optimum temperature range is generally 70 to 90° F. As the temperature decreases, fungal growth is progressively less. When the temperature reaches between 30 to 40° F., most fungi cannot grow, but they do not die. They remain dormant, waiting for better times. Many fungi can withstand freezing temperatures for months or years. In the temperate zone, some parasitic fungi overwinter in a perennial host because they could not survive otherwise. Generally speaking, low temperatures are not effective in eliminating fungi, but high temperatures, 130 to 150° F., will soon kill most of them.

High Temperature Is Lethal

Fungal growth is possible only in moist environments. For example, most fungal molds will grow if the relative humidity is constant and exceeds 70 percent. If the humidity is constant and exceeds 75 percent, mold growth will occur in great abundance. However, this growth will soon cease when the relative humidity falls below 70 percent. In a dry environment, most fungi generally stop growing, produce abundant spores, and mycelia die. However, a dry environment will not kill many types of fungal spores, it simply inhibits their growth.

Fungal Growth Needs Moist Environment

Bacteria Survive:

In Or Near Living Or Dead Tissue

In Hypobiotic State

In Masses of Cells

As Residents

Most plant pathogenic bacteria do not have recognized survival structures. Individual cells of these bacteria not associated with plant tissues probably survive in nature only hours or days. Leben, 1973, suggests that long-term survival of these bacteria--"takes place in or near living or dead plant tissues." He further suggests that masses of hypobiotic bacteria (mature cells with reduced metabolism) are more likely to survive than metabolically active bacteria. Pathogenic bacteria may also survive as residents on or in healthy hosts or non-host plants. Also, bacteria are more likely to survive in masses than as individual cells.

Plant viruses, viroids and mycoplasma-like organisms usually survive in living plant or vector tissue. However, a few viruses, such as tobacco mosaic virus, can survive for long periods in crop debris.

All released inoculum that is within an environment that prohibits its growth, is subject to a death rate that is dependent upon factors, many of which are immediately obvious. The death rate may be rapid or slow; often it is erratic, but never is it uniform. The death rate of a pathogen directly influences the useful control procedures that may be employed against it. For example, crop rotation is useful where soil-borne pathogens survive one year, but fail to survive two or three years.

Death Rate Of Inoculum Is Never Uniform

A number of plant parasitic nematodes are able to survive in soil or plant debris for long periods of time without the presence of host plants, while others are not. Some nematodes possess specialized stages for survival, but no one such stage is common for all pathogenic species. Eggs, either encased in cysts or free in the soil, and second or fourth stage larvae, are most often recognized as resting stages.

Nematode Resting Stages

The golden nematode, *Heterodera rostochiensis*, has survived 8 years as eggs in cysts, in field soil. A related species, the sugar beet nematode, *H. schachtii*, has remained alive as eggs in field soil for 6-7 years. The bulb and stem nematode, *Ditylenchus dipsaci*, has survived 20-23 years as fourth stage larvae in dried plant material. The longevity of root knot nematodes appears to be considerably less; one species was found to survive 6 months as second-stage larvae in field soil.

A major food reserve in infective larvae of some nematodes is fat. This food reserve and the reduced metabolism enable infective larvae to survive extended periods in the advent of unfavorable environment. This is usually a cold, or a warm and dry period; and it brings about a reduced metabolism, movement, growth, and development. This is called "quiescence" by Van Gundy, (1965), who further states this term: "most accurately describes dormancy in nematodes."

The root knot and cyst nematodes are typical sedentary endoparasites in which the second stage larvae are infective and may act as resting forms, also. The migratory endoparasites and ectoparasites are forms in which all stages are infective and they do not appear to have particular resting stages.

The periods of molting when a new cuticle and stylet are being formed, are the weakest links in the life cycle of nematodes. At these times, nematodes appear to be most susceptible to changes in their environment, especially to deficiencies in oxygen.

Even though nematodes may possess resting stages that ensure their survival, one to several years without a host crop will grossly lower the population within a field. An excellent example of this is the golden nematode, where the

Absence Of Host Crops Will Reduce Nematode Population

absence of host crops for one year may reduce the number of viable cysts in the soil by 50-80%. However, after this time, the death rate of cysts continues at a lower rate, and some cysts may remain viable for a number of years.

12-1.3 INOCULUM POTENTIAL

Inoculum Potential Defined

The term inoculum potential has been variously defined. A slight modification of that given by Dimond and Horsfall (1960) is: the amount of tissue invaded per infection and the number of independent infections that may occur in a population of susceptible hosts at any time or place. The amount of tissue invaded per infection may vary from a few cells, as in some local lesion foliar diseases, to the entire plant, as occurs in many vascular diseases.

Optimum Infection Needs Optimum Environment

The word "potential" in the term inoculum potential takes into account the relationship of environment to the amount of infection that can occur. Optimum infection can occur only when the environment is optimum, and this varies for each crop and pathogen.

Rule -- Incomplete Susceptibility

Only a portion of any given mass of inoculum is capable of inciting infection. The reasons are several-fold. First, a natural population of inoculum has non-viable as well as viable entities. Second, some infectious entities may be viable, but lack the vigor to incite infection. Third, during their dispersal, the viability and vigor of some infectious entities are lost. Fourth, even though host plants are susceptible, most if not all, still possess a limited and variable measure of resistance. This resistance prohibits the establishment of infection by some viable inoculum. This concept was illustrated in the writer's laboratory using a single race of *Erysiphe graminis* var. *tritici*, the powdery mildew fungus of wheat, (Ghemawat, 1968). It was found that 37 percent of the conidia failed to infect a susceptible wheat and 89 percent failed to infect a resistant wheat. These data were based upon conidia that had attempted to enter the host plants. In this situation, the two major resistance mechanisms appeared to be similar in both susceptible and resistant wheats. However, both resistance mechanisms were operating at a lower level of efficiency in the susceptible wheat than in the resistant wheat.

Several To Many Spores Per Infection Site

An important concept is that a single spore may incite infection. However, in practice, infection should be thought of as a matter of probability and the number of spores required per infection site will vary according to circumstances. Using the example of powdery mildew of wheat as cited in the previous paragraph, out of 3 spores, 2 infected the susceptible host; out of 10 spores, only one infected the more resistant host. What is important is that this experiment was conducted under conditions that were close to optimum for germination and penetration by the pathogen. How many spores it would take to obtain infection under various suboptimum conditions is a matter of speculation. In inoculation studies, McCallan and Wellman, (1943), found the infection of tomato with spores of the late blight fungus, *Phytophthora infestans*; the early blight fungus, *Alternaria solani*; and leaf spot by *Septoria lycopersici*, to be 6.5, 1.7, and 0.2 percent, respectively. Or in other words, to incite a single lesion it took 15.3 spores of the late blight fungus, 57 spores of the early blight organism, and 526 spores of the septoria leaf spot fungus.

This last set of data should be interpreted in the light that the pathogenicity of the three organisms is unknown, and this data may or may not parallel field conditions.

It would appear that a single infective virus particle could incite infection if placed in the proper susceptible site. However, it is highly improbable. Purified virus suspensions are often applied to local lesion hosts to obtain an approximation of the infective concentration of dilution. The progressive dilution for which no disease symptoms are manifest is called the "dilution end point". However, there are still an enormous number of virus particles in this dilution. Plants and plant parts vary in their susceptibility to infection, inoculation techniques are crude and inefficient, and many particles in the virus suspension are non-infective. These factors obscure the results, but do not explain why 100,000 to 1,000,000 particles of a highly infectious virus, such as TMV, are necessary to produce a single local lesion on a highly susceptible host leaf. Also, it does not explain why an aphid vector must possess about 1,000 stylet-borne particles of cucumber mosaic virus on the distal-end of its stylet in order to be able to infect a tobacco host (Walker and Pirone, 1972). This is several orders of magnitude less than the 100,000,000 particles required for successful manual inoculation per single local lesion by this particular virus (more particles of CMV are required for infection than that of TMV). However, it is indicative of the efficiency of aphid transmission and of the inefficiency of manual inoculation.

Plant infections by bacteria are also more likely to be successful if bacteria are present at infection site in substantial numbers. For example, Crosse et al., 1972, inoculated injured apple shoots with the fire blight pathogen, *Erwinia amylovora*, and found the median shoot infection was caused by an estimated 38 bacterial cells.

Another important concept is that one point of infection seldom results in plant disease. This is most dramatically pronounced in the non-systemic diseases. However, the same concept is true for many vascular wilt diseases, whose pathogens enter their hosts through the root system or stem. The root-infecting parasites enter through young rootlets or roots, through wounds caused by nematodes, or those formed in the cortex by emerging rootlets. Entry through the stems is through wounds. Many vascular disease pathogens will only spread quickly and freely throughout the plant when numerous points of infection have occurred, (Sadasivan, 1961, and Presley et al., 1966). The presence of mass inoculum in some manner overwhelms the limited resistance of the host, which often confines and limits the spread of the pathogen in isolated infections. This may be summarized by the statement: many diseases require numerous points of infection on a susceptible host plant in order to cause disease in that plant.

Rule -- Multiple Infections

12-2 ENTRY INTO PLANTS

Rule -- Pathogen Entry

In respect to the mode of entry into host plants, parasite seed plants, nematodes and most fungi are active, forceful and dynamic; many bacteria enter actively, but not forcefully; viruses, viroids, and mycoplasmas are completely passive. With but a few exceptions, viruses are dependent upon being deposited inside a host plant, usually by a vector or crossing through a living bridge formed by root grafts or through dodder.

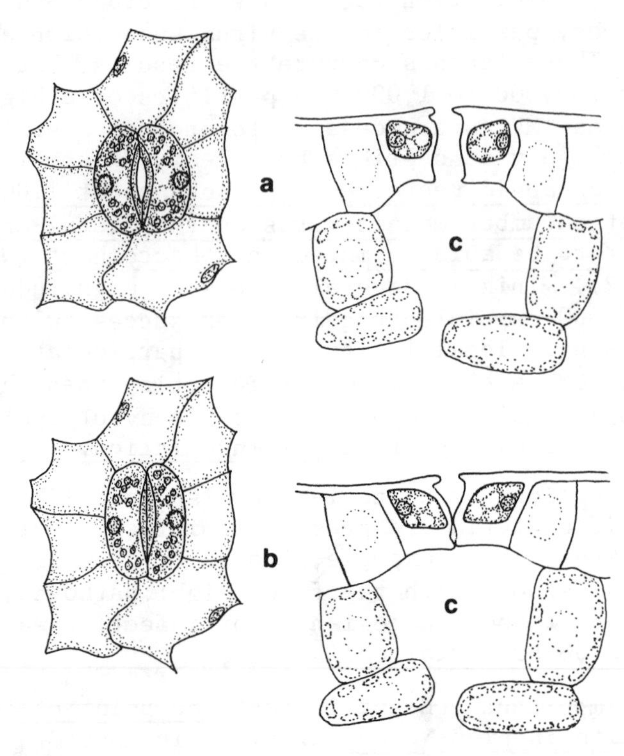

Fig. 12.3 Stomata in top view and cross-section; open (a), closed (b), and substomatal cavity (c).

12-2.1 FUNGI

Some fungi enter plants through natural openings in the host, or through wounds. These portals of entry are utilized by bacteria, as well. However, many fungi penetrate their hosts directly, which bacteria cannot do. In this capacity, fungi perhaps demonstrate one of their most dynamic capabilities. Once fungal spores germinate, they may take only one to two hours to enter their hosts.

A Dynamic Capability

The motile zoospores of *Plasmopara viticola*, the causal agent of downy mildew of grape, are attracted to the host stomata by a chemical stimulus. The zoospores come to rest and germinate. The germ tubes pass through the stomata and into the substomatal cavity. Subsequently mycelia develop intercellularly in the host. *Puccinia graminis* var. *tritici*, the causal agent of stem rust of wheat, also enters through stomata. Furthermore, entry may also occur even when the stomata are closed. A urediospore of the rust fungus germinates in the dark of night to form a germ tube that grows over the stoma. At the tip of the germ tube a swelling forms that is closely appressed to the host surface; this is called the appressorium (pl. appressoria), and much of the protoplasm accumulates here. This leaves the older parts relatively empty. With the advent of light, a tiny projection, called a penetration peg is formed on the underside of the appressorium and passes between the closed guard cells into the substomatal cavity. A substomatal vesicle is formed and infection hyphae grow outward to contact and penetrate into adjacent host cells. Once the thin hyphal strand penetrates into a host cell, a swelling is formed that invaginates the host protoplast. This is an organ of absorption, and is called a haustorium (pl. haustoria).

Entry Through Open Or Closed Stomata

The germinating spores of *Cercospora beticola*, which causes beet leafspot, also enter through stomata, however, entry is only through open stomata. Some fungi are able to enter through stomata and also penetrate directly through the epidermis. One such fungus is the potato late blight organism. However, its usual mode of entry is by direct penetration.

Entry Through Open Stomata

Hydathodes are less often portals of entry for fungi; however, guttation drops at or near the leaf tips allow some fungal spores to germinate and send their penetration hyphae through the hydathodes into the host.

Lenticels serve as avenues of entry for some fungi, most of which also enter through wounds. The fungus *Monilinia fructicola*, which causes brown rot of stone fruits, is one of these. Other fungi belonging to the genera *Penicillium* and *Rhizopus*, are strictly wound parasites, capable of invading their hosts only through fresh or unsuberized wounds.

Lenticels

Wounds

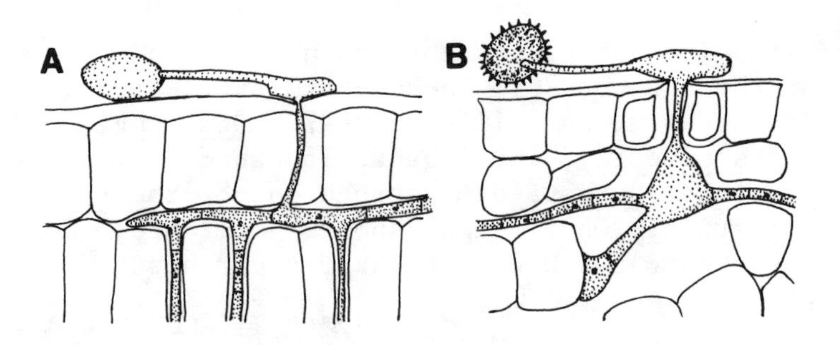

Fig. 12.4 Entry into host plant
A. Direct penetration of a germinating spore. Notice the appressorium, initial intracellular hypha, and the later intercellular hyphae.
B. Entry through a stomata by a germinating spore. Notice the appressorium over the stomatal opening and the enlarged hypha (vesicle) in the substomatal cavity, from which branching hyphae ramify the host.
(After Cummins, et al., 1950)

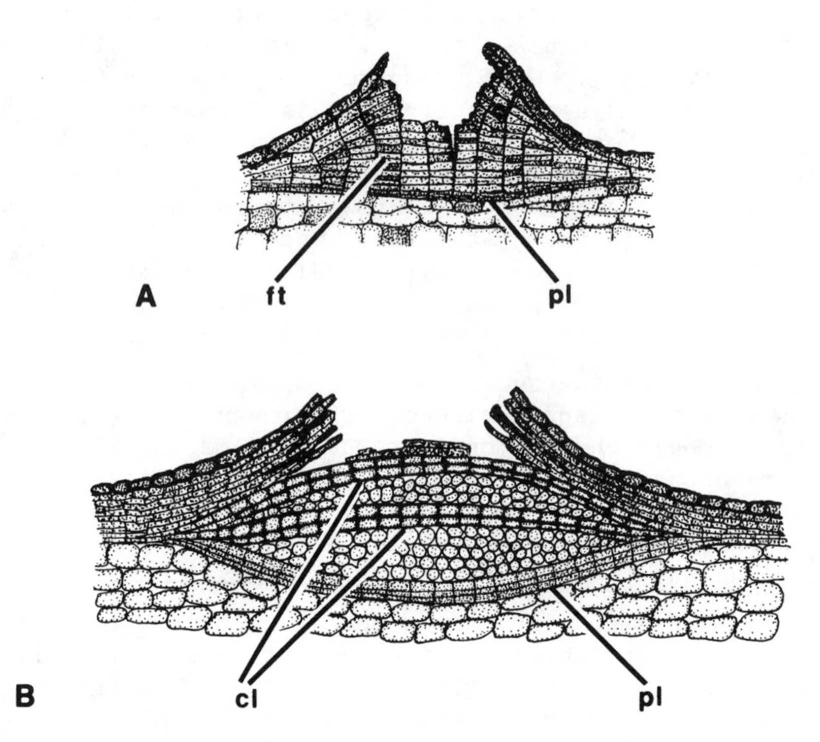

Fig. 12.5 Schematic drawings of two different types of lenticels.
A. Young lenticel with filling tissue (ft), and phellogen (pl); without closing layers.
B. Young lenticel with suberized closing layers (cl), and phellogen (pl).

Direct Penetration -- By Germinating Spores

The direct penetration of the host epidermis layer is often accomplished by a germinating spore that forms an appressorium and sends a penetration peg through the host cuticle and cell wall. The fungus *Colletotrichum lindemuthianum*, which causes bean anthracnose, is an example of this means of entry. The fungi *Venturia inaequalis* and *Diplocarpon earliana*, causal agents of apple scab and black spot of roses, respectively, penetrate directly through the cuticle of their hosts and grow beneath the cuticle, but above the epidermal layer, forming a stromatic mass of hyphae. They do not enter the host cell. The smut fungi usually penetrate directly into their hosts.

The powdery mildew fungus, *Erysiphe graminis* var. *tritici*, which attacks wheat, penetrates directly through the host epidermis and forms a haustorium with several finger-like lobes projecting from opposite sides of a central swelling. An aerial mycelium is then formed that penetrates near-by epidermal cells in which new haustoria quickly form.

Direct Penetration By Mycelium

Penetration of hypocotyl and root tissue by the direct activity of mycelium is commonly accomplished in one of two general ways: by single penetrations from isolated hyphae, or multiple penetrations from a mass of hyphae. Appressoria may or may not be formed, but penetrations are by tiny penetration pegs. The ubiquitous soil-borne *Rhizoctonia solani*, which causes damping-off, root rot, and stem canker on numerous hosts, is known to infect hosts in either manner. However, it appears that various strains of the fungus differ in their preference or ability to use one method or the other. Multiple penetration by this fungus is initiated by knots of hyphae forming on the outer surface of the host tissue. This aggregation of hyphae becomes rather closely knit, and takes on a dome-shaped appearance. It is appropriately termed an infection cushion, which is essentially a multiple appressorium. Various *Fusarium* spp. also form infection cushions.

Infection Cushion

The cotton root rot fungus, *Phymatotrichum omnivorum*, surrounds its host root with a cottony growth of mycelium that proceeds to penetrate the outer walls of root cells or root-hairs around the root.

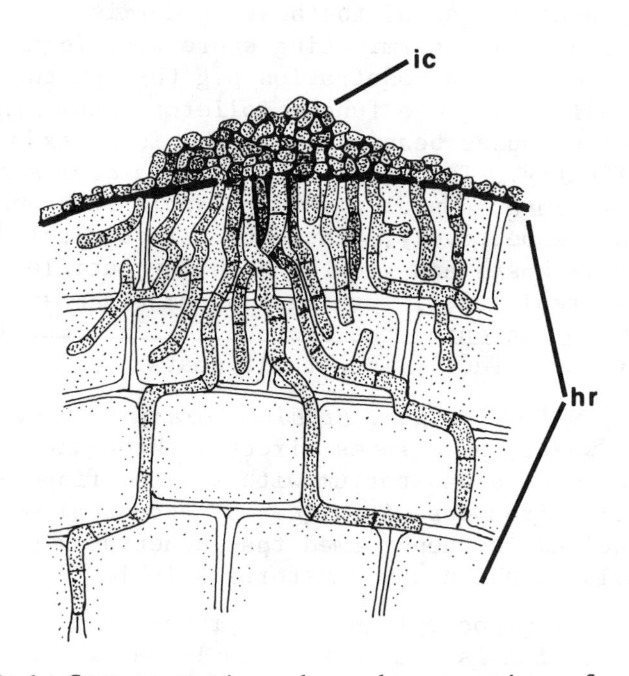

Fig. 12.6 Cross-section through a portion of a young plant root showing a knot of fungal hyphae making an infection cushion (ic) on the outside of the root. Many hyphae penetrate into the cortical cells of the host root (hr). Such an infection cushion is characteristic for a number of soil-borne fungal pathogens.

12-2.2 BACTERIA

Enter Through Natural Openings:

Bacteria primarily gain entry to the host through natural openings, trichomes, nectary cells, and through wounds. The natural openings include lenticels of stems, and stomata and hydathodes of leaves. The usual means of entry of the potato scab organism is through the lenticels in tubers. The causal agent of soft rot of potatoes and many other vegetables, *Erwinia carotovora*, also is capable of penetrating its hosts through lenticels. Both of these bacterial pathogens invade host plants through wounds, also.

Lenticels

The stomatal openings in leaves are avenues of entrance for bacteria that generally cause leafspots; some that use this means of entry include the causal agents of common blight and halo blight of beans, bacterial pustule and bacterial blight of soybeans, and angular leafspot of cotton. The fire blight organism may also enter through stomata.

Panopoulos and Schroth, 1973, studied motile and non-motile (but pathogenically similar) strains of *Pseudomonas phaseolicola*, the causal agent of halo blight of bean. These authors found that motile strains produced from 2 to 10 times as many leaf lesions as the non-motile bacteria. However, when host leaves were water-soaked, the motile bacteria caused 100 to 400 times the lesions as on non-water-soaked leaves. When leaves become water-soaked during a rainy period, there is a continuous column of water from the flooded substomatal cavity up through the stomata to the film of water covering the leaf. Motile bacteria quickly move through stomata into the interior of the leaf. Therefore: motility of bacterial foliar pathogens increases the likelihood of infection. It is also obvious that: water-soaking of host leaves greatly facilitates host entry and subsequent infection by motile pathogenic bacteria. Normally the substomatal cavity contains gasses that pass through the stomata into the outer air. Light rain and dew will often bridge over stomata because of the surface tension of the water and the inner pressure of escaping gasses. Under these conditions, bacteria may not find stomata likely avenues for entry.

Rule -- Water-soaking Of Leaves

Some bacterial pathogens that cause leafspots may sometimes enter hosts through hydathodes. *Xanthomonas campestris*, the incitant of black rot of cabbage and other crucifers, enters the hosts predominantly through hydathodes. Hydathodes are gland cells or pores that release water. They are usually found at leaf tips or margins. Under conditions of high humidity, a drop of water may be seen adhering to the tips or margins of leaves. These drops are excreted from hydathodes and may be resorbed by the leaf when the humidity decreases. When pathogenic bacteria find their way into these drops, they may be carried directly into the leaf and begin to multiply.

Hydathodes

Erwinia amylovora, the incitant of fire blight of pear and apple, may gain entrance through nectary cells while the host is in bloom. The nectary cells are glands that secrete drops of nectar. The nectar is sought by pollinating insects that serve as vectors for the pathogen. The visiting insects deposit the bacteria in the remaining nectar, which is ideally suited for bacterial multiplication. The bacteria soon penetrate down into the ovary, flower pedicel and stem.

Trichomes are foliar hairs that may be single-celled or multiple-celled, with a single basal cell, or a bulbous multicelled base, and they may or may not be glandular. Such trichomes are found on surfaces of leaves, petioles, stems and many kinds of fruits. Lewis and Goodman, (1965), found that the complex glandular trichomes on the midribs and at the tips of serrations of leaves were avenues of entry for fire blight bacteria. Furthermore, these authors found trichomes could be invaded even in the absence of wounding and state: "Leaf infections probably occur through

Entry Through Healthy Trichomes

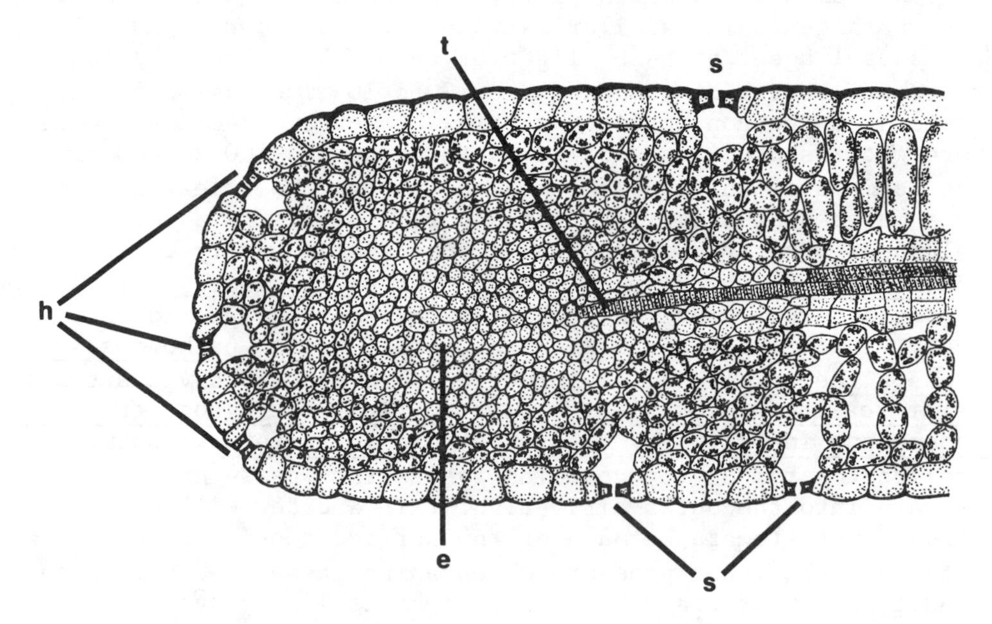

Fig. 12.7 Cross-section of a dicotyledonous leaf tip showing hydathodes (h), tracheids (t), stomata (s), and epithem (e) or modified mesophyll cells. Guttation water passes from the ends of the tracheids through the intercellular spaces of the epithem and through hydathodes to the outside of the leaf. These hydathodes are modified stomata that have a fixed opening.

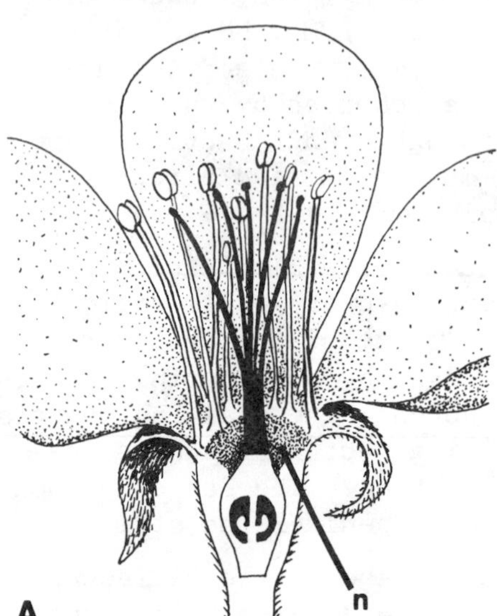

Fig. 12.8A Diagram of a section through an apple blossom. The stippled area (n) shows the location of nectary cells.

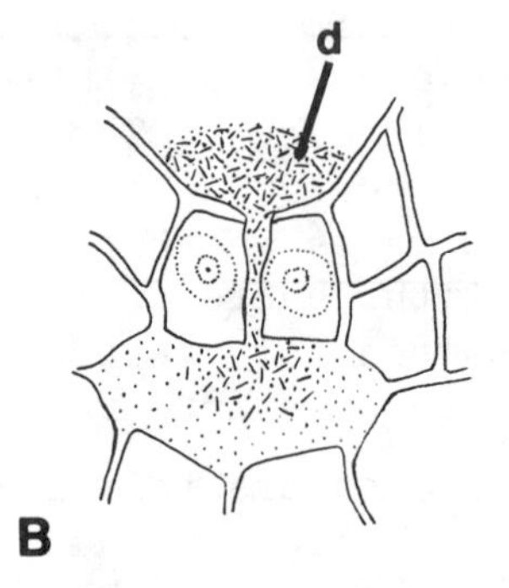

Fig. 12.8B Fire blight bacteria multiplying in a drop of nectar (d) and migrating through the guard cells of the sunken nectary of a pear blossom into the cavity below. *(After Hildebrand, 1937)*

the rupturing cuticle of the glandular trichome where the excretions of the gland have accumulated and permit multiplication of the bacteria."

Leaf trichomes also appear to be infection sites for *Corynebacterium michiganense*, the causal agent of bacterial canker of tomato. This organism appears to be able to enter between the bulbous cells around the basal cell of the leaf hairs without wounding, (Kontaxis, 1962). However, these trichomes are easily wounded by the most gentle touch, and broken hairs increase the over-all incidence and severity of infection, (Layne, 1967). This bacterium also infects tomato fruit, which are without stomata, but possess trichomes. It is thought that fruit infection usually occurs through injured trichomes; and less often, uninjured trichomes.

Increased Entry Through Injured Trichomes

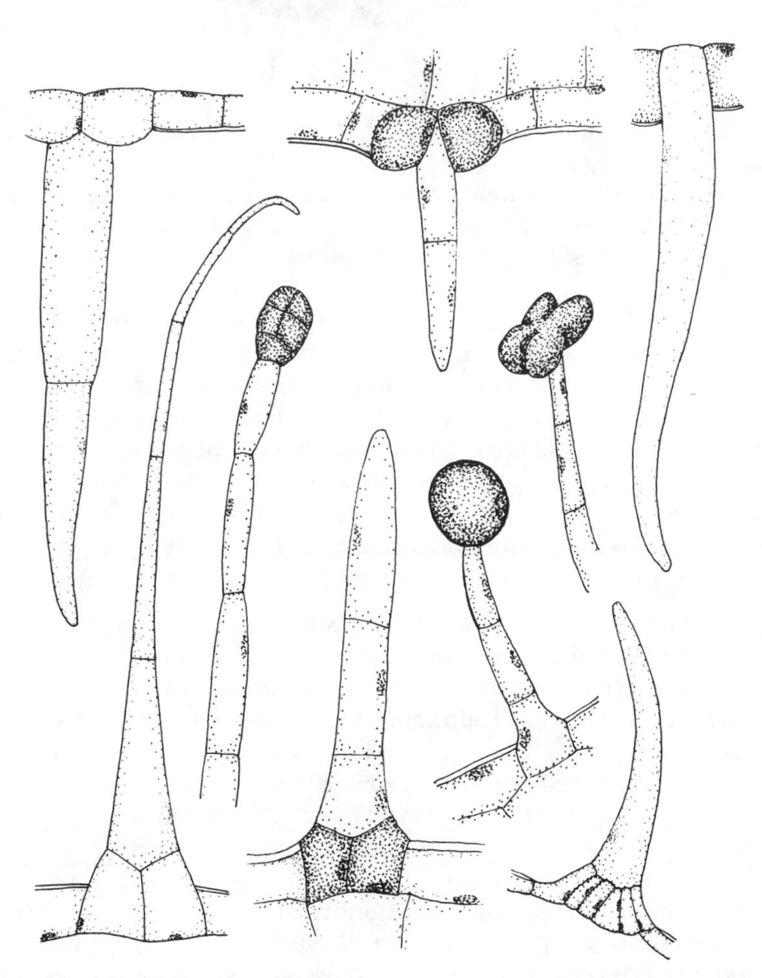

Fig. 12.9 Various types of trichomes (hairs) found on foliage and fruit of plants. The stippled cells are glandular cells of glandular trichomes.

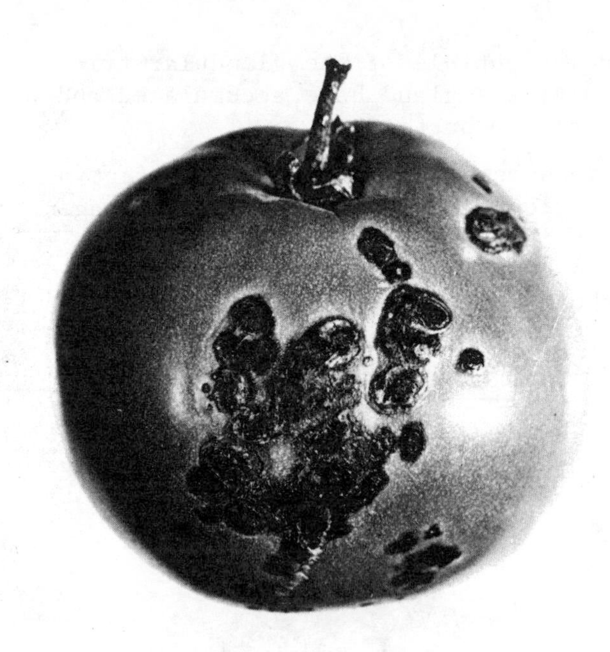

Fig. 12.10 Bacterial spot of tomato, caused by *Xanthomonas vesicatoria*. This pathogen often gains entry to immature fruit through broken trichomes, and less often through uninjured trichomes. There are no stomata on the fruit. *(Courtesy of R. W. Samson)*

Resident Bacteria May Easily Enter Injured Trichomes And Epidermal Cells

Vakili, 1967, found that bacterial spot of tomato was seldom initiated by the entry of *Xanthomonas vesicatoria* through stomata of healthy tissue. Instead, he found that entry was mainly through broken trichomes or epidermal abrasions. Such wounding probably often occurs in the field as a result of windblown sand particles. This author also found the bacterium to be a resident of tomato leaves and believes that infection could take place when the organism, in a film of dew, contacts a freshly broken trichome.

Many bacteria enter the host plant only through wounds. The soft rot organism, *Erwinia carotovora*, and *Agrobacterium tumefaciens*, which causes crown gall, are classic examples. Wounds may be made by windblown sand, hail, branches whipping back and forth, insect feeding wounds, etc. Wounds need not be large to permit entry of bacteria, however, there seems to be a direct relation between the size of the wound and the size of the lesion in many bacterial diseases. The feeding on root systems by soil-borne insect larvae and nematodes causes an enormous amount of root damage. In the case of cucurbit wilt, caused by *Erwinia tracheiphila*, cucumber beetles carry the infectious bacteria and secrete them with their feces. The bacteria enter wherever feeding injuries expose tips of vascular tissues. The corn flea beetle is a vector for *E. stewartii*, the incitant of Stewarts' wilt of corn. When the beetles feed on corn leaves, they introduce the bacteria into the feeding wound.

The relationship of wind-rain storms to the initiation of outbreaks of bacterial blight of soybean caused by *Pseudomonas glycinea*, was studied by Daft and Leben, 1972. These authors found the wind component of storms was necessary for initiation of the disease. The wind caused the foliar injuries that predisposed the soybeans to infection.

Selected References

Dimond, A. E., and J. G. Horsfall. 1960. Inoculum and the diseased population. pp. 1-22, In J. G. Horsfall and A. E. Dimond, (ed.), Plant pathology. Vol. III. Academic Press, New York.

______________ and ______________. 1965. The theory of inoculum, In K. F. Baker and W. C. Snyder, (ed.), Ecology of soil-borne plant pathogens. University of California Press.

Garrett, S. D. 1960. Inoculum potential. pp. 23-56, In J. G. Horsfall and A. E. Dimond, (ed.), Plant pathology. Vol. III. Academic Press, New York.

Sadasivan, T. S. 1961. Physiology of wilt disease. Annual Review of Plant Physiology. 12:449-468.

Stakman, E. C., and J. G. Harrar. 1957. Principles of plant pathology. The Ronald Press Co., New York. pp. 258-300.

Van Gundy, S. D. 1965. Factors in survival of nematodes. Annual Review of Phytopathology. 3:43-68.

CHAPTER 13

THE PATHOGEN--VARIATION

13-1 GENERAL CONCEPTS

One of the properties of living organisms is self-reproduction. A living organism arises from another preexisting living organism. Life is kindled only by life itself. Every form of life today is the very latest member of its kind, evolving through countless generations from the very dawn of life. This is the essence of evolution. While man has pieced together many dramatic changes that both plants and animals have undergone from prehistoric times, it is of the utmost importance to understand that all living forms undergo change continuously and slowly. The forms of life that will appear in tomorrow's world are those that will be most able to survive in the world of tomorrow. Although each kind of life tends to reproduce itself faithfully from generation to generation, changes in heredity do occasionally occur. These changes may in turn be reproduced from one generation to another. When we are dealing with a fungus that produces countless billions upon billions of spores, the number of changes is sizeable. Mother nature ever so slowly but constantly produces changes in every form of life, and those changes that better enable the life form to survive are maintained. The changes that hinder the life form in the race for survival are eliminated. This is the basis for the old saying "survival of the fittest". Most of the changes that life forms undergo are completely undetected, simply because they are small, unobservable, or lethal.

Survival Of The Fittest

Most Changes Undetected

Pathogens Undergo Changes

All pathogens are able to produce variants and thus, are able to evolve new capabilities of growth habits. Among these capabilities are changes in ability to attack different varieties of hosts; loss of virulence (total or partial); or less frequently, an increase of virulence. Such capabilities are particularly essential for pathogens such as the fungal rusts. These pathogens require living tissue to grow on and they have an extremely narrow host range. When a new "resistant" variety of the host is grown, the rust pathogen is usually capable of producing new races that can attack this host variety. For some pathogens, the capacity of variation is handy as a mechanism of survival; for others, it is an absolute necessity.

It is necessary to define our terminology and thereby establish a "common ground" to clearly understand the concept of variation. First of all, we need to understand what is considered to be the basic, or lowest unit of subdivision of any group of organisms. This is the biotype. A biotype is a population of life forms that are identical in all inheritable traits. Therefore, the members of a biotype of a fungal pathogen will be alike in all genetically controlled characters. If the members of this biotype reproduce asexually, they will continue to produce members of the same biotype. However, if a member of one biotype mates with a member of another biotype, hybridization occurs, and the offspring are of new and different biotypes.

Biotype

The next higher step up the classification ladder is the physiologic race, which is composed of one or a number of biotypes of a species that possess similar characteristics of pathogenicity. A physiologic race of a pathogen differs from other physiologic races of the same pathogen by the ability to attack one or several different and particular host varieties and the inability to attack certain other host varieties. Genetically, members of a race are fairly stable, and if presented with suitable hosts to grow on, the race will generally remain intact and identifiable for many years.

Physiologic Race

Race development in pathogens is dependent upon conditions involving both host and pathogen. In respect to the host, the lack of susceptible varieties and the presence of resistant varieties, increase race development in pathogens. By growing resistant crop varieties, we simultaneously and automatically sort out and select new races of pathogens. Conditions of the pathogen that accelerate race development are: obligate parsitism; sexual reproduction; inability to complete its life cycle, or multiply in the field as a saprophyte; and a narrow host range. Of course, the opposite conditions of host and pathogen decrease race development.

Rule -- Host Factors That Increase Race Development

Rule -- Pathogen Factors That Increase Race Development

We are concerned here with only the variations in pathogenicity that occur in fungi, bacteria, viruses and nematodes; even though other variations are known.

13-1.1 ASEXUAL REPRODUCTION

In the fungi, asexual reproduction is common. Essentially, this is cell division without the union of nuclei, and each new cell is capable of growing into an individual that has the same genetic characteristics as its parent. It would seem then, that vegetatively produced hyphal cells and spores do not offer a chance of becoming genetically different from each other. However, we now have undisputable proof of at least two asexual mechanisms that afford the possibility of genetic change. These mechanisms are known as (1) mutation, and (2) heterokaryosis.

Mutation

Mutation is a sudden departure from the parent type, as when an individual differs from its parents in one or more inheritable characteristics. In order to explain how mutation comes about, we must examine mitosis, which is the usual method of cell-division. It is characterized by the resolving of the chromatin of the nucleus into a thread-like form, which separates into segments, called chromosomes. Each chromosome separates longitudinally into two parts, one part being retained in each of two new cells. In the vast majority of mitotic cell divisions, the chromosomes are duplicated exactly so that the daughter cells contain the identical genes possessed by the parent cell. However, occasionally during mitosis, each chromosome is not duplicated exactly; or during the process in which the paired chromosomes separate, bits of a chromosome are broken off. The end result is that a daughter cell may contain different, more, or fewer genes than the parent cell. After mutation has occurred, the change is faithfully reproduced and passed on to the resulting progeny.

Mitosis

Mutation Occurs During Mitosis

Heterokaryosis

Heterokaryosis refers to the multinucleate condition of hyphae or individual cells of a hypha, wherein the nuclei possess different genetic factors. Heterokaryosis can result when hyphal fusion occurs between similar, but genetically different fungi. Such fusion has been described as being roughly comparable to natural grafts in higher plants. This condition is common in many species of the higher fungi. It is not uncommon for fungi with a heterokaryotic condition to have hyphae that possess various combinations of genetically different nuclei. This is evidently a regrouping or disunion of the different kinds of nuclei.

The Result Of Hyphal Fusion

Heterokaryosis should not be confused with dikaryosis, whereby a hyphal cell contains two nuclei of opposite sex, and each of which is reproduced and is passed on to all succeeding cells. The dikaryotic condition is involved in many of the higher fungi where it is a necessary prerequisite to sexual reproduction.

13-1.2 PARASEXUALISM

The prefix para- comes from the Greek word para, meaning beside or near. Thus, parasexualism literally means "near-sexualism", but may be described as: the process of genetic recombination occurring within vegetative heterokaryotic hypae. Parasexualism was first described by Pontecorvo and Roper in 1952, as occurring in a non-pathogenic fungus.

The parasexual cycle has been shown to occur in *Verticillium albo-atrum* (verticillium wilt), *Fusarium oxysporum* (fusarium wilt), *Ustilago maydis* (corn smut), *Puccinia graminis* (stem rust of cereals) and a number of other fungi in the Ascomycetes, Basidiomycetes and Fungi Imperfecti.

The parasexual cycle is initiated within heterokaryotic hyphae. Occasionally, two unlike haploid nuclei within a heterokaryotic cell fuse to form a diploid nucleus. The diploid nucleus typically divides mitotically along with the other haploid nuclei as the hypha grows. Infrequently, the diploid nucleus undergoes "haploidization", crossing-over also occasionally occurs. The haploid nuclei are therefore genotypically different from the parent nuclei. Essentially, this infrequent genetic recombination occurs within vegetative heterokaryotic hyphae. Such genetic recombinations may give rise to new physiologic races possessing new pathogenic capabilities. For further discussion of the parasexual phenomena, the reader is referred to the review by Bradley, 1962.

13-1.3 SEXUAL REPRODUCTION

Hybridization

The end result of sexual reproduction is hybridization, which is brought about by gene recombination and segregation. Hybridization may be defined as the production of offspring from two different parent races, varieties, or species of genera. Hybridization in fungi is initiated by the union of two compatible haploid nuclei, often designated as plus or minus strains, or simply as nuclei of opposite sex. One nucleus comes from each of two parents. Two haploid nuclei unite to form a diploid nucleus. The diploid nucleus then undergoes meiosis. It is during meiosis that segregation and gene recombination occur. Each resulting haploid nucleus may receive a different combination of genes from each parent. Hybridization is generally the most common and frequent source of variation in pathogenicity in those plant pathogens that usually form a sexual reproductive stage.

Common Source Of Variation

13-2 FUNGI

The race concept is most clearly discernible in the cereal rust fungi, where the physiologic races are exceptionally numerous. These pathogens possess a number of genes, each of which may control a specific property of virulence that enables a race to infect only certain host varieties. The host plants of these fungi also possess a number of genes, each of which may control either resistance or susceptibility to one or more specific properties of virulence in the pathogen. For example, in the flax rust disease, it is found that, for each gene in the host capable of mutating to give resistance, there exists a gene in the pathogen capable of mutating to overcome that particular

Theory -- Gene For Gene

resistance. To a lesser extent this theory may apply to some other diseases incited by fungal pathogens, although this remains to be clarified.

Puccinia graminis, the causal agent of stem rust of cereals and grasses, was perhaps the first obligate parasitic fungus for which the concept of races was discovered and developed. It was found that sub-groups of this species could infect only certain hosts. At least six such groups in all have been identified, and they have been designated as varieties or sub-species. Stem rust of wheat, for instance, is caused by *Puccinia graminis* var. *tritici*. A

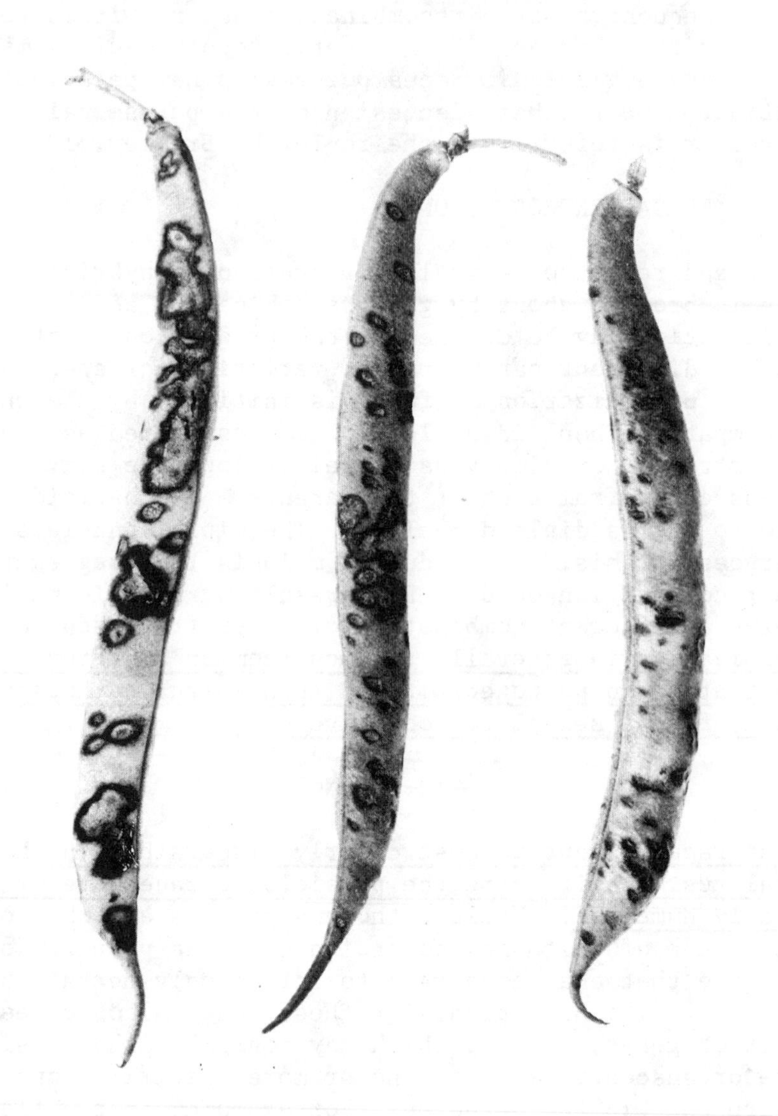

Fig. 13.1 Bean anthracnose, caused by *Colletotrichum lindemuthianum*. Four races of this pathogen are found in the United States. The pathogen is often seed-borne and is frequently a soil invader. *(Courtesy of M. W. Gardner)*

third Latin name is added to the binomial, and it is now called a trinomial. Many races of each of the varieties are known. For example, over 240 races of *P. graminis* var. *tritici* are known. These races, and the races of most fungi, are designated by Arabic numerals. They are distinguished only by differences in pathogenicity on special hosts.

Trinomial

The pathogen *P. coronata*, the causal agent of crown rust of oats, has over 275 physiologic races; and *P. recondita*, the causal agent of leaf rust of wheat, has over 128 physiologic races. *Erysiphe graminis* var. *tritici*, the wheat powdery mildew fungus, has over 50 physiologic races. Other fungi, such as *Phytophthora infestans*, the causal agent of potato and tomato late blight, have a number of races. The causal agent of fusarium wilt of tomato, *Fusarium oxysporum* var. *lycopersici*, has two recognized races. Race 2 is the most virulent, and is capable of attacking all tomato varieties that are resistant to race 1.

One of the first fungi for which the concept of races was developed was the causal agent of bean anthracnose, *Colletotrichum lindemuthianum*. Four races of this fungus are commonly found in this country, and they are designated as alpha, beta, gamma, and delta. The method of designating these races is an exception.

True races are maintained by means of asexual reproduction. For example, a race of the wheat stem rust fungus is perpetuated by means of asexual urediospores. However, in the smut fungi, a slightly different story exists. Smut fungi are single-cycle disease pathogens whose inoculum is in the form of dikaryotic teliospores (chlamydospores), that undergo nuclear fusion prior to germination. When the teliospores germinate, meiosis occurs, and genetic segregation and recombination produce progeny that are genetically diverse. Therefore, for all practical purposes, races of smuts are considered to be masses of teliospores that possess a relatively uniform pattern of pathogenicity and that maintain this behavior in successive generations.

Smut Races

13-3 BACTERIA

It is a well-established fact that many plant pathogenic bacteria have numerous variants. Although sexual reproduction has been proven to occur in some bacteria, it has never been proven to occur in plant pathogenic species. Variations in plant pathogenic bacteria are often explained by processes of mutation and transformation. Mutations may occur naturally at random, or may be induced by chemicals or radiation. Transformation has been proven to function in some plant pathogenic bacteria and may play a role similar

Variation Through Mutation

Variation Through Transformation

to that of sexual recombination. Transformation is the acquisition by a bacterial clone of heritable characteristics of another clone by growing the former in an extract or filtrate of the latter. The agent of transformation appears to be deoxyribonucleic acid (DNA), which is liberated by growing bacteria and confers the heritable characteristic on the recipient bacteria. The extent of transformation and the process of mutation in plant pathogenic bacteria needs a great deal of further clarification.

Variation Through Conjugation

Another means by which bacterial variation may occur is through the phenomenon of conjugation, whereby two compatible bacteria come in contact and a fraction of genetic material is passed from one bacterium to the other. Therefore, both bacteria are changed genetically; both are different than they were before and from each other. As these bacteria multiply, they pass their newly acquired inheritable traits on to their progenies, giving rise to new strains or races.

Shifts In Virulence

Variation in virulence may or may not be associated with such recognizable features as colony size, color, and appearance in culture. The apparent loss of virulence of bacteria in culture is very common, and is often due to the better growth of non-virulent strains in the culture medium as compared to that of the virulent strain. The opposite effect is often observed when bacteria are inoculated into a susceptible host. The virulent strains prosper and the non-virulent strains do not. Sometimes, successive isolations and inoculations into host plants have the effect of increasing the infectivity of a bacterial strain. Such shifts in virulence point to the fact that many so-called "pure" cultures of bacteria are not single biotypes, but are a number of biotypes that differ in certain characters, including pathogenicity.

13-4 VIRUSES

Man has been cognizant for many years of the existence of variants in plant viruses, and more proof exists of the variation than does proof as to exactly how this variation occurs.

In nature, a virus is transferred from one plant to another "en masse", meaning a large number of infective particles. The symptoms that are produced are alike on almost all infected plants, and the bulk of the virus particles produced are the same as their progenitors. However, occasionally a virus particle arises that is unlike its progenitors. The effect of these unlike viruses is masked by the predominance of the like viruses. In order for the unlike viruses to gain the "upper hand", and show their different symptom characteristics, they must somehow be separated, multiplied, and transmitted to a susceptible host. When this happens, the new strain is segregated from the parent virus. Like the mutant strains of microorganisms, mutant strains of viruses arise through selection by the environment.

Variation by Mutation

Over 40 years ago, it was reported that tobacco mosaic virus had a strain variation that procuded chlorotic spots instead of the usual mosaic pattern. Since then, probably more than 400 strains of this virus have been recognized. This is the greatest variation yet reported in plant viruses. Most plant viruses have been reported to possess variants, and the rate of mutation appears to vary with each virus.

Virus Strains Identified By Measurable Changes In Specific Traits

A strain is recognized by a measurable difference in a specific property. Prominent among these and directly related to pathogenicity are: infectivity, symptoms, severity of disease, effect on yield, longevity, optimum temperature for development, rate of movement in host tissues, host range, and specificity to insect vectors. Other detectable variants include: dilution end point, thermal inactivation point, electrophoretic mobility, and serological and immunological reactions.

13-5 NEMATODES

Variation By Hybridization And Mutation

There are numerous examples citing the existence of physiologic races among several genera and species of nematodes, such as cyst nematodes, *Heterodera* spp.; root knot nematodes, *Meloidogyne* spp.; stem and bulb nematodes, *Ditylenchus dipsaci*; and the burrowing nematode, *Radopholus similis*. Races are composed of a number of biotypes, and new races may arise by both hybridization and mutation.

Variation in pathogenicity in nematodes is almost exclusively that of variation in host range. This is to be expected from such animal forms that are obligate parasites.

Selected References

Bradley, S. G. 1962. Parasexual phenomena in microorganisms. Annual Review of Microbiology. 16:35-52.

Buxton, W. W. 1960. Heterokaryosis, saltation, and adaptation. pp. 359-405, In J. G. Horsfall and A. E. Dimond, (ed.), Plant pathology. Vol. II. Academic Press, New York.

Day, P. R. 1960. Variation in phytopathogenic fungi. Annual Review of Microbiology. 14:1-16.

Day, P. R. 1966. Recent developments in the genetics of the host-parasite system. Annual Review of Phytopathology. 4:245-268.

Fincham, J. R. S., and P. R. Day. 1965. Fungal genetics, Ed. 2. Blackwell Scientific Publications, Oxford.

Johnson, T. 1960. Genetics of pathogenicity. pp. 407-459, In J. G. Horsfall and A. E. Dimond, (ed.), Plant pathology. Vol. II. Academic Press, New York.

Kehr, A. E. 1966. Current status and opportunities for the control of nematodes by breeding. pp. 126-138, In Pest control by chemical, biological, genetic, and physical means. U. S. Dept. of Agr. A Symposium. ARS 33-110.

Moseman, J. G. 1966. Genetics of powdery mildews. Annual Review of Phytopathology. 4:269-290.

Parmeter, J. R. Jr., et al. 1963. Heterokaryosis and variability in plant-pathogenic fungi. Annual Review of Phytopathology. 1:51-76.

Price, W. C. 1964. Strains, mutation, acquired immunity, and interference. pp. 93-117, In M. K. Corbett and H. D. Sisler, (ed.), Plant virology. University of Florida Press, Gainesville.

Stakman, E. C., and J. J. Christensen. 1953. Problems of variability in fungi. pp. 35-62, In Plant diseases. U. S. Dept. Agr. Yearbook.

______________, and J. G. Harrar. 1957. Principles of plant pathology. The Ronald Press Co., New York. pp. 121-177.

Starr, M. P. 1959. Bacteria as plant pathogens. Annual Review of Microbiology. 13:231-233.

Thorne, G., and M. W. Allen. 1959. Variation in nematodes. pp. 412-218, In C. S. Holton, et al., (ed.), Plant pathology problems and progress 1908-1958, The University of Wisconsin Press, Madison, Wisconsin.

Wallace, H. R. 1963. The biology of plant parasitic nematodes. Edward Arnold (Publishers) Ltd., London. pp. 186-191.

CHAPTER 14

THE RESISTANT HOST

14-1 INTRODUCTION

Host resistance is one of the most challenging and rewarding areas of study in plant pathology. Host resistance in disease control is of immeasurable value. The planting of resistant varieties is the first line of disease control for most crops. Resistance may be due to morphological or physiological characteristics of the plant. It is common to find that a plant has more than one resistance mechanism that operates against a pathogen. Multiple resistance mechanisms may operate independently, or they may be interrelated in some way.

More Than One Resistance Mechanism Against A Pathogen Is Common

In respect to a given pathogen, plants may be described as immune; or highly, moderately, or slightly resistant; or slightly, moderately, or highly susceptible. While immunity is absolute, without variation, other forms of resistance are relative, and a considerable amount of variation exists within them.

14-2 DISEASE TOLERANCE

Disease tolerance has been defined as--the ability of a host plant to survive and give satisfactory yields at a level of infection that causes economic loss to other varieties of the same host species (National Academy of Science, 1968). In other words, the plant tolerates the presence of the pathogen. Disease tolerance has been termed disease endurance by some authors.

Host Tolerates Pathogen

To illustrate, there is some evidence to indicate that *Cephalosporium gregatum*, the incitant of brown stem rot of soybeans, may invade soybean plants and do little damage while the plants are growing fast. But if the air temperature is decreased to 60°F. for several days, plant growth is retarded, and the pathogen is then able to injure the host. Also, as the soybean plants approach maturity their growth is reduced and the pathogen is then able to inflict damage.

Another example of disease tolerance is the wheat variety that produces a good yield even though it is moderately to severely infected with rust. This is especially noticeable when neighboring wheat varieties produce lower yields even though they are similarly infected. Such examples are not as well documented as they might be, and in some cases it might be argued that these are mechanisms of true resistance. However, in these situations, the pathogens are often abundantly present in the hosts and do not appear to be retarded or suppressed. For a further discussion of tolerance the reader is referred to the review by Schafer, 1971.

Tolerance Most Apparent In Parasitic Nemas

Tolerance is clearly demonstrated in the parasitic nematodes, where certain plant varieties are found to be able to support large numbers of nematodes without being severely injured. The mechanisms involved in tolerance are unknown.

14-3 DISEASE ESCAPE--APPARENT RESISTANCE

Escape Concerns Susceptible Plants

The capacity of a susceptible plant to avoid infectious disease through some character of the plant or other factor that prevents infection, is called escape. It is important to differentiate between what appears to be resistance and true plant resistance. A plant breeder must be aware of this situation and conduct his breeding program accordingly. This does not mean that disease escape as a control measure is valueless; for quite the contrary, any measure which lessens disease, or the chance for disease, is certainly worthwhile.

Disease escape may be accomplished by the planting of an early maturing variety that escapes infection due to its maturing before the environment becomes favorable for infection and disease development. Another example is the planting of a variety capable of producing a suitable crop when planted late. In this situation, the late planting is to avoid the early part of the growing season that is also favorable for the development of some diseases.

A number of plants may be susceptible to a virus disease, yet escape, because the insect vectors prefer to feed on other hosts.

Brome grass mosaic virus (BGMV) can be experimentally transmitted by mechanical means to some corn hybrids, inbreds, and popcorns. Presently, there is no known insect vector that is able to transmit this virus to corn. The corn escapes this disease because of the lack of a vector. However, this virus is a potential threat to corn, since an insect vector may become available at some future time that will carry the virus to corn.

Escape Lessens Disease Severity And Occurrence

The mechanisms of escape are variable and relative. Seldom is the mechanism of escape so efficient as to totally prevent disease. Rather, escape is a tool for lessening the occurrence and severity of disease. Plants possessing the capacity of disease escape, are still susceptible. Therefore, if a change of conditions renders the escape mechanism inoperative, disease can occur.

14-4 IMMUNITY

Immunity is freedom from attack and injury by an infectious pathogen. It means complete disease resistance, the highest level of resistance. So strictly speaking, all plant species that a given pathogen is incapable of infecting are immune to that pathogen. There is no pathogen capable of inciting disease in all species of higher green plants. In fact, there are few pathogens capable of attacking more than a small percentage of species of higher green plants. So from this standpoint, most plants are immune to most pathogens, a point of view that is often overlooked.

Highest Level Of Resistance

Rule -- Immunity

Perhaps most cases of immunity may be described as an incompatibility between the attacking pathogen and the host. Truly, this is a great oversimplification, but it serves a useful purpose here. The "would-be" pathogen fails to become established because of morphological characteristics or biochemical properties of the cells of the plant. Of course, it can also be said that the "would-be" pathogen may lack the necessary enzyme systems to attack, breakdown, and/or utilize the substances of plants as a source of food.

Perhaps it is a strange occurrence that a number of fungi, such as the potato late blight fungus and numerous rusts, enter many immune plants even though they cannot incite disease. Entry is generally made through natural openings such as stomata. The young fungi die when they cannot obtain nourishment from the plant. The host permits the entry of the pathogen, but blocks its further growth.

Rule -- Entrance Into Immune Plants

14-5 RESISTANCE LESS THAN IMMUNITY

14-5.1 STATIC RESISTANCE TO PATHOGEN ENTRY

The term static resistance refers to mechanisms that are continuous on or in the host, regardless of whether the pathogen is present or not. Some of these mechanisms operate to restrict or inhibit the entry of pathogens when the latter attempt to enter the host directly, through natural openings, or through wounds.

Definition

Direct penetration usually takes place through the leaves, roots, or fruiting structures. Penetration is made through the cuticle of the outer epidermal cell walls, and barriers to penetration may be found in both the cuticle and epidermis. Sometimes the epidermis is covered with a

Thickened Cell Walls Form Barriers

thick wax layer, and often the walls of the epidermal cells become thickened and may become toughened by being suberized or lignified. In some cereals, the epidermal walls may undergo a silicification. Each of these may be a barrier which mechanically impedes the pathogen. Some leaves and fruits may be covered with hairs. In some cases, waxy coatings and hairs may cause water droplets to run off quickly and thus do not permit the pathogen a film of water in which to germinate.

Role Of Cuticle Is Not Great

In a review of the role of the cuticle as a barrier to penetration, Martin (1964) concluded that: "its contribution is not great." He based his conclusion upon the fact that the cuticle layer varies in thickness, and is often poorly developed; also, the chemical composition of the cutin is such that it should not be a serious barrier to penetration.

Healing-over Of Wounds

Wounds are very important in the relationship of disease development. They may occur on leaves, flowers, stems, branches, tree trunks, roots and underground storage organs. Wounds provide an open avenue of entrance for many pathogens. Therefore, the ability of the host to heal over the wounded area quickly is of vital importance in the prevention of disease. The wounded surfaces of potato tubers and sweet potatoes are good illustrations of this process. Injured cells liberate wound hormones that evoke cell division in healthy adjacent cells. A thick layer of suberized cells, cork cells, are laid down over the wounded surface.

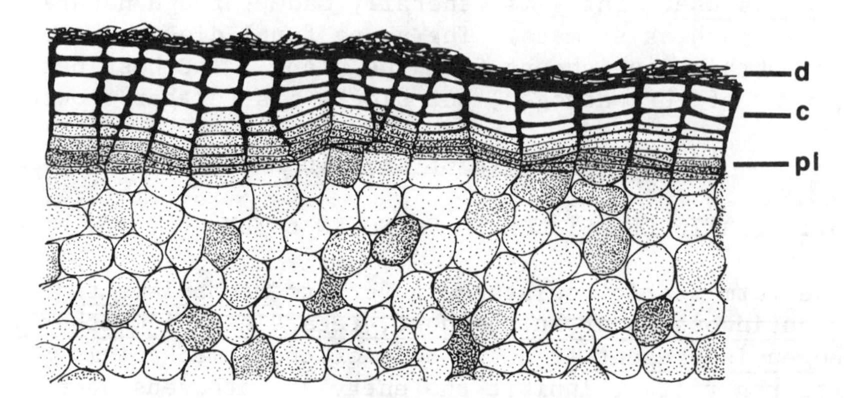

Fig. 14.1 Schematic drawing of a section through a root of sweet potato showing the development of wound periderm. Healthy cells underneath wounded surface cells become meristematic, this is the phellogen (pl) and it produces the cork layer (c) which is heavily suberized. Dead surface cells (d) are outside the cork.

Open stomata are passageways through which many fungi and bacteria enter plants. Some rusts and other fungi, such as *Cercospora beticola*, the incitant of beet leafspot, find that open stomata are necessary in order to gain entry into their hosts. There is limited evidence to show that plants whose stomata remain closed for longer periods than other plants, are less prone to infection.

Chemical defense mechanisms are also known within this category. The outstanding example is the control of onion smudge, caused by *Colletotrichum circinans*. Fungal spores germinate in a film of water on the outer dead, dry scales of the bulb on white onions, and grow saprophytically. The fungus then penetrates directly into the inner, living, fleshy scales and grows as a parasite. The white onions are highly susceptible to attack, whereas, colored onions are resistant. The resistance is due to two phenolic compounds, catechol and protocatechuic acid. These compounds are toxic to germinating spores of the invading fungus. They are produced in colored scales of the bulb, but are independent of the pigments. They diffuse out of the dead scales and inhibit the germinating spores, but are incapable of diffusing out of the living fleshy scales. The live fleshy scales therefore, have no protection against the pathogen.

Chemical Defense

14-5.2 DYNAMIC RESISTANCE TO PATHOGEN ENTRY

The term dynamic resistance refers to mechanisms that are initiated by the host in the presence of the pathogen. An example to illustrate this category is the host response to *Erysiphe graminis* var. *tritici*, the powdery mildew fungus of wheat. The conidia of this fungus germinate on wheat leaves and form germ tubes and appressoria. Host responses in the form of papilla-like projections occur on the inner walls of epidermal cells below the appressoria. These projections, called collars, push into the lumen of the cell. Each collar has a wide, plate-like and circular upper flange attached to the cell wall. The flange may even extend into adjoining cells. The collar is formed immediately below the penetration peg, and fungal germ tubes may or may not be able to penetrate this structure, (see Fig. 14.2, a-f, and x).

Collars Formed On Cell Wall Beneath Appressoria

In the writer's laboratory, two wheat selections, one resistant and the other susceptible, were studied using a single race of the wheat powdery mildew fungus. When inoculated with conidia, both wheats formed collars and flanges at both successful and unsuccessful penetration sites. The resistant wheat prevented 73 percent of attempted penetrations by primary germ tubes versus 36 percent for the

Collars May Be Dynamic Barriers To Fungal Penetration

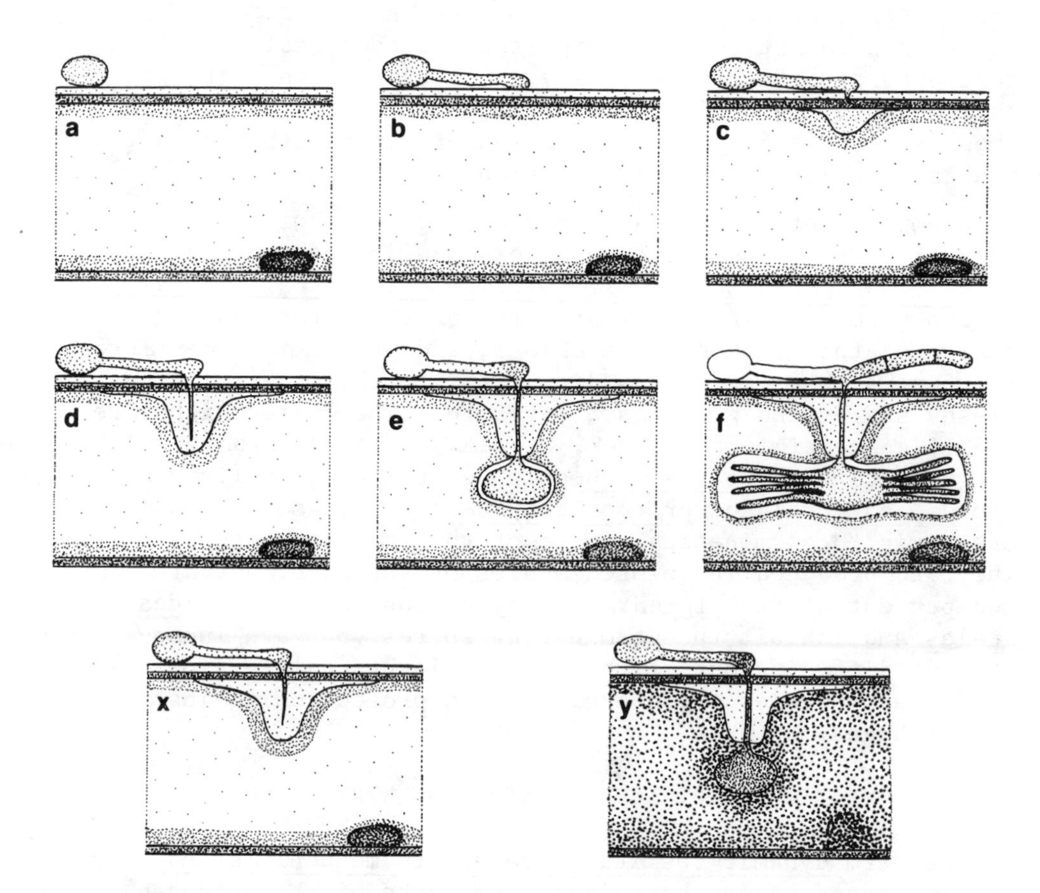

Fig. 14.2 Diagrammatic illustration of host cell response to attempted penetration, and to penetration, by germinated conidia of *Erysiphe graminis*. Susceptible host cell response is shown in (a) through (f). (a) Conidia in contact with host epidermal cell. (b) Germination of conidia to form a primary germ tube with an appressorium at its tip. (c) The host cell forms a thickening (collar) next to the host wall underneath the appressorium. (d) An infection peg penetrates the cuticle, cell wall, and partially through the enlarging collar. (e) Penetration of the collar is complete and the haustorium body begins to form. The fungal protoplasm has migrated out of the conidia. Notice the wide flange of the collar. (f) Fully developed bipolar, lobed haustorium. Protoplasm has migrated out of the germ tube, and an aerial hyphal branch begins to grow.
Resistant host responses are shown in (x) and (y). (x) A dynamic barrier to penetration of the collar by the infection peg; the germ tube begins to die. (y) The hypersensitive cell response; dynamic resistance to establishment of the pathogen. The host cell dies before the haustorium becomes fully developed, which kills the haustorium and germ tube.

susceptible wheat. The same mechanism is present in both the resistant and susceptible wheats; however, they differed in the quantitative response (Ghemawat, 1968). Histochemical tests indicate the presence of lignin in collars and flanges. This lends support to the hypothesis that they are dynamically formed mechanical barriers to fungal penetration (Ghemawat, 1968).

Collar-like Structures Are Common

The presence of similar structures in other plants, that are formed in response to various fungal pathogens has long been known. However, further studies must be made to determine the universality of collars.

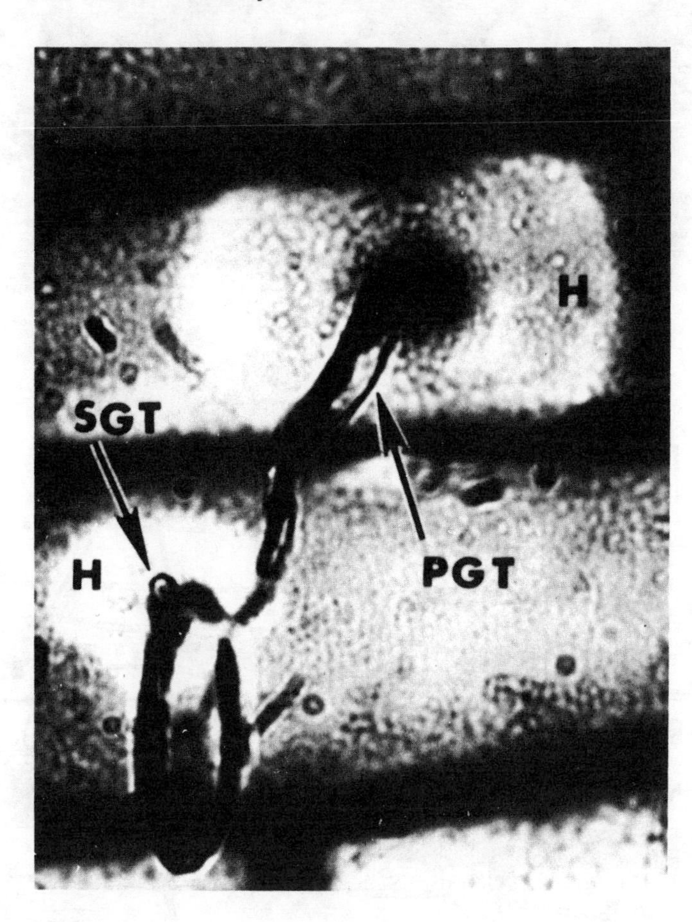

Fig. 14.3 A photomicrograph of a germinated conidium of *Erysiphe graminis* var. *tritici* on stained epidermal cells of a host wheat plant. The conidium has formed a primary germ tube (PGT), and a much smaller secondary germ tube (SGT). The epidermal cells have formed collars with wide circular flanges, that appear as halos (H), in response to attempted penetration by the germ tubes. The epidermal layer was stripped from the plant 36 hours after inoculation by the conidia and stained with 0.05% toluidine blue O (O'Brien, Feder, and McCully, 1965). Epidermal cells are stained pink to red; the flange, light blue. Secondary germ tubes (SGT) appear unable to penetrate into the host. *(Photo by M. S. Ghemawat, 1968)*

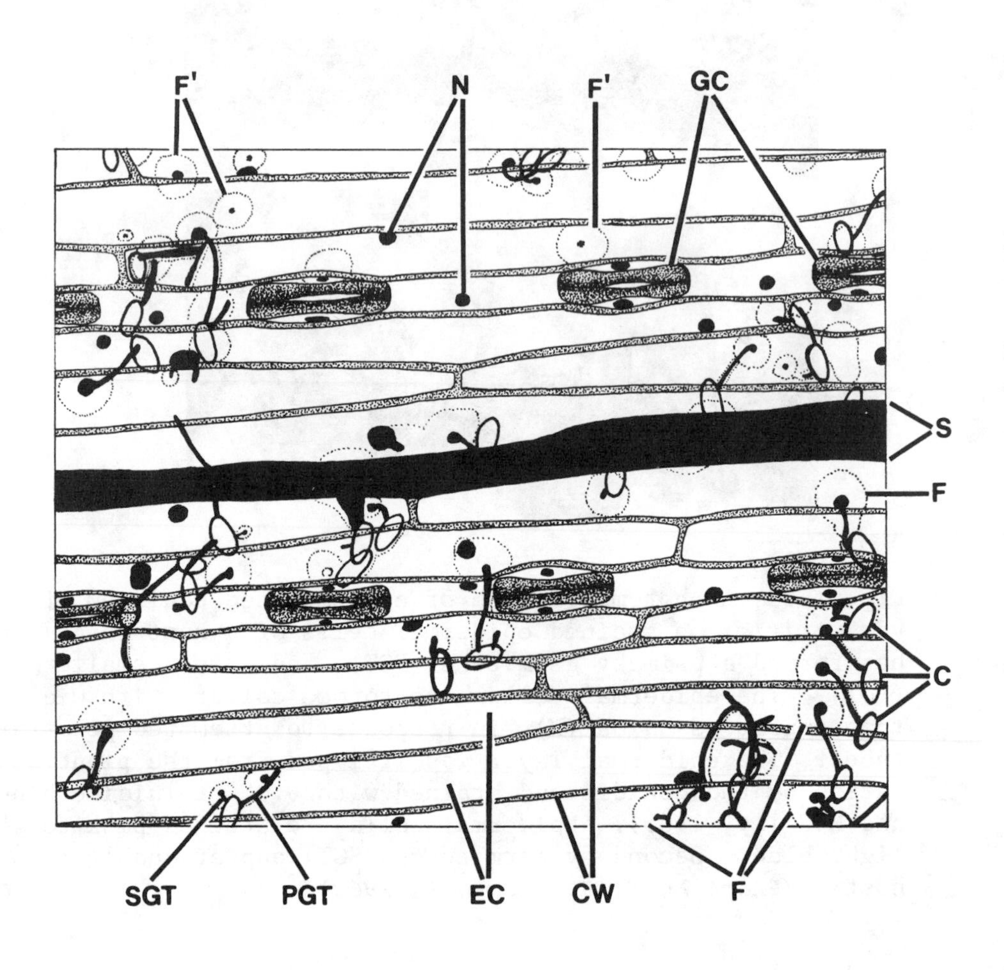
F'
N
F'
GC
S
F
C
SGT
PGT
EC
CW
F

Fig. 14.4 Photomicrograph and drawing of an epidermal layer of a young wheat plant showing host cell responses to attempted penetration by germ tubes of powdery mildew conidia. The epidermal strip with germinated conidia was prepared and stained as in Fig. 14.3. Some conidia and their germ tubes have been dislodged during the preparation of the epidermal strip, leaving halo-appearing flanges (F') as evidence of their previous existance. Conidia (C), cell walls (CW), epidermis cells (EC), collar flanges (F), guard cells (GC), nuclei (N), primary germ tube (PGT), secondary germ tube (SGT), and sclerenchyma fibers (S).
(Photo by M. S. Ghemawat, 1968)

14-5.3 STATIC RESISTANCE TO ESTABLISHMENT OF THE PATHOGEN

These mechanisms are mainly chemical in nature and may be grouped into: (1) intracellular toxic substances, and (2) insufficient or unavailable nutrition.

The evidence that intracellular toxic substances have a hand in the defense mechanisms of plants should not be too startling. Many plant chemicals extracted from numerous plants and plant parts exhibit toxic effects on microorganisms. However, these mechanisms are difficult to study, and the evidence is largely circumstantial.

Intracellular Toxic Substances

Phenolic compounds continue to receive a considerable amount of attention in respect to resistance in a number of diseases. However, because the evidence is so scattered, only one example will be presented.

The resistance of potato plants to verticillium wilt, caused by *Verticillium albo-atrum*, has been attributed to o-dihydroxy-phenols. These substances are found in the underground vascular systems of plants. Furthermore, they are present in higher concentrations in resistant plants than in susceptible plants. The development of disease in the field is thought to be correlated with the amount of phenols in different plant varieties. As the potato plants increase in age, the concentration of phenols is lowered. Again, this is correlated with an increase in susceptibility.

Phenolic compounds such as chlorogenic acid, caffeic acid, hydroquinone, and others have also been studied with respect to resistance. It is probable that these compounds themselves, or compounds derived from them are involved in some mechanisms of plant resistance.

Insufficient Or Unavailable Nutrition

The fact that a host may provide an insufficient or unavailable nutrition for a pathogen also should not be surprising. The plant may be deficient in one or more essential substances required by the pathogen, or the essential substances may be in a form that is unavailable to the pathogen. The latter situation may be characterized also as an enzyme deficiency on the part of the pathogen. The host plant may furnish all the essential nutrition required by the pathogen, but without the necessary enzymes to break down these host substances, they are unavailable to the pathogen.

14-5.4 DYNAMIC RESISTANCE TO ESTABLISHMENT OF THE PATHOGEN

Mechanisms induced by the pathogen may be subdivided into three general areas: (1) mechanical barriers; (2) hypersensitivity; and (3) diffusible toxic substances. The first two areas are the most widely known.

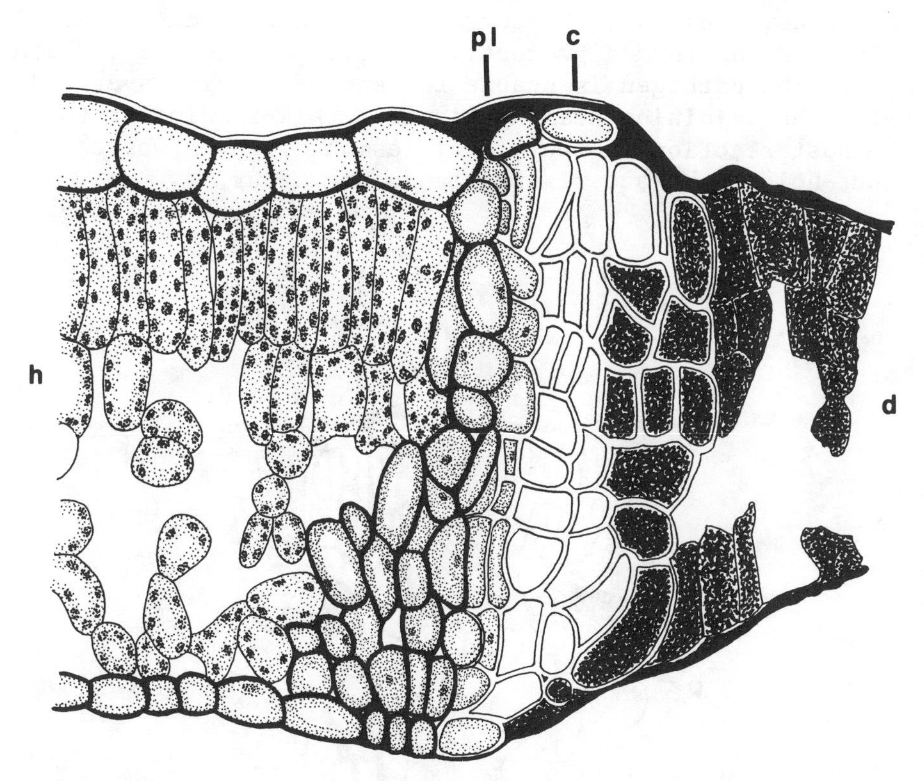

Fig. 14.5 Cross-section through the edge of a wound on a *Prunus* leaf showing bands of corky cells (c) and phellogen (pl) separating healthy tissue (h) from diseased tissue (d). This reaction is similar to the leaf's response to infection by a pathogen. *(After Cunningham, 1928)*

Formation Of Corky Tissue

The formation of mechanical barriers includes the response of the host to repair damage. So called "plant hormones" are involved in this reaction that leads to new cell division and often the formation of corky tissue. The corky tissue walls-off an infected or wounded area, thereby restricting the growth of the pathogen or healing the wound. Sometimes the formation of corky tissue is followed by the dropping-out of infected tissue. An example is cherry leafspot, caused by *Coccomyces hiemalis*. The fungus enters the leaf and nearby cells turn purplish or reddish. A narrow and roughly circular band of cells surrounding the infection point, and a number of cells away from it, become meristematic and produce a compact layer of cells that fill the intercellular spaces. A corky layer is formed around the outer edge of the lesion and extends from the upper to the lower leaf surface, isolating the fungus from the adjoining healthy tissue. The tissue surrounded by the corky layer turns brownish and necrotic; while an abscission layer is formed outside of the corky tissue. The necrotic tissue then falls out leaving a "shot-hole". Cells bordering the shot-hole

Shot-hole

then become corky. In some plants, shot-holes normally occur during moist weather; in dry weather abscission may not occur. Instead, the surrounding occluded cells become suberized and form a protective barrier around the infected area. The pathogen is unable to penetrate this layer and enter the adjoining healthy tissue. This is similar to the host reaction in many local lesion diseases where "shot-holes" characteristically do not occur.

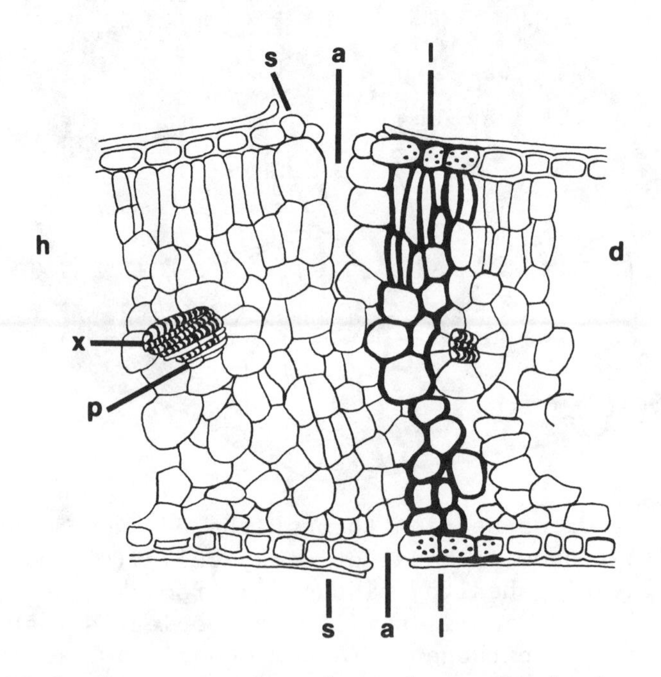

Fig. 14.6 Cross-section through an abscission zone in a cherry leaf.
Healthy palisade and spongy parenchyma cells surrounding a diseased area enlarge and occlude the intercellular spaces. A lignified zone (l) is formed and abscission (a) begins outside of the lignified area. Healthy cells (s) adjacent to the abscission area quickly become suberized. Healthy tissue (h) is on the left, diseased tissue (d) is on the right. Xylem (x) and phloem (p) cells of a vein. When abscission is complete, the infected tissue drops out leaving a shot-hole.
(After Samuel, 1927)

Definition Of Hypersensitivity

Hypersensitivity is the sudden localized host response to the presence of a pathogen that involves the quick death of host cells around the pathogen and the restriction and/or death of the pathogen. Phenolic compounds and their oxidation products are quickly formed after the entry of the pathogen. They are produced in the surrounding host tissue and migrate to the host cells around the pathogen where they accumulate and apparently kill the host cells when toxic levels are reached. This localized death of host tissue also results in the death of an invading obligate parasite. (See Fig. 14.2,y).

Facultative parasites that cause necrosis of infected tissue are not inhibited or killed by the mere death of host cells. These pathogens vary in their response to plant produced phenols, also. Sometimes phenolic compounds and their oxidation products are toxic to the facultative pathogen, in which case they are at least partially responsible for the plant resistance. When the phenols are not toxic to the pathogen, this reaction is not responsible for plant resistance.

Hypersensitivity Is Variable

There are varying degrees of plant resistance due to hypersensitivity. A high level of resistance is when only a few plant cells are affected, resulting in minute localized lesions that are barely visible. A lower level of resistance is when the lesions are much larger and are easily visible. The higher level of hypersensitive resistance appears to occur faster than the lower level.

The sudden collapse of the host cells in the hypersensitive reaction is said by some authors to be a mechanism of extreme susceptibility to the pathogen. In any case, the end result is a containment of the pathogen that prevents injurious colonization and results in plant resistance.

Rule -- Hypersensitivity

Hypersensitivity appears to occur more rapidly in the younger, metabolically active cells than in older, less active cells. The reason for this differential response is still unknown. However, it is well documented that the hypersensitive reaction occurs as a result of infection by fungi, bacteria, viruses, and nematodes. Hypersensitivity is a common defense reaction of green plants to invasion by many pathogens.

14-6 RESISTANCE TO NEMATODES

Nematodes are complex animals, and it is to be expected that plant resistance to nematodes is also complex. It appears that plant resistance to nematodes may involve one or more of the following reasons: (1) plant roots secrete toxic substances; (2) plant roots contain toxic substances and/or do not contain some nutrient necessary for survival; (3) the host fails to respond to the nematode--nurse cells are not formed; and (4) the host is hypersensitive and/or forms toxic substances in the presence of the pathogen.

Plant resistance to nematodes is based on the inability of these parasites to live and multiply on or in the host. In apparent contrast to other plant pathogens, resistance is not necessarily correlated with plant health. Sometimes resistant plants are adversely damaged by their own reaction to the presence of nematodes. This may occur even though attacking nematodes fail to survive and multiply. Furthermore, tolerant plants may support large populations of nematodes without exhibiting severe injury; while susceptible plants may be severely injured by parasitic nematodes that multiply and thrive at the expense of their hosts. Unfortunately, this latter condition is by far the most common plant reaction.

Selected References

Agrios, G. N. 1969. Plant pathology. Academic Press, New York. pp. 105-143.

Akai, S. 1960. Histology of defense in plants. pp. 392-434, In J. G. Horsfall and A. E. Dimond, (ed.), Plant pathology. Vol. I. Academic Press, New York.

________, M. Fukutomi, N. Ishida, and H. Kunoh. 1967. An anatomical approach to the mechanism of fungal infections in plants. pp. 1-20, In C. J. Mirocha and I. Uritani, (ed.), The dynamic role of molecular constituents in plant-parasite interaction. The American Phytopathological Society, Inc., St. Paul.

Allen, P. J. 1960. Physiology and biochemistry of defense. pp. 435-469, In J. G. Horsfall and A. E. Dimond, (ed.), Plant pathology. Vol. I. Academic Press, New York.

Barnett, H. L. 1959. Plant disease resistance. Annual Review of Microbiology. 13:191-210.

Block, R. 1941. Wound healing in higher plants. The Botanical Review. 7:110-146.

________. 1952. Wound healing in higher plants II. The Botanical Review. 18:655-679.

Hare, R. C. 1966. Physiology of resistance to fungal diseases in plants. The Botanical Review. 32:95-137.

Kuc, J. 1966. Resistance of plants to infectious agencies. Annual Review of Microbiology. 20:337-370.

Martin, J. T., 1964. Role of cuticle in the defense against plant disease. Annual Review of Phytopathology. 2:81-100.

Müller, K. O. 1960. Hypersensitivity. pp. 470-520, In J. G. Horsfall and A. E. Dimond, (ed.), Plant pathology. Vol. I. Academic Press, New York.

Rhode, R. A. 1965. The nature of resistance in plants to nematodes. Phytopathology. 55:1159-1162.

_____, R. A. 1972. Expression of resistance in plants to nematodes. Annual Review of Phytopathology. 10:233-252.

Schafer, J. R. 1971. Tolerance to plant disease. Annual Review of Phytopathology. 9:235-252.

Stakman, E. C., and J. G. Harrar. 1957. Principles of plant pathology. The Ronald Press Co., New York. pp. 519-527.

Tomiyama, K. 1963. Physiology and biochemistry of disease resistance in plants. Annual Review of Phytopathology. 1:295-324.

Walker, J. C., and M. A. Stahmann. 1955. Chemical nature of disease resistance in plants. Annual Review of Plant Physiology. 6:351-366.

Wingard, S. A. 1953. The nature of resistance to disease. pp. 165-173, In Plant diseases. U. S. Dept. Agr. Yearbook.

Wood, R. K. S. 1967. Physiological plant pathology. Blackwell Scientific Publications, Oxford and Edinburgh. pp. 399-509.

CHAPTER 15

THE PATHOGEN PRESENT IN THE HOST

15-1 INTRODUCTION

Colonization Of Host

After a pathogen successfully gains entry to a susceptible host and begins to grow, or in the case of viruses takes command of the host cell metabolism, the pathogen becomes established. To place it in another light, colonization has begun and it may be described as local, extensive, unrestricted, or systemic. Of course, very often there is no clear-cut line of demarcation between these arbitrary categories.

Changes Of Host Usually Detrimental To Host

The pathogen may proceed subcuticularly or subepidermally, intracellularly and/or intercellularly. The pathogen may affect the metabolism of host cells a long distance away from the location of the pathogen, or those just adjacent to it. The changes initiated by the pathogen or by the host cells in response to the pathogen are usually detrimental to the host and often disorganize cellular structure resulting in varying amounts of necrosis. It is often extremely difficult to clearly define which metabolic processes directly originate from the pathogen, and which originate from the host in response to the pathogen. Sometimes it is not reasonable or desireable to discuss the relationship of host and pathogen separately. However, the growth of the pathogen and processes that clearly originate from the pathogen are briefly described herein.

15-2 GROWTH AND SPREAD OF THE PATHOGEN

15-2.1 LOCAL LESION

Limited Growth Is Characteristic Of Some Fungi

Local lesion diseases involve a limited amount of host tissue invaded around each infection site. This may be a result of containment of the pathogen by the host and/or a limited growth by the pathogen. The containment of the pathogen by the host is usually a result of a hypersensitive reaction. A limited growth is characteristic of some fungi that invade the host and grow for a short period of time then initiate reproductive growth. These fungi do not resume vegetative growth from old lesions. The stored reservoir of food obtained by the vegetative fungal cells in attack upon the host is utilized by the fungus in the formation of spore-producing structures and spores.

Growth May Be Limited By Environment

Some fungi may be limited in their invasion of the host by an unfavorable change in environment that inhibits the growth of the fungus. This situation is typical for the potato late blight fungus, where growth is checked by warm, dry weather; but may resume when cool, moist weather returns. An extended warm, dry period, or high temperatures, may kill the invading fungus and thus prevent further growth.

Haustoria Invaginate Host Protoplast

The wheat stem rust fungus ramifies the host tissue soon after entering the substomatal cavity. The fungus grows intercellularly and develops haustoria in adjacent host cells. Each haustorium invaginates the host protoplast and is surrounded by the host plasma membrane that does not contact the haustorium. Characteristically, there is a decrease in the size of the chloroplasts of the host cells containing haustoria. This is accompanied by an increase in the size of the nuclei, which later collapse. Four to five days after infection, the fungus forms a subepidermal

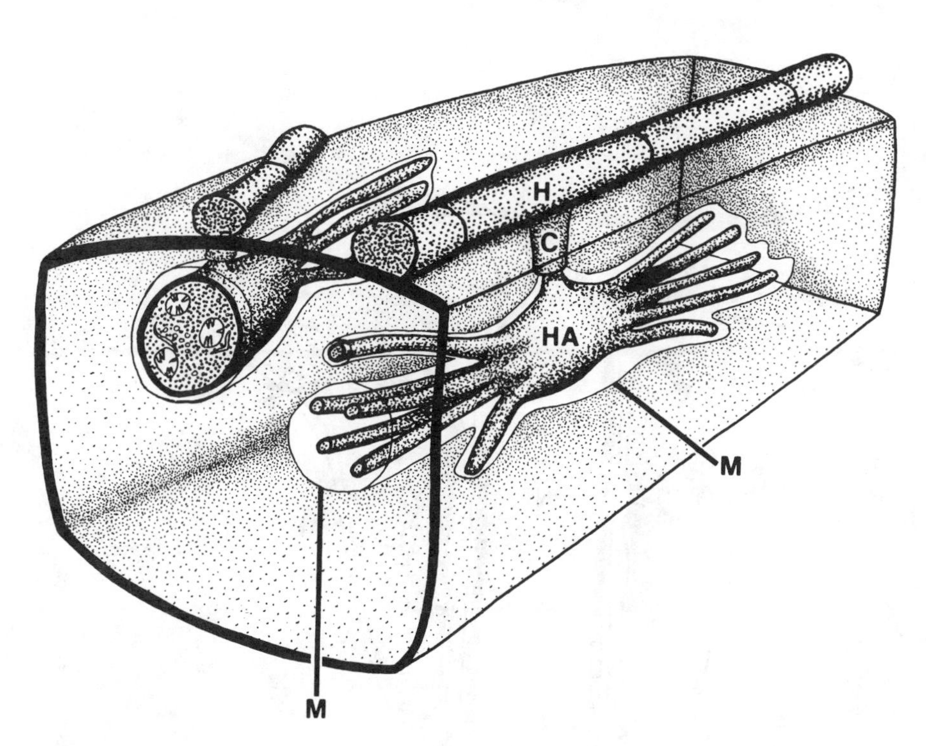

Fig. 15.1 Cutaway three-dimensional diagram of a barley epidermal cell containing two haustoria of the powdery mildew fungus. Haustoria (HA), invaginate the host protoplasm and are surrounded by an extension of the host plasma membrane (M) that does not contact the haustoria. Hypha (H), collar (C). *(After Bracker, 1968)*

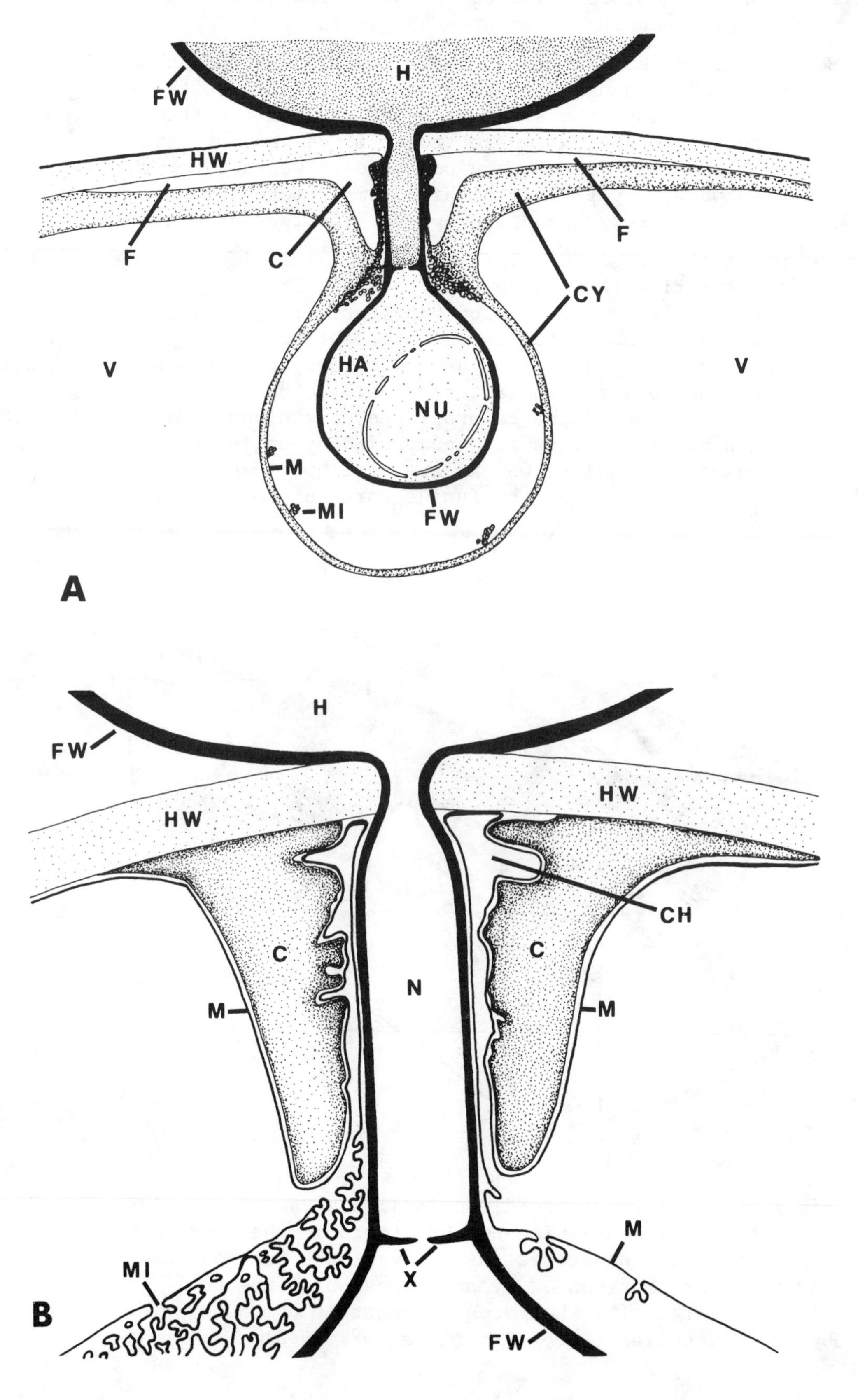
H
FW
HW
F
C
F
CY
V
HA
V
NU
M
MI
FW
A
H
FW
HW
HW
CH
C
C
N
M
M
M
MI
X
B
FW

Fig. 15.2

A. Diagram indicating the physical relationship of *Erysiphe graminis* to the components of the host cell.

B. Diagrammatic interpretation of the penetration region showing the extension of the host plasma membrane and its invaginations.
(Modified from Bracker, 1968)

Key to Labeling

C = collar
CH = channel between haustorial neck and collar
CY = host cytoplasm
F = flange of the collar
FW = fungal wall
H = hypha
HA = haustorium
HW = host wall
N = haustorial neck
NU = nucleus in haustorium
M = membrane
MI = membrane invaginations
V = host vacuole
X = cross wall with central pore

stroma, a uredium, in which urediospores are differentiated. As the developing uredium enlarges, the epidermis is ruptured, and mature urediospores are freed. The individual sizes of the uredia vary according to the frequency of infection and the resistance or susceptibility of the host. In susceptible hosts, the individual uredial pustules are much larger if the frequency of infection is low. This suggests a competition between individual infection centers for essential substances.

Rust Thrives In Susceptible Host

The rust fungus thrives in susceptible hosts. Hyphae are abundant between the host mesophyll cells and haustoria are large and numerous. Host cells show little injury and remain alive until late in the disease cycle, showing mutual compatibility between the fungus and the host. The urediospores are abundant and are quickly formed.

Rust Conflicts With Resistant Host

In resistant hosts, there is ample evidence of conflict between host and the rust. The uredia may vary in size from slightly smaller than normal to nonexistent. Sometimes the uredia are noticeably slow to develop. However, the hypersensitive reaction is the common resistant reaction.

The mycelia of the obligate powdery mildew fungi are entirely outside of the plant, growing profusely over the surface of infected host leaves. Individual hyphae send haustoria into the underlying epidermal cells. The powdery mildew fungus of wheat forms haustoria that possess enlarged central bodies with several long finger-like lobes arranged in a bipolar fashion. As a result of infection by this fungus, there may be anywhere from one to 200 or more haustoria in one host epidermal cell.

There is much yet to be learned about the relationship of a haustorium to a host cell. The extension of the plasma membrane that surrounds but doesn't contact the haustorium undoubtedly plays a very important role in the transfer of metabolites from the host, (see Fig. 15.1 and 15.2).

The apple scab fungus, *Venturia inaequalis*, colonizes its host beneath the cuticle, but above the epidermal cell layer. The lesion tends to be roundish because of the radial growth of the pathogen. The fungus grows at the expense of the underlying host cells, which soon degenerate and die. The fungus produces a stroma and breaks the cuticle allowing the conidia to be released.

Once inside the host cell, the penetrating hyphal strand of *Colletotrichum lindemuthianum* (causal agent of bean anthracnose) enlarges and continues to grow. It penetrates subcuticular cells through small openings probably made by enzymatic activity. This growth continues for several days without killing the host cells. Aggregates of hyphae form beneath the epidermis producing cavities that become acervuli containing masses of conidia borne on conidiophores. A single acervulus may contain 3 to 50 or more conidiophores depending upon the size of the lesion.

A most important observation is that many facultative fungi live in intimate contact within and between living host cells for varying lengths of time, usually one to several days, before the host cells die. This time is substantially shortened if the host exhibits a hypersensitive reaction.

Intimate Relationship Between Host and Facultative Fungus Prior to Death Of Host Cells

15-2.2 EXTENSIVE LESION

Diseases characterized by extensive lesions are commonly incited by a number of bacterial and fungal pathogens. The host tissue affected by a single point of infection varies according to host and pathogen. Many of the so-called "blight" diseases fit this category. Potato late blight is a good illustration of a fungal disease whose pathogen is capable of extensive invasion of the foliage. Typically, however, many lesions are necessary to kill the plant. Two examples of extensive lesion diseases inflicted by bacteria are fire blight of apple and pear, in which blossoms, leaves, fruit, twigs, and large branches may be invaded and killed; and Stewart's bacterial wilt of corn, which has a leaf blight phase with elongated lesions typically being formed.

Blights

15-2.3 UNRESTRICTED LESION

The unrestricted lesion spreads as long as there remains tissue of the type originally attacked. Bacterial and fungal pathogens that are characteristic of this type of lesion are soft rot, dry rot, and root rots. Typical examples include: damping-off of seedlings, *Pythium* spp.; sclerotinia disease of numerous vegetables and field crops, *Sclerotinia sclerotiorum*; rhizopus soft rot of sweet potato, *Rhizopus stolonifer*; stem canker, root rot, leaf blight, and storage rot of many plants, caused by *Rhizoctonia solani*; and bacterial soft rot of vegetables, incited by *Erwinia carotovora*.

Soft Rot
Dry Rot
Root Rot

15-2.4 SYSTEMIC COLONIZATION

The systemic colonization of a plant may or may not lead to wilting. Loose smut of wheat and barley, *Ustilago tritici*, and *U. nuda*, respectively, are illustrations of systemic invasion without wilting. Each of these pathogens

Systemic Growth Without Wilting

resides as dormant mycelium within the seed and grows systemically in the growing point of the shoot. The mycelia in the spike give rise to masses of teliospores that replace the kernels.

Vascular wilt diseases: are a uniquely different, but large and important group of diseases. There is a marked similarity in the symptomology of the various wilt diseases. This is evident even though the virus, fungal, and bacterial pathogens involved are highly dissimilar. The similarity of symptoms of the vascular wilt diseases suggests that a common biochemical phenomenon is involved. What then do these pathogens have in common? The most apparent common relationship is that these pathogens are located in the vascular elements in the early stages of pathogenesis, and the invasion and/or disintegration of other tissues occur later. There is further general agreement that blocking of vascular elements is commonplace and is responsible for most, if not all, of the water deficiency that causes wilting to occur. However, there is disagreement as to what causes the blockage.

Blocked Vascular Elements Cause Wilting

The physical presence of the pathogen in vascular vessels is frequently considered to be a contributing factor of wilting. Saaltink and Diamond (1964) found that mycelium of *Fusarium* formed plugs in the vessels of the first internode of tomato. The mycelial plugging was not continuous longitudinally. They also reported that the reduction in the rate of flow of water in tomato stems was correlated with the number of vessels in which transport was blocked. Waggoner and Dimond (1954) showed that even a few hyphae in a vessel can dramatically reduce the liquid flow through that vessel. There is also evidence that vascular plugging in diseases caused by some bacterial pathogens is at least partially caused by the bacterial cells themselves and the slime they secrete (Husain and Kelman, 1958). However, many fungal vascular pathogens cause the formation of gums and gels that are widely associated with restricted water flow. These gums and gels arise from host tissues, not from the pathogen itself (also see Fig. 16.2 on page 269).

Pathogens In Vessels Contribute To Wilting

A few fungal vascular pathogens were long known to spread throughout their hosts at a rate that was faster than could be accounted for by the mycelial growth. Subsequently, it was found that spores are produced by mycelium in the vessels, and these spores are indeed carried considerable distances in mature vessels.

In the cotton plant, the verticillium wilt fungus penetrates the root directly, or more frequently, through the root tip. The pathogen grows intercellularly and intracellularly until it penetrates the vascular tissue.

Upon reaching the xylem elements, the mycelium proliferates and produces conidia. When the conidia are mature, they are apparently carried rapidly upward. Presley et al., (1966), found that one month old cotton plants contain vessels that are sufficiently large and continuous to allow conidia to move from the base of inoculated stems to the uppermost petiole. This free movement of conidia accounts for the multiple infection sites that are typically found in infected plants.

Spores Transported Through Vessels

A study of bacterial wilt of carnation (*Pseudomonas caryophylli*), has shown that the first evidence of the pathogen in the stem is the presence of a mass of bacterial cells in the lumen of one or more xylem vessels (Nelson and Dickey, 1966). The bacterial mass fills the lumen then moves into the bordered pits of the vessel. Sometimes, when xylem parenchyma cells are adjacent to infected vessels, the mass of bacterial cells in the vessels push through the bordered pits into the neighboring parenchyma cells but appear to be surrounded by a membrane presumably from the vessel pit. The membrane then ruptures, releasing the bacteria into the lumen of the parenchyma cell. The bacteria then multiply and join with masses from adjoining bordered pits forming one large mass. The bacteria multiply at the expense of the parenchyma cells whose walls collapse forming a pocket of bacteria. The bacterial pocket continues to expand and gradually involves large portions of the pith, xylem, phloem, and fiber tissue. The thin-walled cells around the periphery of the pocket all appear to be crushed. Occasionally, a longitudinal crack in the outer stem surface is produced, allowing the bacterial mass to ooze to the stem surface.

Most, if not all, plant pathogenic bacteria are able to multiply in the fluid that is contained in the intercellular spaces and in the conducting vessels of host plants. Not all bacterial pathogens are able to enter the conducting vessels, however. In the resistant host, the growth of the bacteria is checked by various host responses. In the susceptible host, the growth of the bacteria is unchecked, and the bacteria degrade the cell-wall components and intercellular cement, and utilize the breakdown products as nutrients for their continued growth.

Rule -- Bacterial Multiplication

15-2.5 THE VIRUS WITHIN THE PLANT

A number of viruses, including TMV, may be thought of as "nonrestricted", referring to the fact that they are capable of inducing infection when introduced into any type of living host cell. Some viruses, such as beet curly top, are "phloem dependent", meaning they must be deposited into a phloem cell of their host in order to incite infection. Some phloem dependent viruses are capable of spreading into other tissues after initial invasion of the host phloem; while others, including beet curly top virus, appear to be restricted to the phloem.

Since virus infection occurs only when virus is multiplied within a host, it follows that there is a build-up of virus particles within the plant. There is a great variation as to the rate of multiplication and ultimate concentration of virus particles within host cells. It has been estimated that mosaic virus infected cells may contain 100,000 to 10,000,000 virus particles per cell.

Three Probable Synthesis Sites

Virus synthesis sites: there are three sites within a plant host cell where viral synthesis is thought to occur; the nuclei, the chloroplasts, and cytoplasm. The bulk of the present data regarding viral synthesis is based upon studies of tobacco mosaic virus (TMV) and a few other viruses. Most of these studies suggest the host cell nucleus as being involved in viral nucleic acid and protein synthesis. The complete virus particles are then formed by a self-assembly process, and the particles then move out of the nucleus into the cytoplasm. However, in the review of virus synthesis by Schlegel et al., (1967), it is suggested that RNA of TMV is synthesized in the nucleus, and then moves into the cytoplasm where it acquires its protein coat on, or near, the surface of the nucleus. It is further suggested that free movement of cytoplasm in and out of the nucleus occurs, and accounts for intact TMV particles observed in the nucleus.

Chloroplasts of virus infected cells have been shown to undergo structural changes. An involvement of chloroplasts with viral protein has been suggested. However, viruses have been shown to multiply in root tip cells that are free of chloroplasts. It is possible that other plastids could substitute for chloroplasts in these cells. Ushiyama and Matthews, 1970, presented evidence which suggested that turnip yellow mosaic virus (TYMV) multiplies at least in part, in chloroplasts of infected plants. Their studies indicated that RNA of TYMV replicates in peripheral vesicles of the chloroplasts then moves out into the cytoplasm. Synthesis of TYMV coat protein and assembly of the viral particles appears to take place in the cytoplasm.

Some animal viruses containing RNA have been shown to be synthesized in the host cell cytoplasm. Such definitive reports are lacking for plant viruses. However, some studies point to the ribosomes as the sites of synthesis of viral protein of some viruses.

Sites of viral synthesis may be different for different viruses. We should not extrapolate the evidence for synthesis of TMV and TYMV so as to include all plant viruses.

Movement of viruses: the cell to cell movement of a virus in parenchyma tissue apparently involves virus multiplication in each cell into which the virus moves. This type of virus spread is slow and methodic. Most viruses move systemically in the phloem. The majority of the exceptions to this are the few viruses that move systemically in the xylem. Movement in the phloem is accomplished without virus multiplication in each phloem cell. Therefore, virus movement in the phloem is much faster than is the movement into and through parenchyma tissue.

Virus movement in the phloem is usually in the direction in which metabolites move; however, they can also move in the opposite direction. Once viruses enter the phloem, they are rapidly transported to meristematic regions of the shoot and root, and into food storage organs such as rhizomes and tubers. Viral distribution within the host is quite variable. Many viruses become well distributed throughout the living cells of a plant. Other viruses appear to allow the roots, or various sections of tissues to remain virus-free. Sometimes the apical meristems are relatively virus-free. Most plant viruses are not seed transmitted. These

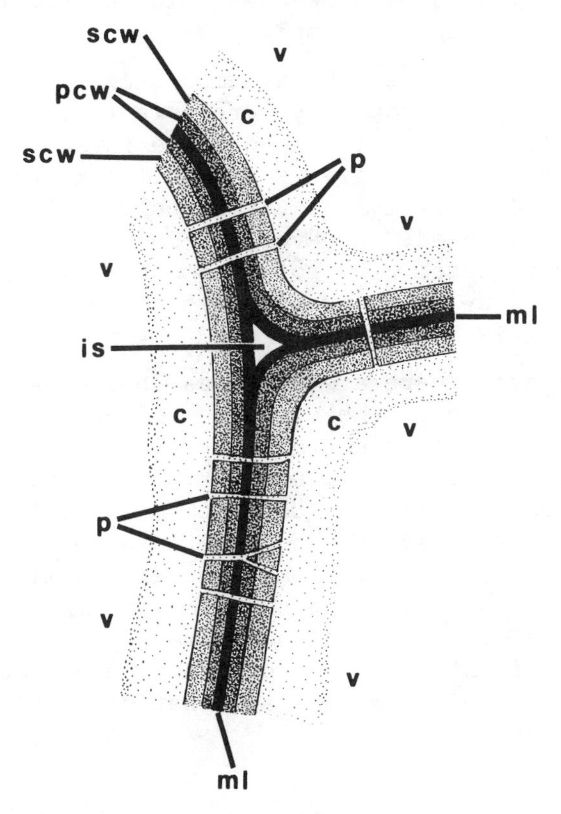

Fig. 15.3 Cross-section of plant cell wall detail showing plasmodesmata (p), middle lamellae (ml), primary cell wall (pcw), secondary cell wall (scw), intercellular space (is), host cytoplasm (c), and vacuole (v).

viruses apparently cannot enter or cause infection in pollen and egg mother cells. The viruses that are capable of being seed transmitted may enter the pollen and/or the ovule.

Plant Viruses Move Through Plasmodesmata

The movement of plant viruses from cell to cell in parenchyma tissue is believed to be through plasmodesmata, (sing. plasmodesma), which are cytoplasmic strands through cell walls interconnecting the living protoplasts, (Fig. 15.3). Complete particles of several viruses have been observed in the plasmodesmata of infected cells with the electron microscope. Virus particles have also been seen in the plasmodesmata joining parenchyma cells with sieve elements. Movement of viruses within sieve elements of the phloem takes place through the sieve-plate pores.

Most virus movement is thought to be by complete virus particles. However, some strains of tobacco rattle virus and tobacco necrosis virus do not produce protein; they exist as RNA. Also, some strains of TMV produce anomalous protein that does not unite with the RNA. Since these viruses do not have protein coats, their movement within hosts is accomplished by naked RNA. Therefore, virus movement is not restricted to complete virus particles.

The viruses that enter the xylem system pose the interesting question as to how they were able to enter and leave these dead cells. It would be possible for a virus to enter an immature xylem cell while it was still living, then when the cell reached maturity the virus would be freed after the cell membrane ruptured. However, it is not known how the virus is able to enter the protoplasm of a living cell from an adjacent dead xylem cell, although there is ample evidence that this occurs.

Virus inclusions: there are numerous instances of virus inclusions occurring in the cytoplasm of host cells. TMV infected plants commonly contain crystalline material that appears as hexogonal plates, needles, and spindles. Crystalline aggregates of viruses often form orderly arrangements within the cell and may be scarce or so numerous as to appear to saturate the cytoplasm. Sometimes crystalline particles are found in the nucleus, and some TMV particles have been found in the chloroplasts, although rarely so. It appears doubtful that virus particles routinely enter vacuoles. Virus inclusions in host cells are characteristic for some viruses, but not others.

15-3 LATENT INFECTION

Many microorganisms enter host plants but are unable to colonize them because the hosts are immune or resistant. In resistant hosts the microorganisms often remain alive. If the host resistance becomes reduced, the microorganisms may resume growth and subsequently colonize the hosts forming necrotic lesions. Such infections are known as latent infections. Latent infections often occur in fruit prior to harvest, and the apparently healthy fruit later becomes diseased in storage. Various species of *Colletotrichum* have been shown to cause latent infections. For example, tomato fruit are commonly attacked in this manner by *C. phomoides* and *C. atramentarium*. Other fungi, such as *Botrytis cinerea* and *Monilinia fructicola*, are known to cause latent infections in various fruit.

Another form of latent infection occurs when facultative parasitic fungi and bacteria enter plants and colonize them to a limited extent, but do not noticeably damage them. Such parasites are often non-pathogenic to the particular hosts. The hosts are said to be resistant or in some cases, tolerant. There are numerous examples of microorganisms that are found to inhabit hosts that are seemingly not affected. Yet these same microorganisms are often pathogenic on appropriate hosts.

Latent infection also occurs in many plants that are able to limit the growth of the pathogen while the plants are in an active stage of growth. Plant resistance is sometimes lowered by maturity or a reduced growth brought about by unfavorable weather conditions, and this enables the parasite to resume growth and colonize the host in a pathogenic nature.

Various vascular fungi, such as *Verticillium albo-atrum*, and several subspecies (forms) of *Fusarium oxysporum* are often found in the vascular systems of seemingly healthy host plants. Again, these microorganisms may be virulent pathogens on other appropriate hosts; other times they are avirulent (non-pathogenic).

Latent Infection Is A Common Occurrence

The phenomenon of latent infection has special implications for many pathogens. These pathogens can survive year after year on apparently unaffected weed and crop plants and then cause severe disease when susceptible hosts are planted. The many references in the literature clearly show that latent infection by various fungi and bacteria is a common occurrence.

Latent infection of plants by viruses is also a common

Three Types Of Latent Virus Infection

occurrence. There are at least three types of "latent virus" infection. First, there are viruses that cause evident symptoms at the time of infection; these symptoms may disappear only to reappear at intervals in response to appropriate stimuli. Such stimuli include wide variations in temperature, severe pruning, or the application of fertilizer that is high in nitrogen; the latter two induce new growth. Secondly, some viruses cause symptoms at the time of infection but the host seemingly recovers; and although it still contains the virus, does not exhibit symptoms again. Thirdly, there are plant viruses that never cause a disease in some of their hosts. The latter situation occurs, for example, with some soil-borne viruses that are known to multiply in weed hosts. These weed hosts grow unimpaired, and provide an important source of inoculum for crop plants.

15-4 TOXINS PRODUCED BY PARASITIC MICROORGANISMS

Defined

Metabolic substances that are toxic to plants are often referred to as "toxins". A number of toxins have been shown to be produced by parasitic microorganisms; some of these play a clear and important role in pathogenesis. In fact, some microorganisms cause disease by means of their toxins. However, most toxins are less important or unimportant in pathogenesis.

Toxins Are Non-enzymatic

Regardless of what the causal agent is, degradation and degeneration of a plant or its parts may produce many toxins that do not function as disease causing entities. This is especially the case in advanced stages of disease. Many of these are undoubtedly produced by the plant and not by the microorganism. Another concept is important--toxins are non-enzymatic and should not be confused with toxic enzymes.

15-4.1 TOXINS THAT INCITE MANY OR ALL OF THE SYMPTOMS INDUCED BY PARASITIC MICROORGANISMS

Victorin: victoria blight of oats is a disease restricted to derivatives of the oat variety Victoria and is caused by the fungus *Helminthosporium victoriae*. This fungus is a seed and soil-borne pathogen that causes severe leaf blight, a basal rot of the stem, and root necrosis. Infection often occurs near the soil line. The first symptoms usually appear as yellow to orange-red stripes on the affected leaves.

Pathogenic strains of *H. victoriae* produce a toxin "victorin" that is capable of producing all of the characteristic symptoms of the disease in the absence of the fungus. Furthermore, victorin is restricted to the exact same host specificity as the fungus and is therefore often termed a host-specific toxin. Non-pathogenic strains of *H. victoriae* do not produce toxin. Pathogenicity is correlated with the amount of toxin produced in culture. Toxic concentrations of victorin have been extracted from *H. victoriae* infected plants and used to produce similar disease symptoms on susceptible oat plants.

Host-specific

Susceptible tissues treated with victorin show an increase in respiratory activity which reach a maximum after 4 to 10 hours. Within limits, the response is directly related to the concentration of victorin. There also appears to be an increase in the permeability in treated susceptible plants that corresponds with plants naturally infected with *H. victoriae*.

Victorin is probably a polypeptide linked to a tricyclic secondary amine, with a molecular weight of between 800 and 2,000. Extremely minute amounts of victorin will kill susceptible tissues. In fact, it is one of the most potent toxins known. It is possible that pure victorin may be active in dilutions above 1 part in 50 billion, (Wheeler and Luke, 1963).

It appears that this fungus causes disease by means of its toxin. This correlation between toxin production and pathogenicity of the fungus has been augmented by data which show that resistance in oats to both *H. victoriae* and victorin are the same genetically. The latter finding proved useful in the screening for resistant oat seedlings in breeding programs. Seedlings were exposed to the toxin rather than to the fungus itself.

Toxin Necessary For Pathogenicity

Periconia toxin: "milo disease" on milo grain sorghums and milo derivitives including the Darso sorghums is caused by the fungus *Periconia circinata*. This fungus also attacks the sweet sorghum "Extra Early Sumac", but all other grain and related sorghums are probably either resistant or immune. The disease is at present only known in the United States.

P. circinata is soil-borne and invades the roots and lower internodes of the culm. The cortex of older roots decays, and the central cylinder turns reddish and dies. This is followed by a reddening and death of the culm base. Although they are not invaded by the fungus, leaves of older plants may roll, wilt, and turn yellow. Stunting and early blooming is common and plants often die prematurely. Young seedlings may take on a scalded appearance.

A toxin has been found to be associated with the fungus and has been isolated from culture extracts. It is known as periconia toxin. It has the same host specificity as the fungus. Filtrates from liquid cultures of some isolates of the fungus have inhibited root growth of susceptible seedlings at dillutions of 1 part in 3,200. Root growth of resistant seedlings was not affected. Cuttings of susceptible plants were killed by culture filtrates at the same high dillutions that inhibited root growth of susceptible seedlings. Again, resistant cuttings were not affected. Filtrates containing this toxin have not produced toxic effects on other species of higher plants, or on bacteria or fungi. Purified periconia toxin will inhibit root growth of susceptible sorghum seedlings at 1.0 - 0.1 μg/ml.

Host-specific

Periconia toxin appears to be a relatively low molecular weight polypeptide that may be in the same general class, chemically speaking, as victorin. However, there is no biological cross-reactivity between the two toxins, and their chemical properties are distinctly different.

Toxin from Alternaria kikuchiana: a host-specific toxin is also produced by the fungus *Alternaria kikuchiana*, which causes black spot of Japanese pears. American and European varieties apparently are resistant to this disease, however, Japanese pear varieties evidently vary from resistant to highly susceptible.

Kills In Advance Of Fungus

The toxin of *A. kikuchiana* kills cells of susceptible plants in advance of fungus growth, which supports the conclusion that it is of real importance in the development of this disease. Non-pathogenic strains of the fungus are known, and these do not produce the toxin.

Altered Cell Permeability Causes Water-soaking

Colletotin: this toxin is produced by the fungus *Colletotrichum fuscum*, which causes anthracnose of *Digitalis* spp., (foxglove). Invaded areas on the stems appear water-soaked at first, then become sunken or depressed, and finally turn necrotic. Lesions on leaves are water-soaked at first. Close lesions often coalesce to form irregular spots which later become necrotic. Colletotin apparently alters cell permeability, and the water-soaked condition of the early stages of infection is due to the movement of cell sap into the intercellular spaces.

Colletotin causes symptoms that are similar to those caused by *C. fuscum*, and resistant hosts are also resistant to the toxin. It is unusual in that it is able to produce water-soaking and necrosis on tomato plants, which are immune to infection by *C. fuscum*.

Wildfire toxin: the wildfire toxin, produced by the bacterium *Pseudomonas tabaci*, which causes wildfire of tobacco, has been studied in some depth. Wildfire disease is characterized by numerous local lesions that are yellowish at first with water-soaked centers. The spots increase in size and the centers turn brown as the host cells die. Individual lesions may finally have a necrotic area of 20 to 30 mm in diameter and are surrounded by a chlorotic halo. The lesions are often numerous and neighboring spots quickly coalesce to form large areas of necrotic tissue. The wildfire toxin apparently invades host tissue in advance of the bacterium. Bacteria cannot be isolated from the "halo", but are readily isolated from the necrotic centers.

P. tabaci produces the toxin in culture. The extracted toxin is capable of producing typical chlorotic symptoms in tobacco plants and in a variety of plants that are not susceptible to *P. tabaci*. Therefore, it is termed non-host specific.

Non-host Specific

Cultures of *P. tabaci* may lose their ability to produce toxin, in which case the bacterium appears to be identical with another bacterial pathogen of tobacco, *P. angulata*. This latter bacterium causes angular leafspot of tobacco.

Wildfire toxin apparently interferes with the metabolism of methionine, an essential amino acid. This belief is supported by the finding that methionine sulphoxime, which is a well known antagonist of methionine in the metabolism of mammals, also induces chlorotic halos around the points of introduction into healthy tobacco leaves. The chemical structure of the wildfire toxin was characterized by Woolley et al., 1955, as having the empirical formula $C_{10}H_{16}O_6N_2$ MW 272. Wildfire toxin is structurally related to methionine. It is labile at room temperature and rapidly looses its activity. Perhaps this is why it has not yet been isolated from diseased tissues. Also, the action of the toxin in the host has not been lessened by the application of excess methionine on affected leaves. It is quite probable that wildfire toxin competes more vigorously than methionine for a reactive site on an enzyme.

Interferes With Metabolism Of Methionine

Other toxins: some other toxins that appear to incite many or all of the symptoms incited by the parasitic microorganisms include the toxin of *Pseudomonas morsprunorum*, a bacterium that causes a canker and leafspot on plum, cherry, and apricot trees. This toxin has only been partially purified, and its chemical nature is thought to be proteinaceous.

The bacterium *P. phaseolicola*, which causes halo-blight of beans, also appears to form a toxin. This toxin incites the halo symptom that is similar to the halo caused by the wildfire toxin.

Toxins are also produced by various vascular wilt fungal pathogens such as *Ceratocystis ulmi*, which causes Dutch elm disease, and *C. fagacearum*, which causes oak wilt. The toxins of both fungi are produced in culture. When applied to cuttings of their respective hosts, they produce symptoms similar to those in infected trees. Little is known regarding their mode of action or chemical composition.

15-4.2 TOXINS THAT INCITE AT LEAST SOME SYMPTOMS THAT ARE INDUCED BY PARASITIC MICROORGANISMS

Piricularin: is produced by the rice blast fungus, *Piricularia oryzae*; and like *P. oryzae*, it is more reactive to susceptible rice varieties than resistant varieties. Perhaps the most interesting feature of this toxin is that it is also toxic to a variety of microorganisms, including *P. oryzae*. In fact it is about ten times more toxic to the parent fungus than it is to a susceptible host plant. At a dilution of 0.25 ppm, piricularin inhibits the germination of *P. oryzae* conidia. The question immediately arises as to how the fungus can survive when the toxin it produces is more toxic to itself than to the host. The answer appears to be that there is a protein produced by the fungus called "piricularin-binding protein" that combines with the toxin rendering it inactive as an anti-fungal agent. However, this does not lessen its toxicity (phytotoxicity) to the host plant.

Piricularin Is Toxic To Its Parent Fungus

Fusarial wilt toxins: are present in a number of diseases incited by various forms of *Fusarium oxysporum*. There appears to be several toxins involved in each of these diseases; and the complete syndrome requires an interaction of at least two, and probably more, of them. Each toxin produces only a portion of the symptom expression. The two most well known toxins that are so involved, are fusaric acid and phytonivein.

Fusaric acid is produced by a number of *Fusarium* spp. It has the empirical formula $C_{10}H_{13}O_2N$, and MW 179. Fusaric acid increases the permeability of cell membranes, but beyond this a definite conclusion concerning its role in pathogenesis is still lacking.

The toxin phytonivein was discovered by Japanese workers and is produced by *F. oxysporum* f. *niveum*, the cause of wilt of watermelon. This is a sterol, with the empirical formula $C_{10}H_{46}O_2$. Phytonivein is thought to affect wilting, since dilute solutions cause irreversible wilting of melon cuttings.

Phytonivein Causes Wilting

15-4.3 TOXINS THAT INCITE DIFFERENT OR ONLY A FEW OF THE SYMPTOMS INDUCED BY PARASITIC MICROORGANISMS

Lycomarasmin: is produced by the fungus *Fusarium oxysporum* f. *lycopersici*, which causes wilt of tomato. Lycomarasmin has been studied extensively. It is a dipeptide with the empirical formual $C_9H_{15}O_7N_3$, MW 277. When tomato cuttings are placed in dilute solutions of lycomarasmin, the leaflets begin to roll upwards, then become desicated without wilting. Necrotic areas then develop in interveinal areas. These are not symptoms that are produced by the fungus.

Alternaric acid: is formed by the fungus *Alternaria solani*. This fungus causes local lesions on leaves and stems of tomato plants. The spots are brown to blackish and often circular with a number of irregular concentric rings. Heavily infected leaves often have a pronounced chlorosis between lesions that often extends into upper uninfected leaves and is more apparent when the lesions are near veins. Leaf abscission is common. The chlorosis and abscission is caused by alternaric acid. This toxin is an unsaturated dibasic acid with the empirical formula $C_{21}H_{30}O_8$, MW 410.

Leaf Abscission And Chlorosis

There appears to be strains of *A. solani* that are virulent, yet lack the ability to produce alternaric acid. Conversely, mildly pathogenic strains have been shown to produce abundant toxin. However, strong circumstantial evidence points to a greater abundance of alternaric acid being produced in the stems and foliage of plants rather than in fruit.

15-5 ENZYMES PRODUCED BY PARASITIC MICROORGANISMS

Fruitful discussion of specific enzymes can only be accomplished when there is a clear understanding of the basic chemical composition of host structures. This is deemed to be beyond the scope of this text. Therefore, enzymes will be dealt with in the most abbreviated and general manner.

All pathogenic microorganisms are at some stage of pathogenesis in contact with the host cell walls. Probably all of these pathogens enzymatically alter at least a portion of the host cell walls. Those parasites that grow intercellularly may alter cells without much obvious and immediate disruption of host cells. Pathogens that cause necrosis and rotting, on the other hand, quickly cause considerable host-cell disruption. Obviously, different types of enzymes are involved in these two different categories of disease.

Various types of pectic enzymes have been shown to be produced by certain pathogens in culture. Probably, most of these enzymes are produced in the parasitized host. These enzymes preferentially attack and cleave pectin at various but highly specific linkages. Such enzymes degrade the pectin of the middle lamellae and that of cell walls. This is a common occurrence in the soft rot diseases and a number of other diseases as well, involving both fungi and bacteria.

Step-wise Breakdown Of Cellulose

The complete degradation of cellulose of cell walls probably requires a step-wise breakdown involving first one enzyme and then another, etc., until the glucose sugar is reached. Other enzymes secreted by particular pathogens include those that attack host proteins, suberin and lignin. A few enzymes secreted by specific pathogens are highly toxic to host cells. The exact mechanism of action of these enzymes is still unknown.

Enzymatic Solubilization

Usually, parasitic fungi and bacteria do not just push their way between and into host cells, but enzymatically solubilize specific host tissues as they grow through the host. All bacteria and some fungi gain access to intercellular passages by the continuous solubilization of pectin of the middle lamellae. Bacteria are generally restricted to intercellular host penetration. However, many fungi penetrate host cell walls, and do so by enzymatic solubilization of pectin and cellulose to form tiny holes through which hyphae quickly grow.

The rate of an enzyme secretion appears to be highly important to pathogenic microorganisms. In a number of studies, the ability to produce ample amounts of an enzyme in culture has been correlated with pathogenicity. For example, pathogenic strains of some fungi produce much more of the pectic enzyme pectinesterase in culture than non-pathogenic strains.

Enzymes are thought to play a role in causing a blockage of xylem elements by the formation of gums and gels in most vascular wilt diseases caused by parasitic fungi and bacteria. Pectic and cellulolytic enzymes produced by pathogens in the vessels, attack the vessel walls and bring about partially hydrolyzed gels, gums, and phenolic compounds, which may contribute to the blocking that leads to the wilting of the host.

Selected References

Braun, A. C., and R. B. Pringle. 1959. Pathogen factors in the physiology of disease - toxins and other metabolites, pp. 88-99, In C. S. Holton, et al. (ed.), Plant pathology problems and progress 1908-1958. University of Wisconsin Press, Madison.

Esau, K. 1968. Viruses in plant hosts. The University of Wisconsin Press, Madison.

Goodman, R. N., Z. Kiraly, and M. Zaitlin. 1967. The biochemistry and physiology of infectious plant disease. D. Van Nostrand Co., Inc., Princeton. pp. 4-5, 19-21, 308-337.

Klement, Z., and R. N. Goodman. 1967. Hypersensitive reaction to infection by bacterial plant pathogens. Annual Review of Phytopathology. 5:17-44.

Misawa, T., and S. Kato. 1967. Changes in nuclear contents of host cells infected by viruses. pp. 256-267, In C. J. Mirocha and I. Uritani (ed.), The dynamic role of molecular constituents in plant-parasite interaction. The American Phytopathological Society, Inc., St. Paul.

Mundry, K. W. 1963. Plant virus-host cell relations. Annual Review of Phytopathology. 1:173-196.

Pringle, R. B., and R. P. Scheffer. 1964. Host-specific plant toxins. Annual Review of Phytopathology. 2:133-156.

Schlegel, D. E., S. H. Smith, and G. A. de Zoeten. 1967. Sites of virus synthesis within cells. Annual Review of Phytopathology. 5:223-246.

Takahashi, W. N. 1967. The biochemistry of plant virus infection. pp. 283-296, In C. J. Mirocha and I. Uritani (ed.), The dynamic role of molecular constituents in plant-parasite interaction. The American Phytopathological Society, Inc. St. Paul.

Wheeler, H., and H. H. Luke. 1963. Microbial toxins in plant disease. Annual Review of Microbiology. 17:223-242

Wood, R. K. S. 1959. Pathogen factors in the physiology of disease - pectic enzymes. pp. 100-109, In. C. S. Holton, et al., (ed.), Plant pathology problems and progress 1908-1958. The University of Wisconsin Press, Madison.

____________. 1967. Physiological plant pathology. Blackwell Scientific Publications, Oxford and Edinburgh. pp. 188-227 and 287-325.

CHAPTER 16

HOST RESPONSES TO INFECTION

16-1 INTRODUCTION

Once a pathogen has entered a host plant and begins to colonize and spread through the host, there is a succession of numerous responses and changes that occur in the host. These responses occur in invaded tissues, adjacent cells, and often in distant cells. Ultimately they may result in visible morphological changes that are regarded as symptoms; i.e., reduction in growth, chlorosis, necrosis, epinasty, leaf abscission, malformation, wilting, etc. These symptoms are brought about by profound physiological alterations of the host in response to the pathogen and products of the pathogen. This discussion deals with the physiological responses of the host.

Physiological Changes In Host In Response To The Pathogen And It's Products

16-2 GENERAL PHYSIOLOGICAL RESPONSES

16-2.1 RESPIRATION

It is generally recognized that most, if not all, infectious diseases increase respiration. This is actually a general reaction of the plant in response to most types of harmful stimuli. Studies concerning respiration are often complicated by the fact that except for viruses, pathogens also undergo respiration. This may lead to an error if the total respiration of the host and pathogen is measured. However, sophisticated studies indicate that most of the increased respiration comes from the affected host tissue. The increase of respiration of wounded plant tissue has long been known. Therefore, it is not surprising that all forms of plant injury result in respiration increase.

Rule -- Respiration Increase

In the susceptible host, affected tissues generally undergo a slight increase in temperature as a result of infection. This is associated with an increased oxygen uptake and an enhanced activity of respiratory enzymes. There is usually an accumulation of compounds around points of infection. This concentration and movement is an active process that requires some energy, and consequently, it is dependent upon respiration. Furthermore, in a number of diseases, especially those involving obligate parasites,

Increase Of Dry Weight Due To Increased Synthesis

there is an increase in dry weight of the host that is correlated with increase in respiration. This occurs in the early stages of infection, prior to the production of symptoms. This increase of dry weight can only be explained by an increase in synthesis by the host, which also requires respiration.

The pattern of increased respiration caused by obligate fungal parasites is different for susceptible hosts than resistant hosts. The obligate rust fungi do not cause a degradation of cells of susceptible hosts until advanced stages of infection. The respiratory response of the susceptible host to the rust fungus begins soon after infection and increases in direct proportion to fungal growth, while the host undergoes an increased synthesis. Respiration then declines when fungal growth ceases and as infected tissue degenerates.

Obligate Rust Fungus In Susceptible Host

When obligate rust fungi attack resistant hosts, a hypersensitive reaction quickly occurs, involving limited tissue necrosis. However, this reaction is preceeded by a sudden rise in respiration that is initiated before the cells degenerate. Respiration does not increase as much in the resistant host as it does in the susceptible host. It appears that host cells not in immediate contact with the pathogen are responsible for the increased respiration. After the abrupt collapse of host cells, the respiration rate declines and soon returns to a normal level. The degree of host resistance appears to be dependent upon the speed of host cell collapse. This rapid cell degeneration is in contrast to the susceptible host which undergoes increased synthesis.

Obligate Rust Fungus In Resistant Host

The infection of susceptible hosts by facultative fungal parasites causes an increase in host respiration that preceedes symptoms. The respiratory peak coincides with the degenerative changes in the host. However, when plant cells undergo the dissorganization that finalizes in necrosis, this involves oxidative reactions that should be distinguished from respiratory responses occurring in the initial stages of infection.

Facultative Fungi In Susceptible Host

Few studies have been made with bacterial pathogens in respect to respiration. However, it appears that the pattern of respiratory effects is similar to that caused by fungi.

In plants <u>systemically</u> infected with a virus, higher respiration usually occurs during the advanced stages of infection. This is in sharp contrast to plants with <u>localized</u> virus lesions, where there is a rapid increase in respiration soon after infection.

The factors controlling respiratory mechanisms in normal

plant cells are still not fully known. Therefore, it is not surprising that we know even less about the respiratory responses of plant cells to infection. However, it is probable that the early rise in respiration after infection is a mechanism of energy production, some of which is used in reactions of synthesis and in the movement and concentration of compounds around points of infection. This respiration increase in the early stages of infection should be distinguished from oxidative metabolism associated with tissue degeneration that occurs in the latter stages of disease.

16-2.2 PHOTOSYNTHESIS

Rule -- Photosynthesis Reduced

Plant diseases cause a reduced photosynthesis. Although there are a few instances of increased photosynthesis immediately after infection, these appear limited to the early stages of disease and/or light infections. The overwhelming bulk of literature cites a reduction of photosynthetic activity.

The amount of photosynthetic reduction by many diseases, and especially the foliar diseases, often has been said to be proportional to the area of tissue invaded. However, some pathogens injure a greater amount of tissue than they actually invade. This is due to translocated effects of the pathogen. There is a destruction of chlorophyll in chlorotic and necrotic regions, which results in a loss of photosynthetic ability and surface area. Often, this appears to be a destruction of the chloroplasts. Under conditions of wilting, the stomata are closed and directly restrict gaseous exchange, which reduces photosynthesis. Of course, indirectly, any pathogen that lowers the vigor of a plant and thereby prevents the growth of potential photosynthetic surface, must also be considered.

In some virus diseases, the host plant has areas of green leaf tissue that are noticeably darker than normal, and these areas contain higher than normal amounts of chlorophyll. Therefore, these areas have a higher than normal photosynthesis. However, the entire plant usually has a reduced photosynthesis because of a reduction of: chlorophyll per unit of leaf, chloroplast efficiency, and photosynthetic surface area.

16-2.3 TRANSPIRATION

Rule -- Transpiration Increased

Although there are a number of exceptions, the transpiration rate of infected tissues is increased in most plant diseases. Some of the exceptions warrant consideration here. For example, it has been shown that bean leaves infected with rust transpire less than normal until the uredial pustules open; but after the pustules open, they transpire more than normal. Vascular wilt fungal diseases usually bring about a decrease in transpiration. Powdery mildew infections of clover and bean also cause a reduced transpiration; while latent infections by viruses usually do not alter transpiration.

It can be reasoned that when rust pustules open, this causes a rupturing of the waterproof covering, the cuticle; therefore, increased transpiration would logically follow. Also, it is logical that wilted plants, with their stomata closed and with a water deficit, would transpire less than normal. Latent virus infections do not cause disease symptoms, so with a normal appearing plant, it is not surprising that there is not a noticeable increase in transpiration. Such reasoning only partially explains the transpiration effects in the above illustrations.

The bulk of water loss through transpiration in the healthy plant occurs in the daylight when stomata are open. A very limited and continuous water loss also occurs through the cuticle. Therefore, the healthy plant undergoes a much higher transpiration in daylight than at night. In wheat plants infected with rust, it has been found that before the uredial pustules open, transpiration is reduced in daylight and increased at night. This is probably because the rust infection causes the stomata to remain closed; but because the disease increases permeability of affected host tissue, there is an increase in water loss through the cuticle. Overall, the water loss before the rust pustules open is below that of the healthy plant.

16-2.4 CELL PERMEABILITY

Rule -- Permeability Increased

Plant cells adjacent to cells infected with either obligate or facultative parasites, have increased permeability to water and solutes, (Thatcher, 1942). Increases in permeability occur in cells beyond the region of rotting in soft-rot diseases, i.e., botrytis rot and some other diseases where toxins are known to play a role in pathogenesis, (i.e., victoria blight of oats, incited by *Helminthosporium victoriae*; and wildfire of tobacco, incited by *Pseudomonas tabaci*).

16-2.5 TRANSLOCATION

Rule -- Normal Translocation Decreased

Normal translocation is usually decreased by infectious diseases that often increase the movement of metabolites towards affected tissues. There appears to be a general concentration of carbohydrates around lesions as well as metabolites; such as phosphorous, sulfur, lipids, amino acids, and phenolic compounds. The nature of many diseases restricts normal translocation, but this does not account for the accumulation of materials around and near diseased tissues, which is an active process, dependent upon respiration.

16-2.6 NITROGEN METABOLISM

Nitrogen Metabolism In Infected Tissues Is Variable

Studies on TMV infected tissues indicate that host deficiencies of nonprotein nitrogen compounds result when virus synthesis is high, especially when conditions of nitrogen deficiency exist. Virus replication requires production of viral protein. Therefore, it is reasonable that the host should suffer abnormal nitrogen metabolism when infected by a virus such as TMV, which reaches a high titer in the host. However, once the period of high virus synthesis has passed, no nitrogen deficiency will occur if ample nitrogen nutrition is available to the host. Viruses that reach only a low titer in their hosts do not appear to cause nitrogen deficiencies during virus synthesis. Also, when a high level of nitrogen nutrition is available, a virus infected plant may contain more total nitrogen than a healthy plant.

A common consequence of virus infection is also an accumulation of nonprotein nitrogen compounds, generally amides, in diseased tissues (Diener, 1963).

Plant tumors, incited by the bacterium *Agrobacterium tumefaciens*, have been found to contain considerably more protein than could be produced by normal plant tissue and cells of the pathogen. Apparently the meristematic activity of the gall draws heavily upon the available nitrogen, often causing a nitrogen deficiency.

In diseases caused by the rust fungi, nitrogenous compounds accumulate around the site of infection. Since there is an increased general synthesis following infection, there is also an increase in total nitrogen. As compared to the tissue of healthy plants, rust infected tissue may have a two-fold increase in insoluble protein nitrogen and a four-fold increase in soluble nitrogen in the free amino acids and amino compounds. The total net synthesis in the infected tissue is a two-fold increase over normal, non-infected tissue. This is typical of what is to be expected in the susceptible host.

16-2.7 GROWTH REGULATING SUBSTANCES

Growth promoting substances, i.e., indole-3-acetic acid (IAA), gibberellins, and others (known collectively as "auxins"), have been found to exert a hormonal control of plant growth in association with disease. Auxins have been found in plant tissues infected with pathogenic bacteria, fungi, viruses, and nematodes. Various plant diseases are characterized by host deformities, i.e., overgrowths, excessive branching of stems or roots, swellings, puckering, and epinasty. All of these are positive growth responses. This growth may vary from simple cellular elongation to cell multiplication without disorganized growth; or there may be highly disorganized growth as characterized by galls and tumors. Probably increased levels of growth regulating substances are associated with many of these host responses.

IAA Increases In Many Diseases

The aecial stage of the rust *Gymnosporangium* was found to promote up to 8-fold increases in growth promoting substances in *Pyrus*. Other rust fungi, such as *Puccinia graminis* var. *tritici* on wheat, do not produce abnormal growths of the host. However, a 24-fold increase in IAA has been found in a highly susceptible wheat variety following infection with stem rust. It should be noted that wheat does not react with hypertrophic or hyperplastic responses with IAA, which is unlike corn and most broadleaved plants that do show such responses. Therefore, it is not surprising that corn smut causes various distorted overgrowths of the host because of increased IAA.

Some plant viruses cause a lower auxin content, and produce growth inhibitors as well. The nature of these growth inhibitors is undetermined as yet. However, they probably affect the host cell response and may be of greater importance than growth promoting substances. Decreases in auxin level apparently occur in advanced stages of infection by many plant viruses.

Vascular parasites, such as *Fusarium* spp., may induce hypertrophy of vascular parenchyma cells in response to metabolites. These growths are called tyloses (sing. tylosis). A tylosis is a cell outgrowth into the cavity of a water conducting (xylem) vessel, which partially or completely plugs the vessel. Tyloses obviously are controlled by growth regulating substances, which have been shown to accumulate following infection, but details of which are still unknown.

16-2.8 PHENOLIC COMPOUNDS

Phenolic substances formed in living plants constitute a large class of compounds that include anthocyanins, anthoxanthins, coumarins, tannins, etc. Anthocyanins are red, blue, and purple pigments. Anthoxanthins are generally yellow pigments. Coumarins are growth substances that are involved in such activities as inhibition of seed germination and leaf expansion and cell enlargement of some plants. Tannins are sometimes found in cell sap, but more often are

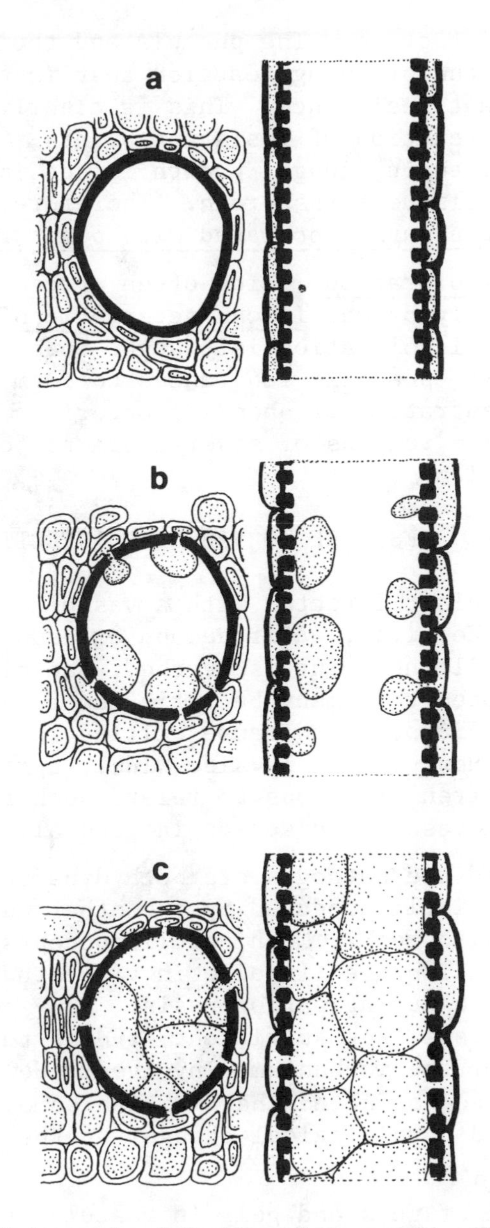

Fig. 16.1 Schematic drawing of cross-sections and longitudinal-sections through vessels to illustrate the formation of tyloses.
a. Normal vessel without tyloses.
b. Partial occlusion by tyloses.
c. Complete occlusion by tyloses.

in the walls of wood and cork cells and in cells of other tissues. Tannins are more recently known as "polyphenols", since they possess more than one hydroxyl group on a benzene ring.

Rule -- Phenol Accumulation

An accumulation of phenolic compounds around centers of infection is a characteristic host response to infection. Phenol accumulation comes about by a migration from uninfected plant parts and the biosynthesis of these compounds that is stimulated by infection. The phenols and their oxidation products cause the browning reaction that is often associated with plant resistance. This is clearly seen in the hypersensitive reaction of resistant plants. When the phenols are not toxic, fungal growth is not inhibited even when the plant tissue turns brown. Therefore, the browning reaction is not always associated with plant resistance.

Rule -- Vascular Discoloration

Vascular discoloration, which often proceeds from yellow to red and finally brown, is characteristic of vascular infection. This discoloration is thought to be caused by the polymerization of phenolic substances to form pigments. Increased concentration of phenolic materials have been shown in vascular regions of several plants following vascular infection.

16-3 RESPONSE TO VASCULAR INFECTION

The wilting plant infected with a vascular pathogen has a water imbalance, i.e., water demand exceeds supply. When the permanent wilting point is reached, the plant dies. It is widely accepted that many bacteria and fungi can produce metabolites *in vitro*, which cause wilting and also chlorosis and necrosis. However, it is also widely agreed that it is difficult and often erroneous to relate such findings to pathogenesis of vascular diseases in general.

The plant and the pathogen are both dynamic systems, each responding to stimuli produced by the other and subject to change with environment. Meaningfull concepts of host-parasite interaction must involve *in vivo* studies of the host and parasite during pathogenesis. Many host changes during pathogenesis lack sufficient studies to provide an understanding of the mechanisms involved. However, recent studies have been conducted that furnish some insight into histological and physiological changes of the host (see Fig. 16.2).

Gums And Gels Contribute To Wilting

The presence of gums and gels in vascular tissues has long been recognized in association with vascular wilt diseases caused by a wide array of pathogens including bacteria, fungi, and viruses. These gums and gels are thought to be partially responsible for blocking the flow of water in infected plants. Gels apparently contain pectinaceous materials and probably other substances as well. It is thought that macerating enzymes, including pectinases, act on the pectin of the vessel walls and middle lamellae, thereby liberating fragments of pectin and perhaps other materials. The gels accumulate in the vessels invaded by the pathogen. Bacterial and fungal pathogens involved in vascular infections are probably able to produce pectic enzymes involved in gel formation. The pectinaceous materials appear to swell (undergo hydration) and are

Fig. 16.2 Possible mechanisms involved in decreased rate of water flow in host xylem--that may lead to wilting. *(After A. Kelman)*

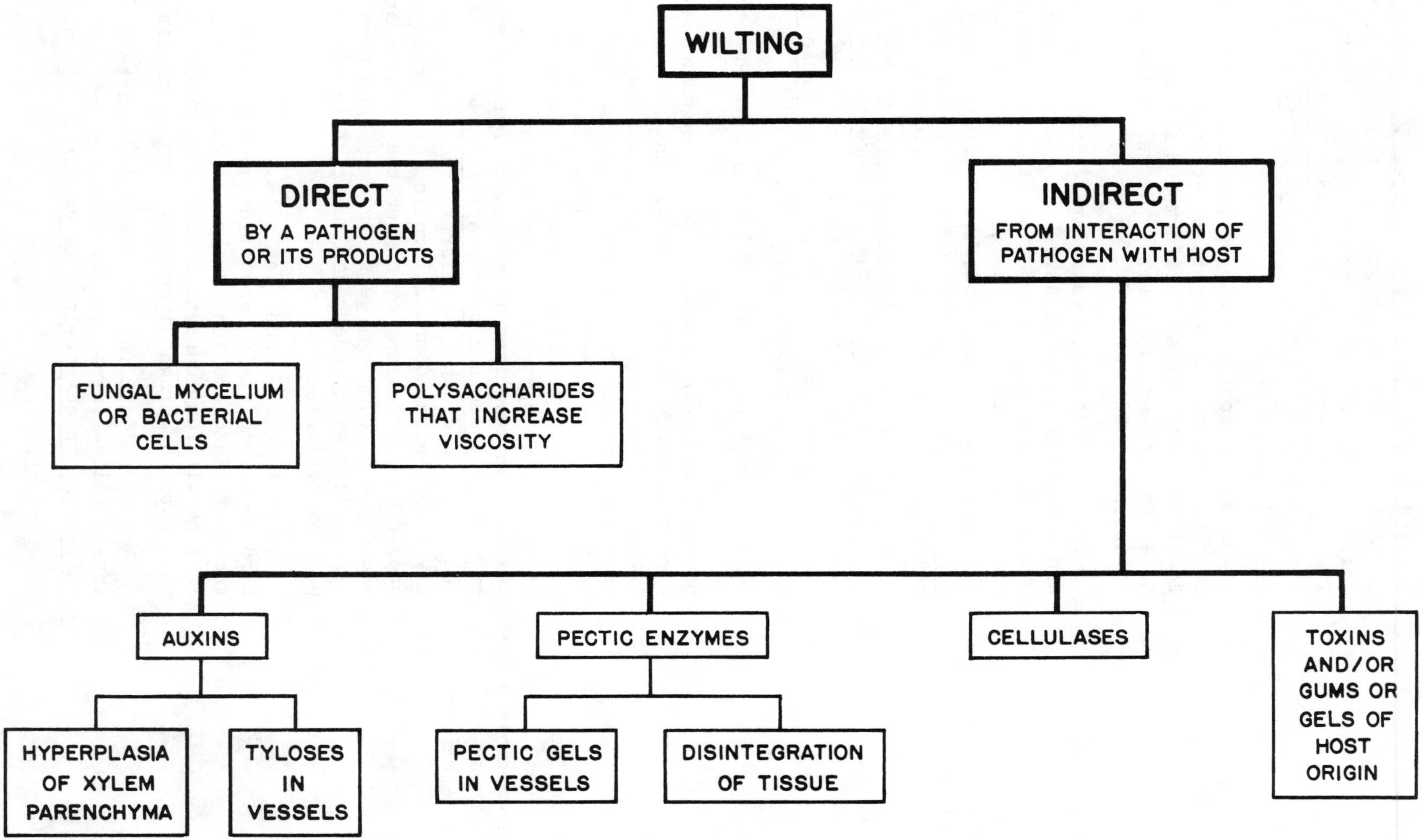

carried in the transpiration stream. They appear to accumulate on the perforation plates or scalariform end walls at the end of the vessels, where they form gel plugs. The gells in some of the vascular diseases are thought to form gums by a reaction with the phenolics associated with vascular discoloration.

Another theory concerning the origin of the gel plugs involves the possibility that pectinases act upon the perforation plates or scalariform end walls. This causes them to soften and swell, forming the gel plugs (Beckman and Zaroogian, 1967).

Tyloses that cause vascular occlusion in both resistant and susceptible hosts, have been found in tomato, banana, broccoli, squash, and sweet potato, in response to fusarium wilt; and in elm trees in response to Dutch elm disease, (Beckman and Halmos, 1962; and Beckman, 1966). In the resistant host, the fungal spores formed by the invading mycelium are trapped at the end of the vessels on the gel plugs. The gel plugs retain the spores until tyloses completely wall-off the further advance of the fungus. This blocking of only a small part of the vascular system by tyloses in the resistant host is insufficient to adversely affect water transport. (see Fig. 16.1).

Limited Tyloses Trap Fungus In Resistant Host

In susceptible hosts, gel plugs form and temporarily contain the spores of the pathogen as in the resistant host. However, the gels appear to become weakened and shear under stress of transpirational pull. The pathogen, its metabolites, and host products surge on to the end of the next vessel before the tyloses can form and block the vessel. As the pathogen moves up the host, it spreads laterally into branching vessels. At successive perforation plates or end walls, spores may germinate and the resulting mycelia cause localized infections. Tyloses slowly occlude, or partially occlude vessel after vessel. When a considerable portion of the vascular system is blocked, the plant may become stunted and chlorotic, or become wilted and perhaps die. Of course, even when the vessels are not completely closed, gums, gels, tyloses, and even the presence of mycelium will impede and restrict the flow of water by partial occlusion.

Extensive Tyloses In Susceptible Host May Cause Wilting

It is further found that tylose formation is retarded or completely absent in the immediate vicinity of vascular discoloration (Beckman, 1966). These findings are in line with our current belief that phenolic substances that form the brown pigments are toxic to living host cells.

Additional studies must be made to determine how widespread tylose formation is, since tyloses are apparently infrequent or absent in some vascular wilt diseases.

The toxin theory and the occlusion theory have long appeared in the literature to account for wilting due to vascular disease. The former theory has been largely discounted and the later is well supported by a wide range of data from numerous host-parasite interactions. The findings may be summarized with the statement: vascular occlusion that leads to wilting appears to be caused by a number of factors including gels, gums, tyloses, and the presence of the pathogen itself.

16-4 PLANT DISEASE REDEFINED

At the beginning of Chapter 2, plant disease was defined as: a harmful alteration of the normal physiological and biochemical development of a plant, (National Academy of Science, 1968). While this is a good and proper definition of plant disease, it is also one quite easily understood by someone new to the study of plant pathology. However, discussions have been presented on the environment, pathogen, and host. This should provide the reader with sufficient background to understand a more complex and complete definition of plant disease which is: an alteration in one or more of the ordered, sequential series of physiological processes, culminating in a loss of coordination of energy utilization in a plant as a result of the continuous irritations from the presence or absence of some factor or agent. (National Academy of Science, 1968).

Plant Disease Redefined

Selected References

Agrios, G. N. 1968. Plant pathology. Academic Press, New York. pp. 83-104.

Beckman, C. H. 1964. Host responses to vascular infection. Annual Review of Phytopathology. 2:231-252.

____________. 1966. Cell irritability and localization of vascular infections in plants. Phytopathology. 56:821-824.

Cruickshank, I. A. M. 1963. Phytoalexins. Annual Review of Phytopathology. 1:351-374.

Dimond, A. E. 1967. Physiology of wilt disease. pp. 100-118, In C. J. Mirocha and I. Uritani (ed.), The dynamic role of molecular constituents in plant-parasitic interaction. The American Phytopathological Society, St. Paul.

Diener, T. O. 1963. Physiology of virus-infected plants. Annual Review of Phytopathology. 1:197-218.

Goodman, R. N., Z. Kiraly, and M. Zaitlin. 1967. The biochemistry and physiology of infectious plant disease. D. Van Nostrand Company, Inc., Princeton.

Husain, A., and A. Kelman. 1958. Relation of slime production to mechanism of wilting and pathogenicity of Pseudomonas solanacearum. Phytopathology. 48:155-165.

Kirkham, S. S. 1959. Host factors in the physiology of disease. pp. 110-118, In C. S. Holton, et al., (ed.), Plant pathology problems and progress 1908-1958. The University of Wisconsin Press, Madison.

Millerd, A., and K. J. Scott. 1962. Respiration of the diseased plant. Annual Review of Plant Physiology. 13:559-574.

Rohringer, R., and D. J. Samborski. 1967. Aromatic compounds in the host-parasite interaction. Annual Review of Phytopathology. 5:77-86.

Rubin, B. A., and E. V. Artsikhovskaya. 1964. Biochemistry of pathological darkening of plant tissues. Annual Review of Phytopathology. 2:157-178.

Sadasivan, T. S. 1961. Physiology of wilt disease. Annual Review of Plant Physiology. 12:449-468.

Sequeira, L. 1963. Growth regulators in plant disease. Annual Review of Phytopathology. 1:5-30.

Shaw, M. 1963. The physiology and host-parasite relations of the rusts. Annual Review of Phytopathology. 1:259-294.

Suzuki, N. 1965. Histochemistry of foliage diseases. Annual Review of Phytopathology. 3:265-286.

Wood, R. K. S. 1967. Physiological plant pathology. Blackwell Scientific Publications, Oxford.

Woolley, D. W. 1959. Metabolic considerations of obligate parasitism. pp. 119-129, In C. S. Holton, et al., (ed.), Plant pathology problems and progress 1908-1958. The University of Wisconsin Press, Madison.

Yarwood, C. E. 1967. Response to parasites. Annual Review of Plant Physiology. 18:419-437.

CHAPTER 17

PLANT DISEASE EPIDEMICS

17-1 GENERAL PRINCIPLES

Epidemic

An epidemic is the extensive development of a disease in a geographical area or community. This definition is imprecise because there is no clear-cut set of conditions as to what constitutes an epidemic. Pathogens vary in their intrinsic ability to cause epidemics. Some have not been known to incite epidemics, although they appear to possess the capacity to do so. Other pathogens are incapable of causing epidemics.

Pandemic

Two other terms should also be defined here. They are pandemic and epidemiology. A pandemic is an epidemic that occurs over an extended geographical area or areas. The most frequently cited plant disease pandemic is the potato late blight pandemic of 1845-46. This disease appeared in devastating form in Ireland, but was also present in several other European countries as well as in the United States.

Epidemiology

Epidemiology is the branch of plant pathology that treats of disease in plant populations.

As popularly used, the term epidemic gives a connotation of destruction and economic loss over a wide area. However, it does not imply the death of host plants, although this may occur. Epidemics may be mild, moderate, or severe. They may cover a small cropping area of less than 100 square miles. They may be devastatingly severe to the crop or mildly destructive.

The Successful Pathogen Is Numerous And Perpetual

The fact that one pathogen is able to infect a wide host range does not necessarily make it more successful than one that can infect only a single host. This is because the pathogen that infects the single host may be in greater abundance than the one with the wide host range. The success of a pathogen must be measured by its population and longevity throughout the years.

Man may rate the success of pathogens according to how they affect his pocketbook. By this measure, the most successful pathogen is the one that causes the greatest annual dollar-loss when averaged over a period of years. Pathogens that incite epidemics may cause a great dollar-loss during the period of the epidemic. However, epidemics of annual tissue generally occur sporadically, often separated by several years. When the average annual dollar-loss of such a disease is compared to that of an endemic disease, which occurs each year, the endemic disease may be found to cause the greatest average dollar-loss.

Rule -- Crop Devastation

Crop devastation is incidental to the pathogen, but paramount to man. Actually, it is often to the advantage of the pathogen not to kill its host, or not to kill it too quickly. The pathogen must have sufficient time for ample growth and reproduction if it is to persist.

Rule -- Multiplication vs. Survival

In order for a pathogen to increase substantially, it must have a survival time that is sufficiently long enough to allow a build-up to occur. The multiplication rate must exceed the death rate. Pathogens vary considerably in their survival time as well as in their rate of multiplication. Therefore, it is useful to note the following general principle: pathogens with a short survival time generally multiply more rapidly than those with a long survival time. Indeed, pathogens with a short survival time must multiply quickly if they are to survive, let alone build up to quantities sufficient to cause an epidemic; whereas those with a long survival time do not need to hurriedly initiate reproduction.

High Population Necessary

When there is a suitable rate of reproduction together with a sufficient survival time to allow an exceptionally high level of inoculum to build up during the hosts' susceptible phase of growth, an epidemic may occur. An epidemic can occur only when there is a high population of the pathogen. Therefore, the factors that promote this high population need to be studied. These are namely, the influence on the pathogen by the host and by the environment, and various properties of the pathogen itself.

Rule -- Epidemic Abatement

For microorganisms in general, there are numerous limiting factors of growth such as nutrition, space, competition from other organisms, environment, and metabolic by-products of the microorganisms themselves. Major factors that often limit the growth of plant pathogens after the onset of disease are temperature, lack of moisture, vectors, and susceptible hosts. This last factor is especially important in the present discussion. If all factors are favorable for the development, spread and continuation of an infectious disease, an epidemic may occur that abates only when susceptible host tissue becomes limiting.

17-2 INFLUENCE OF THE HOST

17-2.1 CROWDING OF PLANTS

Pre-historic man lived a nomadic life, roaming the countryside in the continual search for food. Permanent settlements were possible only when man turned to the cultivation of crops. Bountiful harvests in turn permitted population increases, and increased populations demanded intensified cultivation of crops. Thus today, crops are planted close together with their foliage overlapping so as to utilize the maximum energy of the sun and obtain the greatest possible yield per acre.

In nature, plants of the same species are usually separated from each other by many different plants. Furthermore, on a per acre basis, the population of plants of the same species in a natural setting is usually a fraction of the number of plants of the same species in a cultivated field. The elimination of other plant species and close spacing of desired plants are major requisites in crop production. As necessary as this "unnatural" crowding is, it is not accomplished without some undesireable side-effects, such as frequent damage caused by plant diseases. The crowding of plants in a field facilitates the movement of many pathogens from plant to plant. This is true for pathogens whose inoculum is locally disseminated by wind and splashing rain. The diseases they cause are usually foliar leafspots and blights. Massive amounts of inoculum may be quickly produced and lead to the devastation of whole fields and beyond.

Rule -- Crowding Of Plants

17-2.2 MONOCULTURES AND UNIFORMITY

Coincidental with other technological advances has been the rise of monocultures and increasing uniformity. A monoculture is defined as the growing of a single crop. Many of the farms of today are using the monoculture system, or nearly so. This is in sharp contrast to the farms of yesteryear, when it was common for even small farms to raise a number of different crops. Today, farms have little crop diversity even though the acreage per farm has sharply increased. For example, in the Midwest, many growers predominately raise corn, or perhaps corn and soybeans; for others the major crop is soybeans. In much of the wheat-belt in the United States and Canada, wheat is often grown in field after field as far as the eye can see.

Monoculture Defined

The economic production of crops has demanded ever larger fields and has resulted in the creation of monocultures. In respect to the prevention of plant diseases, large fields of single crops are both beneficial and detrimental. From a beneficial standpoint, disease may be lessened where the primary inoculum is produced on scattered hosts near the fields, such as in fence rows, roadways and adjoining fields. In such cases, disease percentage is reduced by combining smaller fields into larger fields and thereby reducing the overall perimeter. Disease reduction occurs because there is a reduction of inoculum bordering the fields. On the detrimental side, large fields of single crops are frequently destroyed by fast spreading infectious foliar diseases. Therefore, monocultures and crowding of plants are both factors relating to the fast spread of foliar disease pathogens.

Reduced Inoculum Reduces Disease

Another factor, perhaps the most important one, is that of crop uniformity. Crop uniformity is a result of increased technology, much of which has resulted in increased production. This uniformity has also resulted in keeping food prices down. Farmers, housewives, food processors and grocers have all demanded uniformity. The housewife wants uniform size peas, apples of a high fruit finish, and smooth skinned potatoes. Farmers want their wheat, corn and soybeans to be of uniform height and with uniform ripening. Processors seek high levels of protein in wheat corn and soybean, etc., while stringbeans must be stringless. High quality cotton lint is sought by the cotton industry. The grocer wants his fruits and vegetables to be unblemished and of particular sizes. The list goes on and on. However, to meet the demand for uniformity there is often a penalty to be paid. That penalty is often plant disease on a large scale. For example, if airborne inoculum of an infectious pathogen infects several plants within a field, and if all plants within that field are alike in factors governing susceptibility to that pathogen, the latter may multiply repeatedly and infect and destroy the entire field of plants.

Uniformity A Product Of Demand

The most striking illustration of plant disease affecting man-made uniformity in crops is that of the 1970 epidemic of southern corn leaf blight. Eighty-five percent of the hybrid corn in the United States possessed Texas male-sterile cytoplasm. This man-made uniformity set the stage for the most devastating loss ever suffered by the corn industry.

In order to understand why so much of the corn possessed the same cytoplasm, it is necessary to explain briefly how and why hybrid seed corn is produced. Hybrid corn is planted because of the increased yield obtained as a result of "hybrid" vigor. It is obtained by crossing inbred lines. Therefore, in order to produce hybrid seed,

Hybrid Vigor

it is necessary to detassel female parent plants before they shed their pollen. Two rows of male parent plants are alternated with six rows of female plants. The pollen from the male plants pollinate the female, or seed-parent plants, and insure the formation of the desired cross. The detasseling operation is done by hand and is both tedious and expensive.

In Texas, in 1931, a male-sterile plant was found in which the sterility factor was governed by factors within the cytoplasm rather than by genes on chromosomes of the muclei. This cytoplasm which conferred male-sterility was designated as Texas cytoplasm male sterility (*Tcms*). *Tcms* was transferred into numerous inbred corn lines, which of course made them all male-sterile. If these inbreds are used as female parents and pollinated by male inbred lines to form hybrid seed, such seed will produce male-sterile plants. Therefore, if a grower were to plant such hybrid seed, there would be no pollen for fertilization and no crop to harvest. However, subsequently a gene was found that would restore male-fertility. This gene was placed in the male parent inbred lines that were crossed with the *Tcms* female parent. The resultant hybrid seed corn was planted by growers who knew it was both fertile and high-yielding. This new method of producing hybrid corn was considerably less expensive since it eliminated the expense of hand detasseling; it subsequently became adopted throughout the United States and in many other parts of the world.

In 1962 and 1965, two small articles were published which indicated that southern corn leaf blight was prevalent on several corn lines and hybrids carrying *Tcms* in the Philippines. This disease is caused by the fungus *Helminthosporium maydis*. Of course these reports did not warn of an impending epidemic in the United States, nor could one be extrapolated from them. It was not until a new race of *H. maydis* was detected in the United States in the early months of 1970 that concern was raised. The new fungus strain was called race T, because it was capable of attacking all corn possessing *Tcms*. Furthermore, race T was more virulent than the normal race which is called race O. Race O only attacks corn leaves, but race T can attack the leaf, stalk, husk, shank, and ear of susceptible hybrids.

Southern Corn Leaf Blight Epidemic Of 1970

The epidemic of southern corn leaf blight in 1970 is a classic example for studying epidemiology. In the early months of 1970, the disease was present in severe form in the Belle Glade region of Florida. It was next found in southeast Georgia and along the Gulf Coast from Florida to Texas, then up the entire Mississippi valley and along the

Ohio River. By harvest time it was present from Florida to Ontario, from Maine to Minnesota. There was a 15 percent loss of the corn crop; the monetary loss was placed at 1 billion dollars.

In defense of the corn breeders, until the discovery of race T of *H. maydis*, cytoplasmic factors had not shown any influence over resistance or susceptibility to infectious plant pathogens. Breeders now know differently; cytoplasmic factors must be carefully controlled as to their widespread use, just as genetic factors carried on chromosomes. Uniformity of inheritable traits, especially in crops planted over vast acreages, endanger those crops to disease of epidemic proportions.

Rule -- Crop Uniformity

For excellent in-depth discussions of the dangers of genetic uniformity in major crops, the reader is referred to the review by Day, 1973, and the book by the N. A. S., 1972, entitled Genetic vulnerability of major crops.

17-2.3 DISTANCE BETWEEN INOCULUM SOURCE AND HOST

Another factor of considerable importance is the distance between the susceptible host plants and the inoculum source. The probability of infection is highest when the source of inoculum is close to the host, and becomes less likely as the distance increases. For some diseases the removal of the alternate hosts from the immediate vicinity of the primary hosts will provide complete control. Such is the case for both white pine blister rust and cedar-apple rust on apple. Few diseases are afforded complete control by such procedures.

Rule -- Infection Varies Inversely To Distance Between Host And Inoculum

17-3 INFLUENCE OF THE ENVIRONMENT

Many pathogens possess a narrow range of environmental conditions under which they infect and incite disease in their hosts. Other pathogens are able to infect and incite disease in their hosts under a wider range of environmental conditions. It naturally follows that the pathogen that is less restricted by environment will be consistently more prevalent within a given locality, than the restricted pathogen. This may be restated: the epidemic potential of a pathogen is increased if it can incite disease under a wide range of environmental conditions.

Rule -- Environmental Conditions

An unusual situation exists for *Erwinia stewartii*, and *E. tracheiphila*, incitants of bacterial wilt of corn and cucurbits, respectively. Both of these species of bacteria appear to overwinter in their insect vectors, and these insects serve as the major source of inoculum. Mild winters allow a high number of vectors to survive and inadvertantly permit the survival of a large amount of inoculum. Also, these pathogens are completely dependent upon insect vectors for their spread. Therefore, a large population of both bacteria and insects is necessary to bring about wide scale disease. Once infection has taken place, the bacteria multiply to fantastic numbers, but the limiting factor in the increase of disease is the number of insect vectors.

Perhaps one of the most important aspects of the study of epidemiology has been the development of forecasting the impending occurrences of a number of plant diseases. Such forecasts are based upon the relation between disease development and weather. This service is useful for certain important diseases that may be controlled by chemical means, but where application of chemicals is prohibitively expensive and wasteful unless disease of considerable proportions is imminent. Once it is ascertained that conditions favor the development of a given disease, the growers within the region are quickly notified.

Forecasts, Based On Relation Of Weather To Development Of Disease

The major plant diseases for which forecasting has become established as an acceptable procedure in reducing losses in various regions of the United States include: potato late blight, *Phytophthora infestans*; tobacco blue mold, *Peronospora tabacina*; downy mildew of lima bean, *Phytophthora phaseoli*; and downy mildew of cucurbits, *Pseudoperonospora cubensis*.

17-4 PROPERTIES OF THE PATHOGEN

17-4.1 MULTIPLE-CYCLE DISEASES

A multiple-cycle disease pathogen may increase according to sigmoid curve, where the early part of the curve is logarithmic in character (see Chapter 12). The lower portion of the curve is relatively flat and represents low amounts of inoculum. It has often been called the "lag" phase of growth. The middle portion of the curve is rather steep and straight, and represents exceptionally large increases of population in a short period of time. It is commonly referred to as the "log" phase of growth even though it is only at first logarithmic. As the curve continues upward, it becomes less and less logarithmic. The upper portion of the curve is again relatively flat, which often reflects a lack of susceptible healthy host tissue upon which the pathogen can continue to grow.

Lag Phase

Log Phase

Multiple-cycle disease pathogens that can repeatedly attack the crop all through the growing season possess great explosive potential. Some pathogens are able to attack the host only at specific growing periods and are therefore, somewhat restricted. The fungus that causes brown rot of stone fruits, *Monilinia fructicola*, is a good example of this situation. This pathogen can attack the host early in the growing season while it is in flower, then later in the growing season it is able to attack the fruit as it matures. Green fruit is usually attacked only through wounds caused by insects, etc. How fortunate this situation is, because even with such imposed restrictions of periodic host susceptibility, this pathogen can almost destroy a maturing peach crop in a matter of days!

The time between infection and subsequent production and release of inoculum is also important. The time required for the disease cycle directly controls the possible number of repeated infections that can occur during a growing season. From this discussion it follows that: multiple-cycle pathogens that can repeatedly attack the host throughout its growing season, and have short disease cycles, possess an explosive epidemic potential.

Rule -- Explosive Potential

Van der Plank, 1959, estimates that within a single growing season, local lesions (multiple-cycle diseases) may be multiplied as much as 1 billion times. This emphasizes the explosive capabilities of these diseases.

For multiple-cycle diseases, the relationship of percent infection of available susceptible plants to time may be plotted graphically; and it also forms a sigmoid curve (Fig. 17.1). The faster a disease spreads, the steeper its

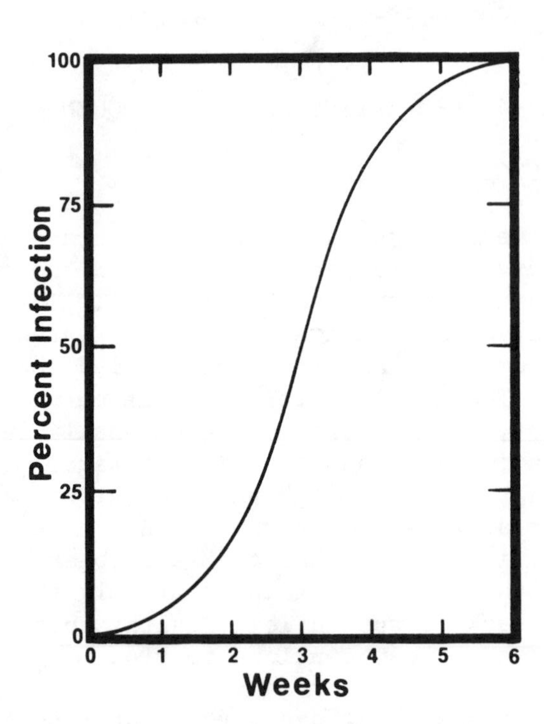

Fig. 17.1 Hypothetical sigmoid curve for multiple-cycle diseases, relating the percent of infection of available plants to time.

curve will be. However, since multiple reinfection of previously infected leaves regularly occurs in multiple-cycle diseases, an increase of inoculum seldom causes a proportional increase of newly infected plants. In multiple-cycle diseases, the proportion between inoculum-increase and increase of infected plants is most nearly alike at disease onset and becomes progressively dissimilar with the increase of infected plants. In a small field of 1 million wheat plants where 500 plants infected with stem rust serve as the source of inoculum that infects 100,000 neighboring healthy plants, there is a 200-fold increase of disease incidence. If these 100,000 infected plants produce inoculum that infects the remaining 900,000 healthy plants, there would be only a 9-fold increase of disease incidence.

Rule -- Inoculum vs. Disease Incidence

17-4.2 SINGLE-CYCLE DISEASE

As stated in Chapter 12, single-cycle diseases are incited by pathogens that lack a repeating stage and therefore increase intermittently during the growing season. This is in contrast to the logarithmic increase of multiple-cycle disease pathogens. Smut diseases and vascular wilt diseases of annual plants, and diseases such as cedar-apple rust are typical of single-cycle diseases. By the very nature of their inoculum liberation, systemic diseases possess single-cycles.

Systemic Diseases Are Also Single-cycle Diseases

Single-cycle diseases increase in a manner similar to money invested at a simple rate of interest. This means a rather uniform increase, but definitely lacking an explosive capacity. In contrast to multiple-cycle diseases that may increase up to 1 billion times in a single growing season, systemic diseases of small herbaceous plants increase only 10,000 times; and systemic diseases of trees increase 10 times, (Van der Plank, 1959). In this evaluation, each systemically infected plant is considered to be a single lesion. Of course, a great deal of variation exists between the pathogens of these three groups as to their ability to multiply. However, this leads us directly to a modification of a rule by Van der Plank (1959): the rate of increase and spread of single-cycle disease pathogens is relatively slow.

Rule -- Increase Of Single-cycle Diseases

Since these diseases spread slowly, why are they of such importance? The point to observe here is that many of them are systemic diseases caused by fungal and bacterial pathogens that are often lethal--totally destructive to their hosts. In the case of the cereal smuts where the diseased plant itself is not destroyed, the grain is. However, some systemic diseases, such as brown stem rot of soybean, may not always reduce the yield. Another important feature of systemic diseases is that in as much as their spread is often insidiously slow, many times they are overlooked in their early stages.

Some of the most prominent single-cycle diseases are the numerous fusarium and verticillium wilts that infect so many different host plants; the many smut diseases; ergot of cereals; and diplodia and gibberella ear and stalk rots of corn. Also included in the single-cycle disease category are many bacterial wilts, with notable exceptions of bacterial (Stewart's) wilt of corn and bacterial wilt of cucurbits. Both of the latter diseases are spread by insect vectors and therefore, are considered to be multiple-cycle diseases.

Except for some fungi, the inoculum of all pathogens is capable of reinfecting the same host plant. The notable exceptions are the heteroecious rust fungi, which have different spore forms that are not always capable of reinfecting the same host plant. For example, *Gymnosporangium juniperi-virginianae*, the causal agent of cedar-apple rust, produces basidiospores that infect only apple hosts, and aeciospores that infect only cedar hosts. The fungus must alternate its infection and growth between apple and juniper and back to apple again. An additional damper on the spread of this pathogen is that it takes two years to develop on juniper before it can sporulate and infect apple.

Poor Dispersal And Lack Of Host Tissue Restrict Nema Population

Although plant parasitic nematodes have the capacity to produce a number of generations in a single growing season, they are restricted in population growth because of limitations of dispersal and susceptible host tissue. Under these restrictions, the golden nematode in the Long Island area increases about tenfold each year in fields under continuous potato production (Fig. 12.2). This population increase of the golden nematode fits that of single-cycle diseases.

17-4.3 DISEASE AGENTS OF PERENNIAL TISSUE

Perennial Tissue Diseases Are Long-lived; Their Spread Is Slow And Steady

Epidemics of perennial tissue are often caused by introduced foreign pathogens of the systemic type. Most pathogens of perennial tissue are long-lived; therefore the inoculum is present annually and often in relatively stable amounts. Disease spread is generally slow but steady. The major epidemics of forest trees in the United States have not been influenced by narrow limits of environment.

White-pine blister rust and chestnut blight are two diseases that took a number of years to run their epidemic courses in the continental United States. At this writing, two epidemics, Dutch elm disease and oak wilt, are still developing. Both diseases started many years ago and have yet to subside.

17-4.4 INOCULUM POTENTIAL

The inoculum potential at the onset of disease is of greater importance for the slow spreading pathogen than for one that is quick to multiply and spread. For example, most plant parasitic nematodes are slow spreading and must be present in substantial numbers and be widespread throughout a field prior to the onset of the growing season, in order to devastate the crop.

High Inoculum Level At Disease Onset is Important For Slow Spreading Pathogens

The explosive multiple-cycle disease pathogens may often be little affected by the amount of available infective inoculum, because a low level of inoculum may be sufficient to enable them to increase to the point of causing epidemics. For example, the two fungi that incite potato late blight and apple scab, have ample inoculum to initiate epidemics each and every year.

Pathogens of annual tissue (this includes annuals, crops grown as annuals, and annual tissue of perennial plants), must be present in large populations during the susceptible phase of growth of the host in order to cause an epidemic. In contrast, pathogens of perennial tissue often have several to many years of longevity and, therefore, need only to multiply relatively slowly to build up to high populations capable of causing epidemics. From this discussion it is obvious that: <u>to incite epidemics of annual tissue, slow-spreading pathogens require a large amount of inoculum at the onset of disease, while fast-spreading pathogens do not</u>.

Rule -- Inoculum Requirement

A brief examination of the sigmoid curve of a disease shows <u>the earlier an epidemic of annual tissue starts, the more severe it is likely to be</u>. Reducing inoculum prior to the onset of disease will be especially effective in delaying severe damage by a slow-spreading pathogen. Such inoculum reduction may involve crop rotation, removal of weed hosts, or perhaps soil fumigation. The reduction of inoculum is best accomplished when it is at its lowest level, which is prior to planting time. Also, the planting of resistant crop varieties; the use of insecticides to reduce insect vectors; and the use of fungicides to reduce foliar infection will retard the build-up of inoculum. <u>Any procedure that delays the onset of disease or retards the build-up of inoculum increases the chances of escaping from damage by an epidemic</u>. From this discussion it also follows that the use of early maturing plant varieties may allow a crop to escape harm wrought by late season diseases.

Rule -- Early Start Of Epidemics

Rule -- Delay Of Epidemics

Selected References

Day, P. R. 1973. Genetic Variability of Crops. Annual Review of Phytopathology. 11:293-312.

Miller, P. R. 1959. Plant disease forecasting. pp. 557-565, In C. S. Holton, et al., (ed.), Plant pathology problems and progress 1908-1958. The University of Wisconsin Press, Madison.

____________, and M. J. O'Brien. 1957. Prediction of plant disease epidemics. Annual Review of Microbiology. 2:77-110.

National Academy of Sciences, 1972. Genetic Vulnerability of Major Crops. National Academy of Sciences, Washington, D. C.

Stackman, E. C., and J. G. Harrar. 1957. Principles of plant pathology. pp. 217-236;332-340. The Ronald Press Co., New York.

Van der Plank, J. E. 1959. Some epidemiological consequences of systemic infection. pp. 566-573, In C. S. Holton, et al., (ed.), Plant pathology problems and progress 1908-1958. The University of Wisconsin Press, Madison.

____________. 1960. Analysis of epidemics. pp. 229-289, In J. G. Horsfall and A. E. Dimond, (ed.), Plant pathology. Vol. III. Academic Press, New York.

____________. 1963. Plant diseases: epidemics and control. Academic Press, New York.

____________. 1968. Disease resistance in plants. Academic Press, New York.

CHAPTER 18

DISEASE CONTROL--BREEDING FOR RESISTANCE

18-1 INTRODUCTION

One of the most important methods of plant disease control is the development and utilization of disease resistant crop varieties. For the "low cash value" crops, such as most field crops, the planting of resistant varieties is the major means of disease control. However, resistant varieties are highly important in the growing of most other crops as well. Of course, growers would like a high level of resistance for every disease of a crop; unfortunately, this is not possible and never will be. Nevertheless, in the future, more and more emphasis will have to be placed upon breeding for resistance.

Resistant Crop Varieties, A Major Means Of Disease Control

Most foliar diseases of fruit and vegetable crops are controlled through the use of chemical pesticides. However, even for these diseases, growing the most resistant varieties together with a chemical control spray program results in a better disease control. There are some other crop diseases for which there is little or no available pesticide control, and resistant varieties are demanded in order to profitably raise these crops.

The frequent and often disastrous outbreaks of diseases in previously resistant plants show all too well how a resistant plant variety is often only temporarily victorious. The battle between host and pathogen is a constant one. The past failure of crop varieties to remain disease free even when bred for resistance, has dramatically demonstrated the importance of variation of pathogens (Chapter 13). Our breeding programs must continuously produce new resistant varieties if we are to tip the scales of disease control to our advantage.

In breeding for disease resistance, one must be cognizant of two series of heritable factors; (1) those of the host plant, and (2) those of the pathogen. Past experience has told us that the more we know about the heritable characteristics of the pathogen, the easier it is to breed better lines of resistance into the host plant.

Heritable Factors In Both Host And Pathogen

18-2 BASIC CONCEPTS AND TERMINOLOGY

Higher green plants possess at least two sets of chromosomes resulting from the union of male and female reproductive cells. Genes are the units of inheritance and are located linearly in the chromosomes. A gene interacts with other genes and the environment to control the development of a character. The gene exists in at least two separate functional forms that are known as alleles or allelomorphs, which segregate at meiosis and recombine during fertilization.

Genes Control Development Of A Character

Genes may be dominant or recessive. The term dominant is applied to one member of an allelic pair of genes that has the quality of manifesting itself wholly or largely to the exclusion of the other member. Recessive is the term applied to one member of an allelic pair that lacks the ability to manifest itself wholly or in part when the other or dominant member is present.

The exchange of corresponding segments between each member of paired (homologous) chromosomes is called crossing-over. When genes that determine characters are located in the same chromosomes, the genes are said to be linked, and the group of genes is called a linkage group. Sometimes it is highly desirable to separate linked genes in order to incorporate one of the linked genes in a new crop variety. This is usually accomplished by crossing-over or by translocation. Translocation refers to a segmental exchange between non-paired (nonhomologous) chromosomes to form two new chromosomes. Genes that are not linked, segregate independently and are inherited in a random nature. However, linked genes are not inherited in a random nature, but occur together much more often than they do apart. The frequency of association of inherited characters indicates whether or not the characters are linked. An individual possessing a desired gene from a broken linkage group will only be obtained after a great number of crosses are made and the progeny examined. The closer together the genes are in a linkage group, the less frequent they are separated by translocation.

Translocation Separates Linked Genes

Single gene inheritance for resistance may be due to; (1) a single completely dominant gene, (2) a single incompletely dominant gene, or (3) a single recessive gene. The single dominant gene is the simplest form of inheritance, and is the easiest form of plant resistance to incorporate into a new variety.

Sometimes two or more genes may interact to provide resistance, and often several genes act independently to provide the desired resistance. These genes may be dominant,

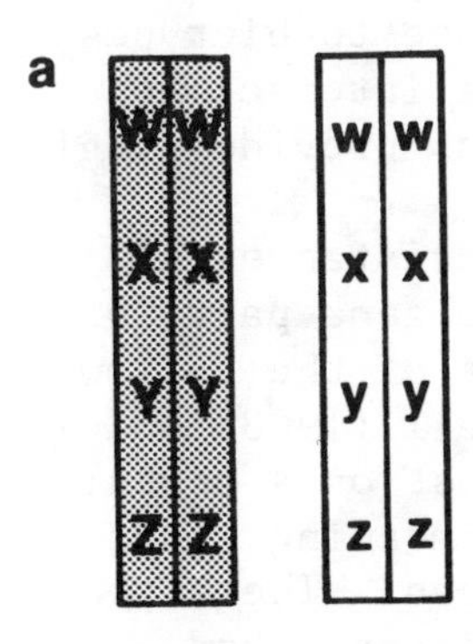

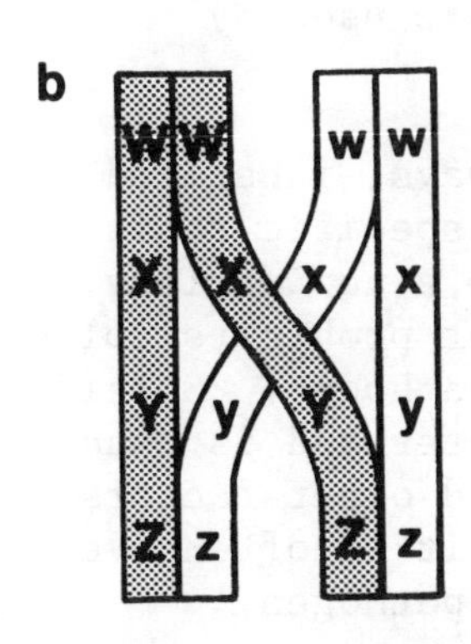

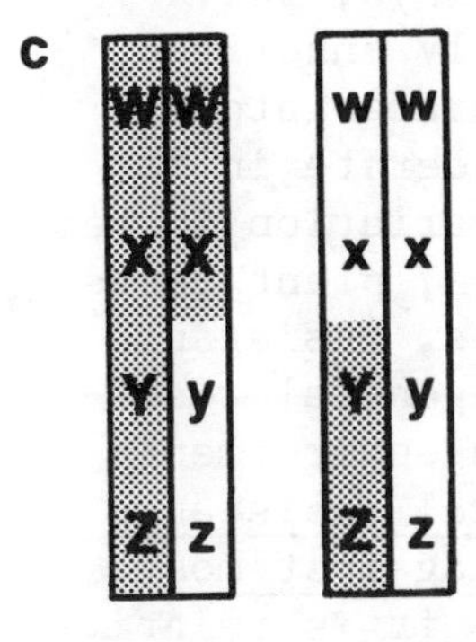

Fig. 18.1 Diagram of crossing-over.
a. Pairing of paternal and maternal chromosomes to form a tetrad.
b. A crossing of paternal and maternal chromatids.
c. The results of crossing-over.

recessive, or in any combination. Many genes are modified to a varied extent by the action of other genes. A gene that affects the expression of another nonallelic gene is usually called a modifying gene. Such genes usually bring about a relatively small change in the effect of another gene. Modifier genes may enhance resistance imparted by another gene.

Another type of gene interaction necessary for the ultimate expression of a character is complementary gene action, which means two or more genes acting together. Again, these genes may be dominant, recessive or in various combinations. Sometimes, two or more genes are incorporated into a plant for resistance where each gene provides a range of resistance to different biotypes of a pathogen. In such situations, the sum of the ranges of resistance to biotypes is usually obtained. Occasionally, the genes interact to provide a higher level of resistance than each provide singly.

Like many factors of inheritance, plant resistance is usually controlled monogenically, by a single gene pair, or polygenically by several gene pairs. The use of the following descriptive terms is strongly urged by Caldwell (1968): specific resistance, meaning resistance against only certain races of a pathogen; and general resistance, meaning a uniform resistance against all races of a pathogen. These two terms are similar in meaning to vertical resistance and horizontal resistance, respectively, which are used by Van der Plank (1963).

Specific Resistance Is Often Unstable

Specific resistance is often, but not always, inherited monogenically. When this is the situation, specific resistance is relatively easy to incorporate into a new variety. It is usually a high level of resistance that prohibits colonization and reproduction of the particular pathogen. Specific resistance is often hypersensitive in character and is usually unaffected by variation in temperature and other factors of environment. Its major drawback is that it is often overcome by the appearance of a new race of the pathogen.

General Resistance Is Relatively Stable

General resistance is usually, but not always, polygenically inherited. When it is polygenically inherited, it is usually extremely difficult to incorporate into a new variety. General resistance is often moderate in its level of resistance, and may be subject to variation in temperatures. For instance, at low temperatures, plant resistance may be good; but at higher temperatures, resistance may be lowered or lacking. At other times, general resistance may be of high level and unaffected by environment. Regardless of its mechanism of action, general resistance is highly desirable because it is effective against most, if not all, races of a pathogen. Therefore, it is relatively long-lasting, or stable.

Rule -- Gene For Gene

In order to better understand the possible effective life-span of a resistant plant variety, the rule of gene for gene as stated in Chapter 13 should be re-examined. This is: for each gene for resistance in the host there is a corresponding gene in the pathogen capable of overcoming that resistance. On the basis of this, a new physiologic race of the flax rust fungus may appear and may differ in only a single gene. However, it may be able to attack a host that previously possessed resistance. This is especially likely to occur when the resistance is specific. A host possessing general resistance may have considerable permanence, because the rust pathogen may have to produce a polygenic change in virulence in order to overcome the host resistance.

Fig. 18.2 Loss of yield and quality of wheat seed sults from infection by stem rust.
Left: 100 shriveled kernels from plants heavily infected with stem rust.
Right: 100 plump kernels of wheat from healthy plants. Control of stem rust is achieved through development of resistant varieties of wheat. *(Courtesy of R. W. Samson)*

18-3 SUCCESSFUL BREEDING VENTURES

Nature Developed Corn That Is Resistant To Leaf Rust

One of nature's foremost breeding accomplishments for disease control has been the development of corn that is resistant to leaf rust, caused by *Puccinia sorghi*. Caldwell (1966) describes this as being: "one really smashing success--". There have never been any accounts of epidemics by this rust, even though open-pollinated and hybrid corn are known to be at least partially susceptible as proven by inoculation. This resistance has proven to be exceptionally stable.

Resistance To Fusarium Wilt

Some remarkable successes brought about by man's breeding programs have been the development of peas, cabbages, and melons that are resistant to fusarium wilt. The development of tomato varieties resistant to race 1 of *Fusarium oxysporum* var. *lycopersici*, has also been a most remarkable accomplishment. The production of these four crops would be commercially unfeasible in most localities of the United States without the use of resistant varieties.

Resistance To Race 15B Of Wheat Stem Rust

Perhaps the most outstanding example of successful breeding in the cereal crops was the development and release of Selkirk wheat. This spring wheat variety was developed by the Canadian Department of Agriculture for resistance against race 15B of stem rust. Selkirk wheat was widely planted in the United States and Canada until the mid-1960's when it was subject to attack by a race of leaf rust. However, the resistance to stem rust was still intact. The financial success of this variety is valued at several hundred million dollars. A success undreamed of at the time of its release. A probable reason for the long tenure of Selkirk is that it possessed general resistance to stem rust when in the mature plant stage.

Resistance To Sugarcane Mosaic Virus

In 1926, varieties of sugar cane that were resistant to sugar cane mosaic virus were released to growers in Louisiana. These and subsequent resistant varieties were responsible for saving the sugarcane industry in the southern United States. The industry had been near bankruptcy since the introduction of sugarcane mosaic virus some 10 to 12 years earlier.

Resistance To Curly Top Of Sugarbeet

Another virus disease, curly top, threatened to ruin the sugar beet crop throughout the irrigated areas of the western United States. In 1929, the first resistant variety of sugar beet was produced. This variety was only moderately resistant, but nonetheless, it was the first hope that the crop could be saved. Subsequent varieties possess much higher levels of resistance.

18-4 BREEDING FOR CONTROL OF NEMATODES

Kehr (1966), lists 22 crops for which resistance to one or more parasitic nematodes is known. This is only meant to be a partial list; however, it serves to establish the fact that host resistance is a proven method of nematode disease control.

Rule -- Resistance To Nematodes

In respect to plant resistance to nematode pathogens, Rohde (1965), states: "Resistance becomes more common as complexity of host-parasite interaction increases." A slight modification of this is: plant resistance to parasitic nematodes becomes more common as the complexity of the host-parasite interaction increases. For example, many plant varieties possess resistance to nematode species in the genera *Heterodera* or *Meloidogyne*, the cyst and root knot nematodes, respectively. These nematodes have an intimate host-parasite relationship that requires host cell modifications in the form of nurse cells. On the other hand, plant resistance is less commonly found against nematodes such as the lesion nematodes, *Pratylenchus* spp., which have a much less intimate relationship with their hosts.

An interesting aspect of nematode pathogens, is that by their very nature they spread very slowly through the soil and from field to field. This automatically prohibits a fast build-up of physiologic races that break resistance. This is in great contrast to the bacteria, fungi, and viruses that spread so quickly.

Slow Spread Of Nemas Restricts Build-up Of Races

18-5 LEARNING FROM PAST EXPERIENCES

Ever since man has cultivated crops for his own benefit, he has selected the seed from the best plants to sow the next season. Without realizing it, he chose seed from plants that had undergone natural, or field selection.

The Irish famine in the middle 1840's, which was the aftermath of the potato late blight pandemic, stimulated a great deal of interest in hybridizing to improve the potato crop. This program was the first concerted effort by man to breed for disease control. A potato variety Magnum Bonum that had a considerable tolerance to late blight was released in England in 1876. It quickly found wide acceptance, but was discarded by the turn of the century for lack of adequate control.

In 1900, the rediscovery of Mendel's laws of inheritance spurred on plant breeding, and programs based on scientific principles were soon initiated.

Mendel's Laws

In the United States in 1904, there was a widespread and destructive epidemic of stem rust of wheat, caused by *Puccinia graminis* var. *tritici*. This led to the initiation of wheat breeding programs to obtain resistance to stem rust. It was soon found that factors for resistance were lacking in the bread wheats, and attempts were made to cross the bread wheats with durum and other wheats. However, it then became apparent that genes governing resistance in durum and other wheats were linked to several undesirable genes. After several thousand crosses, three wheat varieties resulted that appeared to have the desired resistance. This was the first indication that linkage could be broken. We now know that linkage can be broken, but is unlikely unless a large number of crosses are made.

First Time Linkage Is Broken

About 1916, physiological races of stem rust were discovered, but the importance of the races was not fully realized for some time. Then, when the complexity of breeding resistant wheat against the continuously hybridizing pathogen became understood, the "hue and cry" was raised for the eradication of the alternate host, the barberry bush. For it is only on barberry bushes that the sexual stage of the rust takes place, and therefore, the only place for hybridization to occur. The elimination of barberry bushes removed the largest source of the production of new physiologic races.

Discovery Of Physiological Races

Minor Disease Became Major

A lesson was learned from the wheat variety, Marquis. This was an early ripening variety that enabled it to escape severe stem rust damage. Also, it was highly resistant to a major disease, stinking smut, caused by *Tilletia* spp. Shortly after Marquis was released, a new disease appeared. The new disease was called head blight, which is caused by *Gibberella zeae*. Until this time, head blight of wheat was a relatively minor disease, and was scarcely noticed. However, with the widespread planting of Marquis, head blight became a disease of major importance. The genes for early ripening and resistance to stinking smut also brought along genes for susceptibility to head blight! A disease of minor importance quickly became a disease of major importance!

Appearance Of New Races

The sudden rise of new physiologic races is amply and vividly seen in *P. graminis* var. *tritici*, which causes stem rust of wheat. The bread wheat variety Ceres possessed stem rust resistance when it was widely distributed in 1926. Two years later the race 56 was identified. This race was capable of attacking Ceres. By 1934, race 56 was abundant, and in 1935, it became severely destructive over much of the range where Ceres was grown. Again, in 1937, race 56 reached epidemic proportions and mainly on Ceres.

Rise Of Race 15B

Race 15 of wheat stem rust was found to be a composite mixture containing several sub-races, each of which would attack specific wheat varieties. The race was subdivided into three groups on this basis, and these were called race 15, 15A, and 15B. Although the differences between the sub-races were consistant, they were not sufficient to justify the consideration of each as a separate race. Race 15B became the most notorious. Even though it was found only occasionally in the United States, breeding against this race was begun about 1939. This was shortly after its discovery and the realization of its potential. In 1950, this race suddenly appeared in Minnesota and the Dakotas. In 1953 and 1954, race 15B appeared in epidemic form and destroyed most of the durum crop and a quarter of the bread wheat crop! It then became obvious that 15B was composed of biotypes that displayed variation in host varieties they could infect. Wheat varieties that were resistant to some biotypes were susceptible to other biotypes! Our criterion for establishing races is dependent upon host varieties that exhibit resistance or susceptibility. It is not until new host varieties are bred that physiologic races of pathogens may become apparent.

Rule -- Race Detection

The oat variety Victoria and a number of varieties derived from Victoria, were highly satisfactory in resistance to *P. graminis* var. *avenae*, the oat stem rust and *P. coronata*, the crown rust pathogen. However, *Helminthosporium victoriae*, the causal agent of a minor disease called blight, suddenly blossomed into a full-fledged disease of major importance as it severely attacked Victoria oats and all varieties in which Victoria was one of the parental lines. Today, many newer oat varieties contain a single dominant gene that imparts a very high level of general resistance against *H. victoriae*. This disease is once again a pathogen of little economic significance. So it is obvious that minor diseases may become major diseases if susceptible hosts are grown on large acreages.

Rule -- Minor Disease

The epidemic of southern corn leaf blight in 1970 (see Chapter 17) was a bitter lesson of crop uniformity. A highly virulent race of *Helminthosporium maydis*, race T, appeared and destroyed 15 percent of the corn crop. This race was capable of attacking corn that contained Texas cytoplasm which conferred male sterility. Since about 85 percent of the corn contained *Tcms* in 1970, we were fortunate that more corn was not destroyed. This epidemic highlighted the dangers inherent in crop uniformity as no other incident has before it.

18-6 STRATEGIES OF BREEDING

18-6.1 USE OF SPECIFIC RESISTANCE

To date most breeding programs have incorporated specific resistance in their varieties. However, from the lessons of the past, it is at once clear that specific resistance has often failed because of the ability of pathogens, especially fungal pathogens, to overcome such resistance by the shift of a single gene. Occasionally, however, specific resistance has imparted a high level of resistance which the pathogen has been unable to overcome.

Robinson, 1971, has analyzed the use of specific resistance and stated 14 rules concerning its use. The following 3 principles are modifications of his rules. Specific resistance is likely to be of greater value against a single-cycle disease than against a multiple-cycle disease. This principle is based on the fact that a single-cycle disease pathogen undergoes fewer reproductive cycles per growing season than a multiple-cycle disease pathogen. With fewer reproductive cycles, the single-cycle disease pathogen has less opportunity to produce a new virulent race than the multiple-cycle disease pathogen.

Rule -- S.R. And Single-cycle Disease

Rule -- S.R. Most Valuable In Easy To Breed Crops

Specific resistance is likely to be of greater value in crops that are easy to breed than in perennial crops or annual crops that are hard to breed. When specific resistance within a host variety is overcome by a pathogen, another gene for specific resistance must be incorporated in the host. The ease and speed with which this is accomplished is vital. In certain annual crops this may be a relatively easy task, which makes the use of specific resistance of low risk. This is in sharp contrast to crops that are difficult to breed or that are perennials, because it takes a longer time to breed new varieties. Also, in many perennials, such as fruit trees, it takes several years of growth to reach bearing age. The use of specific resistance in the latter crops is of greater risk than in the former.

Rule -- S.R. Plus G.R.

Specific resistance is likely to be of greater value when combined with useful levels of general resistance than when used alone. As is pointed out by Robinson, 1971,: "There is good reason to believe that every host plant possesses some horizontal resistance to every disease." The level of this resistance is highly variable, of course. Often it is so low it is of little benefit. However, if the level of horizontal resistance is high enough to be useful, it should be incorporated into varieties along with specific resistance. The reason for this is that if the specific resistance is overcome by the pathogen, the crop loss would be minimized by the presence of the lower but useful level of general resistance.

18-6.2 USE OF GENERAL RESISTANCE

High Levels Of G.R. Not Yet Available

As stated before, general resistance is usually conferred by multiple genes. It is often inherited quantitatively; the more genes for general resistance in a given variety, the higher the level of resistance. The great difficulty in breeding for general resistance is that the individual genes which govern it are difficult to detect. However, breeders are searching for general resistance in every crop breeding program. When high levels of general resistance can be found it can be augmented by combining it with specific resistance as stated in the rule in the previous paragraph. Of course, if the level of general resistance is sufficient, it may be used alone; in fact, its use is preferred above other forms of resistance. However, for most crops, the use of general resistance is in the future, because it is yet to be identified in sufficiently high levels to provide the needed protection.

G.R. - The Preferred Resistance

18-6.3 USE OF TOLERANCE

There are probably few crop breeding programs seeking tolerance as an objective. However, a strong case for the use of tolerance is presented in the review by Schafer, 1971.

A tolerant variety allows the full growth and reproduction of the pathogen and therefore does not sort out for particular races of the pathogen. The widespread use of tolerant varieties would thus tend to stabilize pathogen populations. This falls in line with the concept of "managing a pest" or "living with a pest", instead of attempting to eliminate it.

Tolerance Stabilizes Pathogen Populations

The major drawback of growing tolerant varieties is that when they become diseased they still produce less than a highly resistant variety. Another factor is that they allow an abundance of inoculum to be produced that may attack other host varieties.

Perhaps in the future tolerance can be combined with specific resistance. The tolerance can provide a back-up system if the specific resistance is overcome.

18-6.4 USE OF VARIETAL BLENDS

A varietal blend is composed of a mixture of a number of genetically different varieties that resemble each other in agronomic traits. A varietal blend is therefore, a deliberate diversification in respect to resistance. It would probably be easier for these blends to be accepted by underdeveloped countries whose markets are not as conscious of product uniformity as is the case in most of the developed countries.

In the future, our standards for crop uniformity may have to be re-examined and allowances made which will trade crop uniformity (with its subsequent high disease-risk and hence high production risk) for crop "blends" (which carry the penalty of non-uniform crop products, but have much less disease risk). Such a trade-off may have to be made under the pressure of a higher human population.

18-6.5 USE OF MULTILINE VARIETIES

A multiline variety is diverse in factors governing disease resistance, but is uniform (or nearly so) in all other inheritable characteristics. Essentially, this is a refined blend of lines that differ only in their resistance to various races of a pathogen.

To obtain a multiline variety, breeders start with a single cultivar (a variety, strain or breeding line) and add a single gene for specific resistance. This is increased as an isoline (all members with identical factors of inheritance). Another, but different, gene for specific resistance is added to the cultivar and it is increased as a second isoline. In this manner, a number of isolines are prepared in which each contains a different specific resistance. The seed of each isoline is increased separately. The seeds of all isolines are then blended together and the mixture is sold as a multiline variety.

A major feature of the use of multiline varieties is that their composition can be varied from year to year in response to race changes in the pathogen. The proportion of each isoline in the mixture is determined by the relative resistance of each isoline to each race in the inoculum population and the prevalence of the different races.

The concept of multiline varieties has been developed during this century amid some controversy. However, such varieties are useful especially in the self-pollinated grain crops, against various rust pathogens, because these pathogens have the capacity to form new races quickly and hence threaten small grain production. Several multiline oat varieties composed of 7 to 11 isolines have been released in Iowa. These varieties have composite resistance to crown rust, caused by *Puccinia coronata*.

An examination of how a multiline variety may buffer against disease loss is in order. The pathogen increases from the initial inoculum (x_o) at a certain rate (r) over a length of time (t), and results in a proportion of susceptible tissue becoming infected (x), (Van der Plank, 1963). The initial inoculum prevalent in a region is often composed of mixed races. This is especially true with cereal rusts. A host variety with specific resistance is selectively resistant to various races, and therefore allows only certain other races of the initial inoculum to grow, which reduces the quantity of the initial inoculum (x_o). However, the rate of reproduction (r) of a pathogen growing on a host with specific resistance is usually so high that much of the susceptible tissue is infected by the end of the disease season. A host with specific resistance is, therefore, only valuable when it confers resistance to all prevalent races--essentially reducing the initial inoculum (x_o) to a very low level.

A host variety with general resistance that confers a moderate level of resistance to all races of a pathogen does not reduce the quantity of initial inoculum (x_o). Instead, it reduces the rate of pathogen increase (r) allowing the host to mature without much loss. Stated another way, by the end of the disease season, the proportion of diseased susceptible tissue (x) is held to a minimum, whereby little crop loss occurs.

A multiline variety possesses a number of genes for specific resistance--supposedly effective against all prevalent races of the fungus. Therefore, because a multiline variety contains specific resistance, the initial inoculum (x_o) is reduced--each isoline usually allowing only one or two of the prevalent races to grow on it. For example, assume a multiline variety is composed of 10 isolines and that one of the prevalent races is capable of infecting 2 of the isolines. There is a maximum chance of 1 out of 5 that a viable spore of that race which lands on a plant of the multiline variety can cause infection. If the spore lands on a susceptible host, causes infection, colonization and then sporulates, the subsequent spores have a maximum chance of 1 out of 5 that neighboring plants will be susceptible. In other words, the rate of pathogen increase (r) is reduced by at least 80%. Therefore, a multiline variety reduces initial inoculum (x_o), a characteristic of specific resistance. It also reduces the rate of pathogen increase (r), a characteristic of general resistance, and it therefore follows that the proportion of diseased susceptible tissue (x) is held to a low level. The example above is adopted from one given by Browning and Frey, 1969.

Multiline varieties are like tolerant varieties in that they permit all races of the fungus population to grow, thus stabilizing the pathogen population. The multiline variety does not "sort out" for new races, which allows breeders to concentrate on agronomic improvements in the original cultivar. To summarize, multiline varieties have characteristics in common with tolerant varieties and those possessing specific resistance and those with general resistance.

Dynamic, Natural, Biological System

Browning and Frey, 1969, have this to say about multiline cultivars: "Being easy to develop, they hold great promise as a dynamic, natural, biological system of effectively buffering the host population against the rust population."

Two major arguments against the development and use of multiline varieties are: (1) the time and effort required in producing and testing the component isolines are so great, that improvement of the basic cultivar is considerably retarded. (2) A multiline variety has resistance that is effective against many races of one pathogen, but each isoline is uniform in possessing the same specific resistance against other pathogens. For example, a multiline variety of oats may have resistance against many races of crown rust, *Puccinia coronata*, but possesses only specific resistance to stem rust, *P. graminis* var. *avenae*. If the latter fungus forms a new physiologic race, the multiline variety may be jeopardized.

18-7 ESSENTIALS OF DEVELOPING RESISTANT VARIETIES

Mode Of Reproduction Affects Type Of Breeding Program

Each crop species must be studied with respect to the mode of reproduction to determine the type of breeding program to be employed. For instance, some crops are reproduced by clonal means and may need special procedures to be induced to produce seeds. Some crops must be reproduced clonally to obtain the uniformity of the variety. Also, there is considerable variation in the crop plants that normally reproduce by seeds. Some plants may be normally self-fertile, which leads to in-breeding and a considerable amount of homozygosity. Other plants normally are crossed, and some are dioecious, both of which lead to a high level of heterozygosity. Still others may be both self-fertile or crossed. A few plants may lend themselves to grafting procedures and the desired resistance may be achieved by grafting them to resistant rootstocks.

Obtain Resistant Plants

Grow In Disease Nursery

Select Plants

Cross

Grow In Disease Nursery

Select

Repeat Until Resistant And Desireable Plants Are Obtained

When setting up a breeding program for disease resistance, the first task is to search for individual plants that show evidence of resistance to the specific pathogen. This may be a long drawn-out procedure involving collecting and selecting plants from all over the world, or it may be a process of selecting local plants. In some cases, satisfactory resistance has not been found even after years of searching and screening. After plant collections are procured, the plants are grown in a "disease nursery", where the plants are subjected to severe attacks by the pathogen or pathogens involved. Resistant plants are selected and the next step is to cross these plants with the desirable agronomic plants. The offspring of these crosses are also grown in the disease nursery, and again the resistant plants are selected. Subsequent crosses must be manipulated until a resistant variety of good agronomic quality is obtained.

It is essential that a breeding program for disease control be a continuous process that is coordinated with the general improvement of the crop. To be effective, resistance must be combined with the best agronomic or horticultural characteristics.

Selected References

Browning, J. A., and K. J. Frey. 1969. Multiline cultivars as a means of disease control. Annual Review of Phytopathology. 7:355-382.

Caldwell, R. M. 1966. Advances and challenges in the control of plant diseases through breeding. pp. 117-126, In Pest control by chemical, biological, genetic, and physical means. A Symposium. ARS 33-110, U. S. Dept. Agr.

Coons, G. H. 1953. Breeding for resistance to disease. pp. 174-191, In Plant diseases. U. S. Dept. Agr. Yearbook.

Hare, W. W. 1965. The inheritance of resistance of plants to nematodes. Phytopathology. 55:1162-1167.

Hooker, A. L. 1967. The genetics and expression of resistance in plants to rusts of the genus Puccinia. Annual Review of Phytopathology. 5:163-182.

Kehr, A. E. 1966. Current status and opportunities for the control of nematodes by plant breeding. pp. 126-138, In Pest control by chemical, biological, genetic, and physical means. A Symposium. ARS 33-110 U. S. Dept. Agr.

Moore, E. L. 1960. Some problems and progress in the breeding and selection of plants for nematode resistance. pp. 454-460, In Sasser, J. N., and W. R. Jenkins, (ed.), Nematology fundamentals and recent advances with emphasis on plant parasitic and soil forms. The University of North Carolina Press, Chapel Hill.

National Academy of Sciences, 1972. Genetic vulnerability of major crops. National Academy of Sciences, Washington, D. C.

Roane, C. W. 1973. Trends in breeding for disease resistance in crops. Annual Review of Phytopathology. 11:463-486.

Rohde, R. A. 1965. The nature of resistance in plants to nematodes. Phytopathology. 55:1159-1162.

Schafer, J. F. 1971. Tolerance to plant disease. Annual Review of Phytopathology. 9:235-252.

Stakman, E. C., and J. J. Christensen. 1960. The problem of breeding resistant varieties. pp. 567-624, In J. G. Horsfall and A. E. Dimond, (ed.), Plant pathology. Vol. III. Academic Press, New York.

____________, and J. G. Harrar. 1957. Principles of plant pathology. The Ronald Press Co., New York pp. 489-540.

Walker, J. C. 1959. Progress and problems in controlling plant diseases by host resistance. pp. 32-41, In C. S. Holton, et al., (ed.), _Plant pathology problems and progress 1908-1958_. The University of Wisconsin Press, Madison.

___________. 1969. _Plant pathology_. Ed. 3. McGraw-Hill Book Co., New York. pp. 786-802.

CHAPTER 19

DISEASE CONTROL--INOCULUM REDUCTION

19-1 INTRODUCTION

This chapter contains a heterogenous grouping of various practices and procedures specifically designed to reduce the amount of inoculum. Plant debris, soil, and various living plants harbor great quantities of inoculum; and it is against these sources of inoculum that many practices and procedures are directed. For the most part, inoculum is not completely eliminated. Generally, the goal is to reduce the inoculum to the point where a crop can be economically grown and harvested. Often, it is possible to utilize two or more procedures to hold the level of the inoculum down and/or reduce it.

19-2 SANITATION AND CULTURAL PROCEDURES

19-2.1 REMOVAL OF INFECTED HOSTS

A number of diseases caused by various pathogens may be partially controlled, or less frequently, completely controlled by the removal and burning of infected host plants. The exact control procedure must be based upon scientific studies, because eradication of diseased plants is usually extremely expensive. There must be substantial evidence that the amount of control achieved will offset the initial costs and those of the subsequent surveillance.

Citrus Canker

One of the very few diseases that lend themselves to complete control by eradication of infected hosts is citrus canker, caused by the bacterium *Xanthomonas citri*. In 1915, both Federal and Florida State officials began a program of inspecting all citrus groves and condemning each grove found to contain any citrus canker. Each condemned grove was burned to the ground with kerosene torches. One-quarter of a million grove trees and over three million nursery trees were destroyed. No infected trees have been reported since 1926. The disease was conquered eleven years after the initiation of the program. Citrus canker was characterized as one disease that "threatened to be the worst catastrophe in the history of the state's agriculture". Control measures now rest with the officials in charge of the quarantine program, for this disease is intercepted many times each year at ports of entry. If citrus canker is ever again introduced, prompt eradication procedures will be certain to follow.

Another major citrus disease in which the removal of infected hosts is a prime factor in control is spreading decline. This disease, first observed in the mid-1920's, spread slowly into many citrus groves, and the threat to the entire industry was clear. In 1953, the nematode *Radopholus similis* was implicated as the major causal agent. The disease first appears as groups of stunted, unthrifty trees bearing sparse leaves and small fruit. Such infected trees may live in this condition for a number of years. According to Suit and Ford (1950), the disease spreads slowly at the average rate of 1.6 trees per year. The disease spread is by movement of the nematodes from the roots of an infected tree to those of an adjoining healthy tree. The foliage symptoms reflect the root deterioration caused by the nematode and perhaps by fungi as well.

Spreading Decline Of Citrus

Eradication of *R. similis* was subsequently attempted by a state enforced "pull-and-treat" method. This program involved locating diseased groves, removing infected trees and their roots, and burning them. A buffer zone of four rows of healthy trees beyond the diseased trees were also removed and burned. The land was then prepared for soil fumigation and the nematicidal soil fumigant D-D was applied at the high rate of 60 gallons per acre. The treated areas were allowed to be planted with nonsusceptible crops for a two-year period; then citrus could be replanted if the soil was found to be free of the nematode.

Host Removal Plus Soil Treatment

It is not reasonable to believe that *R. similis* can be eradicated by this procedure, especially when the nematode can live on over 237 different host plants. However, <u>the pull-and-treat method is the only known effective control procedure</u>. Spreading decline and its control has cost the Florida citrus growers tens of millions of dollars; however, its containment has justified the program.

A number of other diseases could be cited here for which the removal of infected hosts provides at least partial disease control. In some cases this is highly effective control, while in others it is much less effective.

19-2.2 REMOVAL OF ALTERNATE HOSTS

A few of the destructive diseases incited by heteroecious rust fungi are partially controlled by the destruction of the alternate hosts. There are two such "classic" eradication programs in the United States centering around the barberry and *Ribes*. The barberry is the alternate host for

Puccinia graminis, the incitant of stem rust of cereals. *Ribes* spp. are the alternate hosts for white pine blister rust caused by *Cronartium ribicola*.

Shortly after World War I, an eradication program was set up to destroy the barberry plants in the northern half of the United States. This was established at the Federal level. Most of the states involved also passed barberry-eradication laws that are still in effect. Consequently, a rigorous program was carried out in Montana, Wyoming, Colorado, North Dakota, South Dakota, Nebraska, Minnesota, Iowa, Wisconsin, Michigan, Illinois, Indiana, and Ohio. Later the program was carried over to include Missouri, Pennsylvania, Virginia, and West Virginia. Canada carried out a similar program in Manitoba, Saskatchewan, and Ontario.

Stem Rust Of Cereals

Barberry Eradication

The eradication program proved to be immense and costly. The common barberry was often grown by the early settlers. The plant grew and propagated itself well in the wild. It became widely disseminated by birds that fed upon the berries. By 1942, over 300 million bushes had been removed and about 60 percent of the area was practically free from the barberry. It was estimated that the damage to the wheat crop was reduced correspondingly by 60 percent in the area freed of barberry.

It was realized that the eradication of the barberry would not begin to eliminate the disease because urediospores survive in the southern winter wheat belt and in Mexico. These spores infect the locally grown wheat in the spring and are blown northward sometime after the wheat emerges. The removal of the early spring inoculum delays the build-up of massive amounts of inoculum and reduces the likelihood of epidemics. It was also known that the rust fungus only hybridizes on the barberry, and the removal of these bushes therefore, prevents such hybridization. This is an application of the rule: any factor that reduces hybridization of a pathogen will also reduce its race development. The purpose of the barberry eradication was therefore, two-fold--reduce spring inoculum and race development.

Rule -- Reducing Race Formation

Epidemics of stem rust can still occur, because the summer repeating stage, the urediospores, can reinfect the wheat again and again and again, every 8 to 14 days. If weather conditions are favorable for disease development and spread, a massive quantity of urediospores are formed and descend upon the already stricken crop just as it is in the early stages of heading-out.

Although epidemics of stem rust have occurred since 1942, they could have been far more severe if the barberry bushes had not been removed. The program is considered to be successful. However, continued surveillance is maintained in order to keep the population of barberry plants at an absolute minimum.

White Pine Blister Rust

In order to control the white pine blister rust, an immense eradication program was undertaken to remove *Ribes* spp., (gooseberries and currants). Basidiospores, (sporidia) of *Cronartium ribicola* infect the white pine by entering through the stomata of the needles. Mycelia grow down the needles to the twig or branch where a perennial canker is formed. Each spring aeciospores are formed on the surface of the canker until the enlarging canker finally girdles the stem and kills the branch. The aeciospores are airborne and may be viable after being carried for 100 miles or more. These spores infect the *Ribes* spp., by penetrating through the stomata. In 7 to 10 days, urediospores are formed on the *Ribes* spp. Several generations of urediospores may be produced, and they can reinfect *Ribes* spp., but not pine!

Basidiospores Are Weak Link In Life-Cycle

Teliospores are formed in late summer and germinate to form basidiospores. The basidiospores are the weak link in the life-cycle chain of this fungus. They are disseminated by air currents, but because they are thin-walled they are desiccated quickly and are not long-lived. They are killed before they can be carried 1,000 feet. This weak link has made the *Ribes* spp. eradication program an effective control procedure. The program was set up to remove the *Ribes* spp. from the white pine forest areas and in a belt 1,000 feet wide around each area. About 28 million acres of land in 20 states were included in the control areas. Repeated combing of the areas was necessary for the eradication, and constant surveillance is still required.

Ribes Eradication

The success of this program is best illustrated by the fact that regeneration of the white pine forests is taking place wherever the *Ribes* spp. is under control. Thirty years ago, it was thought that the white pine was doomed to extinction because of this disease.

19-2.3 PRUNING OF HOST

Costs Limit Pruning

Removal of infected limbs and leaves, at best affords only a partial disease control. However, in some diseases it is a practical measure. Pruning is generally limited to perennial plants and essentially involves the removal of an inoculum source. It also may prohibit the advancement of established cankers. Such procedures are costly, but nonetheless have value in orchard and shade tree plantings. The removal of fire blight cankers is common in many apple orchards. The pruning of twigs and branches of plum and cherry trees that are infected with the black knot fungus, *Dibotryon morbosum*, is also a widely accepted control procedure.

The removal of heart rot in valuable shade trees is often practiced to extend the useful life of the tree. Also, it is often worthwhile for the home gardener to practice the removal and burning of infected leaves and blighted buds and stems from diseased ornamental plants to reduce the inoculum in his garden. This is commonly recommended for such diseases as botrytis blight of peony and other plants, caused by *Botrytis cinerea*.

19-2.4 SANITATION

Sanitation Is Important For Control Of Some Diseases

The knowledge of the pathogen with respect to the disease cycle is all important. Sanitary measures are only feasible when it is worth the cost and effort. The destruction of potato refuse piles for control of potato late blight is a good example of a sound sanitary control procedure. The infected tubers harbor the pathogen *Phytophthora infestans*, and allow it to overwinter. In the spring, the infected tubers sprout, and the primary inoculum is formed and quickly disseminated to healthy plants. The elimination of the potato refuse piles also aids in control of potato blackleg, caused by the bacterium *Erwinia atroseptica*. Several species of maggot flies are vectors for this disease, and they are attracted to the refuse piles and lay eggs in the rotting tubers. In the spring, adult females spread the pathogen at the time they lay their eggs in the soil just above newly planted potato seed pieces. The young larvae hatch out and feed on the potato seed pieces and the bacteria are deposited in the feeding wound.

Another sanitary procedure is the cleansing of pruning implements with bactericides between cuts, as is recommended for control of fire blight of apple. For instance, a pruning implement contaminated with the fire blight organism could result in the spread of the bacterial pathogen, rather than its control.

19-2.5 CROP ROTATION

There is an old belief--continued culture of one crop in a given soil is detrimental to the soil. However, if this were true, why are we able to grow forest tree crops on land for centuries? Why can we grow apple trees and citrus trees for a period of decades in the same soil? In some regions, certain annual crops produce as well or better in continuous production. In the past, crop rotation has often resulted in the practice being abused, since it will not always reduce the amount of inoculum in the soil. Therefore, for some diseases crop rotation may be useless. On the other hand, continued cropping may increase the severity of other diseases, and crop rotation may well prove of value by decreasing the inoculum. Obviously, what is needed is a thorough knowledge of the pathogen. Will the inoculum remain viable for years in the soil without a host crop or not? Will there be enough reduction of inoculum to allow a crop rotation program to be workable?

Pathogens that are soil inhabitants, such as *Pythium* spp., *Rhizoctonia* spp., and *Fusarium* spp., may grow on a wide variety of crops and can survive indefinitely in the soil, or can at least survive for a long enough period to make crop rotation impractical.

Crop Rotation Useful For Control Of Many Soil Invaders

Soil invaders are usually killed soon after the crop refuse is decomposed. For example, *Colletotrichum lindemuthianum*, the causal agent of bean anthracnose, and *Xanthomonas phaseoli*, which causes bacterial blight of beans, may persist in the soil for two years, but not much beyond. Therefore, a 3-4 year crop rotation for control of such diseases is a practical procedure, and one that is scientifically sound.

19-2.6 CHANGING CULTURAL PRACTICES

Plowing Down Crop Residue Destroys Much Fungal Inoculum

No Plow Tillage

In conventional tillage systems, a crop residue is plowed-down sometime after fall harvesting, but before spring planting. The "clean" plowing of fields has the marked advantage of rotting the crop residue and thereby destroying much fungal inoculum residing in it. In the last few years, there has been an ever-increasing number of growers in the mid-west using "no plow" tillage practices in the growing of row crops, especially corn. Appreciable crop residues are left on the soil surface by many of these tillage practices. The soil is either partially tilled at planting time, or the seed is drilled into the field behind coulters, etc. The benefits of such tillage systems are considerable. The crop residue reduces soil erosion by wind and rain throughout the winter and spring; soil moisture is conserved, fertilizer run-off is reduced and the costs of plowing are saved. No plow tillage systems depend upon the proper use of herbicides to obtain necessary weed control which eliminates the need for cultivation. It is likely that such tillage practices will be expanded and extended to other crops such as soybeans.

Inoculum May Survive In Undecomposed Crop Residue

However, the crop residue left on the soil surface may contain and allow to overwinter the inoculum of seedling, stalk and ear rotting, and foliar pathogens. The problem inherent in leaving exposed crop residue on the soil surface is the almost certain increase of particular plant diseases. Perhaps plowing-down the crop residue every few years when the incidence of diseases becomes too high, or where possible, rotating to a non-susceptible crop will provide the needed control.

In some regions and on certain soils, double cropping is possible. For example, winter wheat may be planted in the fall, harvested in early summer and the field then planted to a short-season crop of soybeans. A full-season soybean crop may be grown in alternate years. Various types of such cropping systems may be utilized in the future where there is sufficient soil moisture to permit it. The growing of different crops is extremely useful in building-up a diverse soil microflora that effectively reduces many soil-borne pathogens.

19-2.7 VARIETY DEPLOYMENT

Within Regions

The risk of plant disease is reduced by the growing of two or more varieties of the same plant, each with different genes governing resistance, in the same general area or geographical location. Of course, when a plant pathogen enters one field, the subsequent spread of the pathogen within the field is not reduced, but usually there is a reduction of the spread of the pathogen between fields.

Between Regions

In many parts of the world, certain plant diseases become extremely widespread during the course of a growing season. In North America, stem rust of wheat is an example of such a disease. The rust fungus, *Puccinia graminis* var. *tritici*, overwinters in northern Mexico and southern Texas on winter wheats, spreads northward with the growing season, and terminates in the spring wheats of the prairie provinces of Canada. To block the northward spread of the rust, different genes for specific resistance could be assigned to particular regions along the course of spread of the rust. The race or races of *P. graminis* var. *tritici* that attack wheat in a southern region would most likely be unable to attack wheats of an adjoining northern region. The risk of disease epidemics in northern regions is thus reduced because inoculum from southern regions is unlikely to be virulent.

Such a plan is entirely feasible, but would require the cooperation of researchers and growers and may need to be backed-up with federal regulations.

19-2.8 DISEASE-FREE SEED AND PROPAGATING MATERIAL

Inspection And Certification

In the United States, state laws generally require compulsory inspection and certification of trees, shrubs, and bulbs before they are moved. However, seed stocks, tubers, and transplants usually do not come under state compulsory inspection laws; instead, systems have evolved for their production, certification, and sale, entirely on a voluntary basis. The procedure for the voluntary certification program varies from state to state, but is also governed by state law. In some states separate seed-inspection agencies control and police the program. Sometimes, state agencies that supervise the compulsory program also administer and enforce the voluntary program. Certification agencies develop, administer, and police the programs, but do not sell the certified commodities. They are sold through commercial nurseries, seed firms, brokers, retailers, etc. In the case of some crops (small grains, soybeans, and hybrid corn), certified seed may be sold by the growers to other farmers.

Seed Certification

The major purpose of seed certification is to maintain the identity of a variety after it has left the plant breeder. Pedigree records for crop varieties are kept; fields from which the seed is to be certified, are inspected; and inspections are conducted on the harvesting and cleaning of each seed lot. Without such a system, seed of varieties tend to become mixed and to lose identity.

Secondary purposes of seed certification are to assure the buyer of certified seed that it has a high rate of germination; and that it meets certain minimum standards as to amount of weed seeds, other crop seed, inert matter, and disease. It is the latter requirement that we are especially concerned with.

The recognition that trees, shrubs, and bulbs were important carriers of pathogens prompted the early establishment of state laws governing their compulsory inspection and certification. Now it is also realized that seeds, tubers, and transplants are also important carriers of pathogens. The planting of pathogen-free seed and planting stock is a first line of disease control; it is the cheapest form of crop insurance a grower can buy.

Disease-Free Seed Produced In Disease-Free Region

Bean seed is generally produced in semi-arid regions of the western United States. This has proved effective in the production of seed that is generally free from anthracnose and common bacterial blight. These are wet-weather diseases that are spread by splashing rain, and therefore, are essentially absent in arid regions where the crop is irrigated in furrows.

The history of peppermint culture in the United States is one in which verticillium wilt has played a dominant role. The mint industry was once centered in Michigan, but due to an increasing infestation of the fields with *Verticillium albo-atrum*, the industry in this state has been reduced to a fraction of what it once was. *V. albo-atrum* is a fungus that spreads slowly in the soil and is capable of existing there for many years. This disease became widespread by planting fields with stolons that were infected.

Seed, dormant root stock, and other propagating materials are often dipped into hot water to free them of pathogenic fungi, bacteria, viruses, and nematodes. The temper-

Fig. 19.1 Infected propagating stock. These tomato transplants were shipped from the southern United States to Indiana for early spring planting. They are heavily infected with *Sclerotium rolfsii*. Note the dark necrotic leaves and petioles and innumerous masses of white mycelium. About natural size. *(Courtesy of R. W. Samson)*

ature and the length of treatment is variable depending upon the material to be treated. Often this heat treatment is followed by the application of a protective fungicide.

Virus-free propagating stock: the distribution and planting of "virus-free" propagating material is essential for the control of many virus diseases. There are at present, no practical treatments to cure virus diseased plants in the field, and the strategy of control requires the removal of infected plants as soon as they become noticeably deteriorated.

To obtain virus-free stock it is first necessary to obtain disease-free "mother" plants. Starting with diseased plants, there are two major methods to obtain plants that are free of the virus. The first technique makes use of erratic virus distribution within the plant. For example, a proportion of young buds of a plant may be virus-free, and these may be grafted onto healthy rootstocks giving a proportion of healthy plants. A variation of this is the "meristem tip culture". It involves the aseptic culture of small pieces of meristematic tissue from about 0.1 to 0.5 mm in length, and of similar width, cut from shoot tips of plants. It appears that most plants that can be propagated by rooting cuttings can be successfully cultured as meristem-tips. The time needed for roots to develop varies greatly among different species and within a group of meristem-tips taken from similar plants; it may take as little as 10 days or as long as 2 to 3 months.

Erratic Virus Distribution In The Plant

Bud-grafts

Meristem-tips

The second method of obtaining virus-free plants is through the use of heat. The treatment of dormant plant material depends upon the destruction of the viruses by the heat, and the ability of the plant tissue to withstand the heat. Hot water ranging from 35 to 54° C., and immersion times of a few minutes to several hours are used. The hot water is less destructive to dormant tissues than to actively growing ones. Ratoon stunt virus of sugar cane is routinely prevented by soaking the planting canes for two hours in hot water at 50° C.

Hot Water Dip

The treatment of actively growing plants with hot air for 20-50 days or more is a widely used method of obtaining virus-free plants. The hot air generally gives better survival of the plant than does hot water. The critical temperature range is 35 to 40° C. Extended exposure above 40° C. is lethal to most plants, and temperatures below 35° C., only lessen the virus concentration within the plant. Virus degradation in plants held at these temperatures is not a direct effect of temperature on the virus; instead, virus multiplication is inhibited and the plant slowly inactivates the virus, thereby freeing itself.

Hot Air Treatment

Unable To Multiply, Virus Is Inactivated By Host

All young plants must be examined to determine if they are virus-free. This is done in several ways, and is called "indexing". The basic technique is to determine if the young plant contains an infective virus by grafting a bud from it to a healthy indicator plant. Young plants found to be healthy are then grown as "mother" plants from which virus-free propagation stock is obtained.

Indexing

19-2.7 STEAM TREATMENT OF SOIL

The steaming of soil is expensive and laborious; therefore, it is profitable only on high cash crops in the greenhouse trade. Generally, rows of tiles are laid about one foot deep in the ground beds of greenhouses. The ground beds are generally covered with canvas, and low pressure steam is passed through the tiles until the soil is pasteurized. The soil is steamed at least once each season.

The close planting and intensive cropping in greenhouses can quickly result in exceptionally large amounts of soil-borne inoculum. This amount of inoculum is often many times as large as that found in field plantings.

19-3 DISEASE CONTROL THROUGH CHEMICAL TREATMENT

The growers in the United States today rely heavily on disease control through the use of various pesticides. All people have benefited considerably from the use of these chemicals. However, many growers have failed to realize the benefits that can be obtained by using other control methods in conjunction with pesticides, because the use of chemicals seldom, if ever, provides complete disease control. There is ample evidence that the effectiveness of pesticides increases when there is a decrease of the inoculum potential. A partial reduction of inoculum by any method and the use of resistant varieties, in combination with pesticides, will provide the optimum control. The use of pesticides is an absolute necessity to control many diseases, and without their use the commercial production of many crops would not be feasible.

Rule -- Pesticide Effectiveness

Pesticides commonly include fungicides, bactericides, nematicides, insecticides, miticides, and herbicides. Each pesticide group is essential in the raising of commercial crops. However, this discussion is limited to disease control, and as such, emphasizes fungicides and to lesser extent, nematicides. Although some fungicides also possess bactericidal activity, there are few bactericidal and no antiviral compounds commercially available for use on edible crops. The antibiotic streptomycin, which is translocated systemically, has substantial antibacterial effectiveness, but its use on edible crops is limited.

Pesticides:
Reduce Inoculum,
Protect Plants,
Provide Therapy

Pesticides may either (1) reduce inoculum (2) provide protection before inoculum is able to contact host plants, or (3) provide therapy after the pathogen has invaded the host. Soil fumigant nematicides are examples of pesticides that reduce inoculum. Most fungicides provide a protective chemical barrier between the plant and the pathogen. The use of therapy to provide plant disease control has received a great deal of research attention, but substantial commercial application is still lacking.

Most foliar fungicides serve to protect plants by forming a chemical barrier between the inoculum and the plant. Hence, they are often referred to as protectant or protective fungicides. Some fungicides are active against a large number of fungi and are described as possessing a "broad-spectrum" of activity. Others have a "narrow-spectrum" of activity and are only effective against a few fungi. Most fungicides are non-systemic, meaning that they are not absorbed by the plant and translocated in the vascular system; however, a few are systemic, or partically so. The activity of most fungicides is of a lethal nature, which is called "fungicidal". A few others exhibit an inhibitory effect on fungi and are termed "fungistatic".

Differential
Toxicity: An
Essential
Property Of
Pesticides

Fungicides often are broad-spectrum "biocides" that are lethal or inhibitory to all forms of life. However, these chemicals are toxic to fungi at very low concentrations that are not toxic to the plants or plant parts on which they are applied. The differential toxicity to various groups of organisms, (humans, domesticated animals, plants, and pathogens), is an essential property of many pesticides. Chemical fungicides must be toxic to fungal pathogens, but much less toxic to the plants on which they are applied and present no hazard to man when properly used.

Retention

An important property of a protectant fungicide is its retention, or its effective residual life on foliage and fruit surfaces. To be effective, a protectant fungicide must resist being washed-off by rains in order to provide protection during prolonged rainy periods. A second important property of a protectant fungicide is its redistribution, which refers to the movement or transfer of residue upon a treated leaf and from leaf to leaf by the action of rains so as to evenly coat all exposed surfaces. This is important in order to protect new growth. A good protectant fungicide must possess a balance of retention and redistribution. It must adhere well, yet not so much as to prevent its being spread by rain.

Redistribution

Protectant fungicides are usually effective for 7 to 10 days, an occasional fungicide may be effective for 3 weeks if it is applied at a dosage rate of 2 to 3 times the normal application rate. The length of time a fungicide remains effective is usually dependent upon the weather. Frequent and hard rains reduce the concentration of even the best fungicides. Also, loss of fungicide is often due to oxidation caused by ultraviolet radiation of sunlight.

Some protectant fungicides also possess "eradicative" properties, especially against superficial fungal pathogens, of which the most notable example is the apple scab fungus *Venturia inaequalis*. For control of this pathogen, an eradicative fungicide must be applied within 18 to 48 hours after infection has occurred. Since the apple scab fungus is capable of infecting the host only during rainy periods, it is often possible to use fewer fungicide applications by spraying after each rain rather than on a regular schedule. Many growers use a combination of protective and eradicative fungicides for control of this disease, or use eradicative fungicides after protracted rainy periods.

19-3. TOXICITY RATINGS OF PESTICIDES

Pesticides are rated as to the mammalian toxicity of their active ingredients. The usual toxicity ratings are given as LD_{50} (lethal dosage for 50 percent of test animals) and expressed as mg/kg (milligrams of active ingredients per kilogram) of body weight of the test animal. Two types of toxicity tests are most frequently used, "oral" and "dermal". Oral toxicity is determined by a single dosage administered orally in solution. Dermal toxicity is determined by a dosage administered to the skin with a 24 hour exposure. The oral and dermal LD_{50} values are usually obtained using test animals such as mice, rats, rabbits, or dogs. Since toxicity is determined on lower animals, LD_{50} values cannot be directly extrapolated to humans. However, LD_{50} values are good approximations of relative toxicities to man and show relative toxicities between chemicals. The following table of classes of mammalian toxicities of pesticides is generally used:

Toxicity Tests -- Oral And Dermal

LD_{50}------in mg/kg

		Oral	Dermal
Class 1	Highly toxic	1-50	1-200
Class 2	Moderately toxic	50-500	200-2,000
Class 3	Slightly toxic	500-5,000	2,000-20,000
Class 4	Relatively nontoxic	over 5,000	over 20,000

Toxicity Classes

The lower the LD_{50} value, the more toxic, or dangerous, the material is. Classes 3 and 4 are generally considered to present little hazard in use. Graphs showing the relative toxicities on a number of common insecticides, herbicides and fungicides are presented on the next two pages using LD_{50} data on rats.

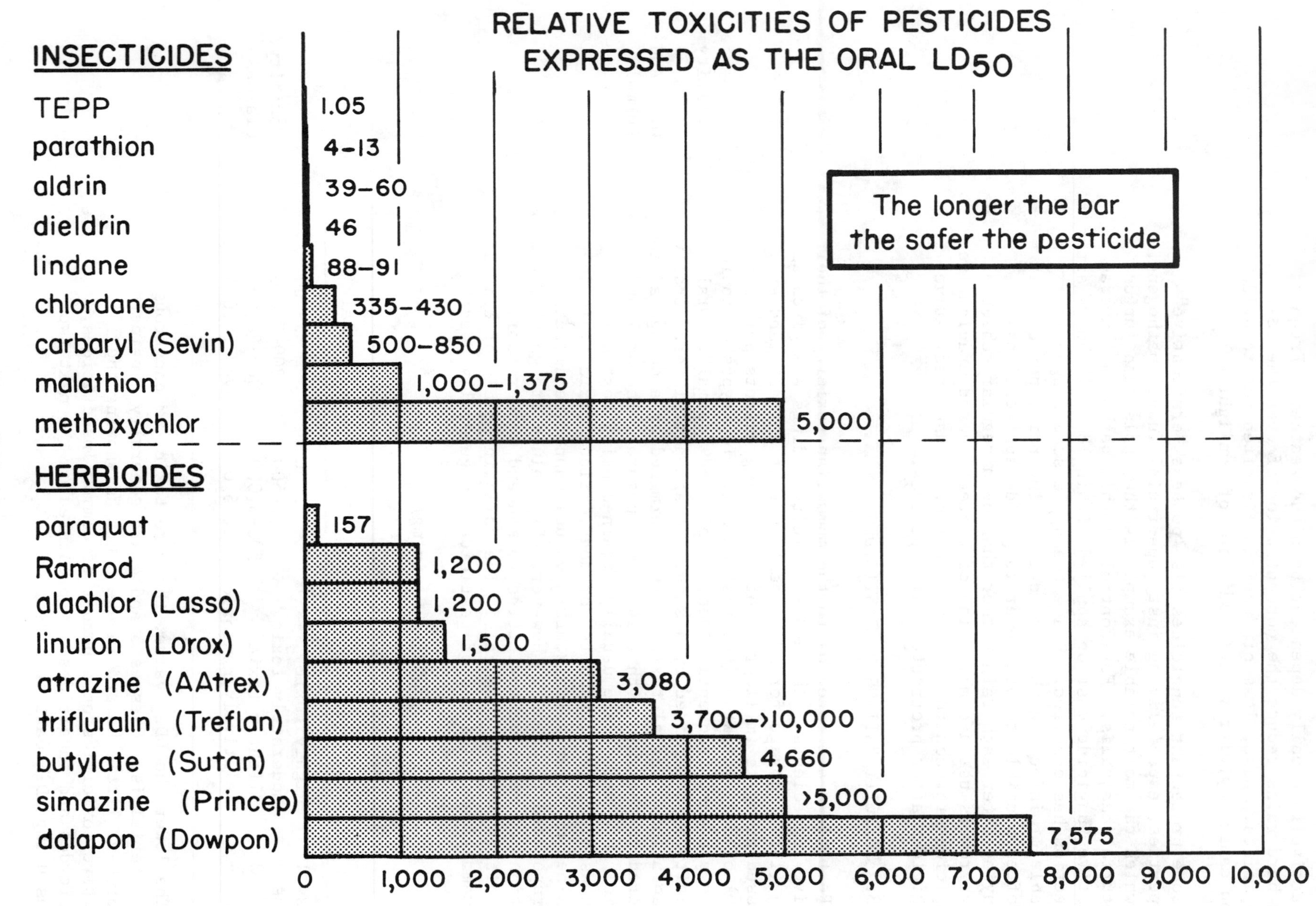

RELATIVE TOXICITIES OF PESTICIDES
EXPRESSED AS THE ORAL LD50
The longer the bar
the safer the pesticide
INSECTICIDES
TEPP
parathion
aldrin
dieldrin
lindane
chlordane
carbaryl (Sevin)
malathion
methoxychlor
HERBICIDES
paraquat
Ramrod
alachlor (Lasso)
linuron (Lorox)
atrazine (AAtrex)
trifluralin (Treflan)
butylate (Sutan)
simazine (Princep)
dalapon (Dowpon)
1.05
4–13
39–60
46
88–91
335–430
500–850
1,000–1,375
5,000
157
1,200
1,200
1,500
3,080
3,700–>10,000
4,660
>5,000
7,575
0
1,000
2,000
3,000
4,000
5,000
6,000
7,000
8,000
9,000
10,000

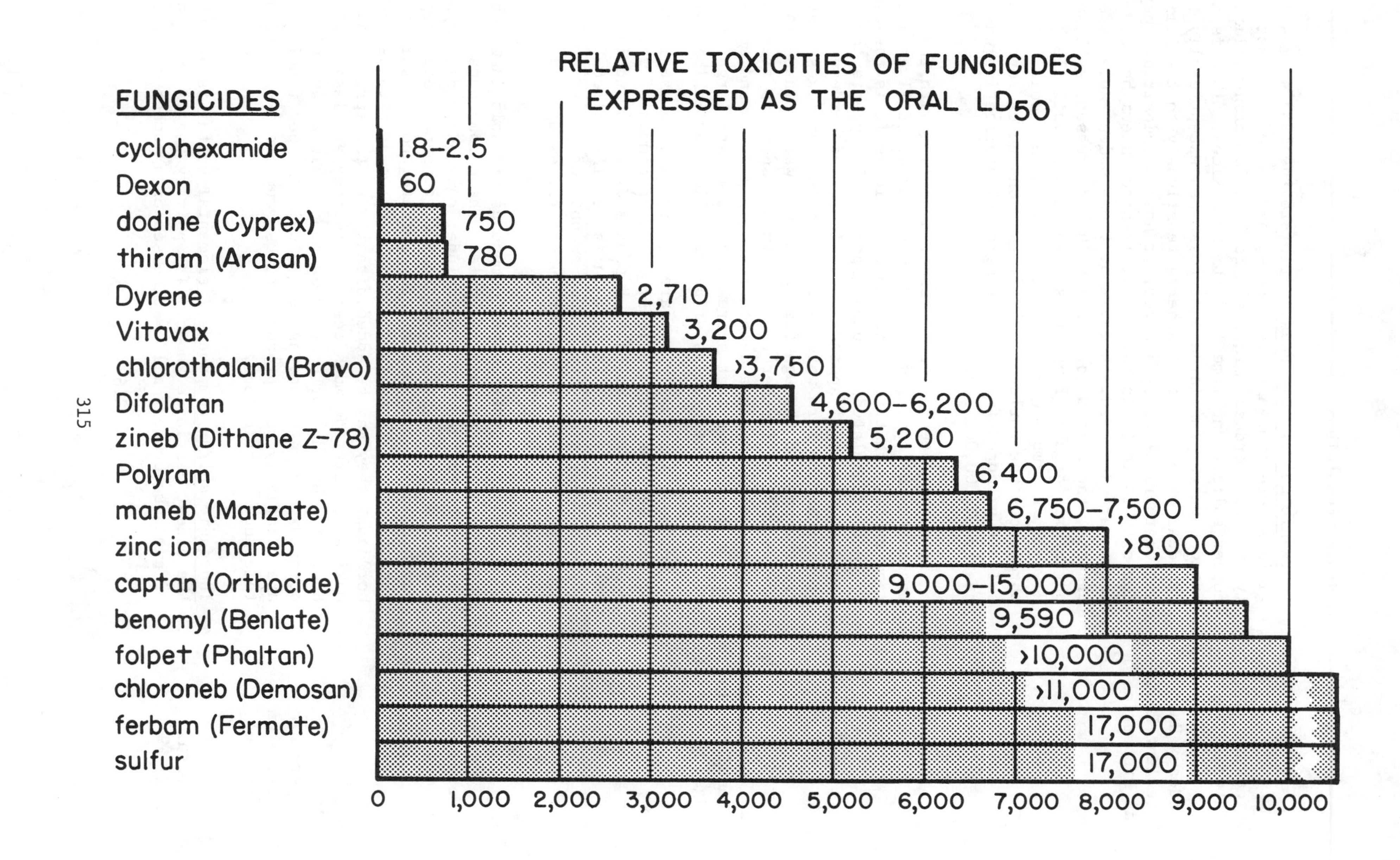
RELATIVE TOXICITIES OF FUNGICIDES
EXPRESSED AS THE ORAL LD_{50}
FUNGICIDES
cyclohexamide
Dexon
dodine (Cyprex)
thiram (Arasan)
Dyrene
Vitavax
chlorothalanil (Bravo)
Difolatan
zineb (Dithane Z-78)
Polyram
maneb (Manzate)
zinc ion maneb
captan (Orthocide)
benomyl (Benlate)
folpet (Phaltan)
chloroneb (Demosan)
ferbam (Fermate)
sulfur
1.8–2.5
60
750
780
2,710
3,200
›3,750
4,600–6,200
5,200
6,400
6,750–7,500
›8,000
9,000–15,000
9,590
›10,000
›11,000
17,000
17,000
0
1,000
2,000
3,000
4,000
5,000
6,000
7,000
8,000
9,000
10,000

19-3.2 FUNGICIDES

Bordeaux Mixture

Bordeaux Has Excellent Residual Properties

Isphytotoxic

Inorganic copper compounds: prior to the 1940's, most fungicides were inorganic in nature, and one of the most widely used was bordeaux mixture. It is a reaction product of copper sulfate and lime (calcium hydroxide). This is a broad-spectrum fungicide that possesses exceptionally fine residual properties; it adheres tenatiously to the foliage and weathers slowly. However, it has an undesirable phytotoxicity to most plants. Fruit is often blemished and leaves are scorched. Because of these undesirable side-effects, the demand for bordeaux has decreased, as less phytotoxic organic fungicides have become available. However, enormous quantities of bordeaux are sold throughout the world. In 1967, the tonnage of copper sulfate produced for use in fungicidal application was nearly 34,000,000 lbs.

The Toxic Copper Accumulates In Fungal Spores

There has been a lot of research dealing with the mechanisms of toxicity of bordeaux, but we still do not know exactly what form the copper is in as it is taken up by, and accumulated in fungal spores. It appears that exudates of fungal spores react with dried bordeaux precipitate to form water soluble copper complexes that are transported through the spore cell wall. These complexes undergo further reaction in the spore to free the copper which causes a denaturation of protein and enzymes. It has been shown that spores may accumulate large amounts of copper in the process of being killed, or inhibited.

Bordeaux is usually prepared just prior to use in one of many various formulations; the most common of which is 8 pounds of copper sulfate, 8 pounds of lime, and 100 gallons of water. This formula is usually called 8-8-100. To lessen phytotoxicity, lower amounts of copper sulfate and/or higher amounts of lime are used; for example, 4-8-100, 6-12-100, 8-24-100, etc.

Fixed Coppers

Coppers Are Exempt From Tolerances

A number of other copper compounds are sold in which the copper ion is less water-soluble. These compounds are called "fixed" coppers and include; cuprous oxide, copper carbonate (basic), copper hydroxide, copper oleate, copper oxychloride, copper oxychloride sulfate, copper sulfate (basic), etc. These fungicides have two advantages over bordeaux, they may be added directly to the spray tank, making preparation much easier, and they are less phytotoxic. However, they do not possess as good residual activity as bordeaux, nor are they as fungicidal. Copper fungicides are relatively safe products to use and are exempt from tolerance restrictions.

Elemental Sulfur

Inorganic sulfur compounds: elemental sulfur is probably the oldest pesticide known. Its primary use is for control of numerous powdery mildew diseases on a number of fruit,

Put under light bulb + it vaporizes

and ornamental diseases. However, it may also control black spot of roses, apple scab, peach scab, brown rot of stone fruits, and several other foliar diseases. It also has insecticidal and acaricidal (miticidal) properties.

Number One In Tonnage

The world consumption of sulfur has long exceeded that of any other single fungicide. The tonnage of sulfur used in disease control in the United States in 1964 was 136,823,000 lbs. On a tonnage basis, it still ranks number one in the United States. However, its use has been declining because of the development of organic fungicides.

May Cause Phytotoxicity, Especially At High Temperatures

Elemental sulfur fungicides are formulated either as dusts or wettable powders. The major drawback to the use of sulfur is the severe phytotoxicity which may occur; scorching, burning, dwarfing, reduction in yield, and sometimes early defoliation. This is more likely to occur at temperatures of 80° F., or higher.

Inexpensive And Safe

On the "plus" side of the ledger, sulfur is inexpensive and relatively safe to man and wildlife; it is classified as a nonpoison or "safe chemical". Tweedy (1969), gives a good review of the mechanism of fungicidal activity of sulfur, much of which is still unknown.

Lime-sulfur

Lime-sulfur (calcium polysulfide) is formed by the reaction of lime with sulfur. It too has a long history of use as a fungicide. In the United States, it has been primarily utilized as a tree fruit fungicide, for use as both dormant and foliage sprays.

Carbamate compounds: these fungicides had their origin in a patent by Tisdale and Williams in 1934, issued to the du Pont Company. This fungicide, tetramethylthiuramdisulfide, with the common name of thiram, is formed by joining two molecules of dithiocarbamic acid through the sulfur atoms.

Thiram, A Seed And Turf Fungicide

Thiram is mainly used as either a seed treatment or turf fungicide; with a limited application as a foliage protectant for control of several fruit diseases. It is formulated as either a wettable powder or dust. Thiram is sold under various trade names such as Arasan, Tersan, and Thylate.

Numerous metal dialkyldithiocarbamates have been tested for fungicidal effectiveness, but only ferric dimethyldithiocarbamate (ferbam) and zinc dimethyldithiocarbamate (ziram) have been commercially developed. Ferbam has a wide spectrum of antifungal activity, with little or no phytotoxicity when applied as directed. It quickly became widely accepted as "the fungicide" for control of apple scab, and was also used for other foliar diseases of fruits, ornamentals, and vegetables. Ferbam is a wettable powder with good residual properties. Its major drawback is the unsightly black residue left on the foliage, flowers and

Ferbam, First Organic Chemical For Apple Scab Control

fruit. It has been largely replaced by newer organic fungicides. Ferbam is sold under trade names of Fermate, Carbamate and Karbam.

Ziram, Has Wide Spectrum

Ziram, formulated as a wettable powder, also has a wide spectrum of activity. It has been used to control numerous foliar diseases of vegetables, cucurbits, fruits, and ornamentals. However, like ferbam, it has been largely replaced by newer organic fungicides. It is sold under the trade name of Zerlate.

The ethylenebisdithiocarbamates were developed by uniting two molecules of methyldithiocarbamic acid, that were previously reacted with a metal, through the methyl groups. The first compound of this family to be developed was the water soluble disodium ethylenebisdithiocarbamate; later

Nabam

assigned the common name nabam, and sold under trade names of Dithane D-14 and Ortho-nabam. This compound is markedly improved by reacting it with zinc sulfate in the presence of carbon dioxide. One zinc atom replaces two sodium atoms

Zineb

resulting in zinc ethylenebisdithiocarbamate (zineb). Zineb, sold under trade names of Dithane Z-78 and Ortho-zineb, has found wide usage against foliar diseases of fruits, vegetables, cucurbits, and ornamentals; and a number of fruit rots.

Maneb, A Standard Protectant Fungicide

In 1950, maganese ethylenebisdithiocarbamate (maneb) was introduced. Maneb, sold under trade names of Manzate, Dithane M-22, etc., was quickly acclaimed for its broad spectrum of antifungal activity and its effectiveness in control of such diseases as late blight of potato and tomato. It is widely used for control of many foliar diseases of vegetables and ornamentals, and is also used as a seed treatment. Since its introduction, maneb has been a "standard" in the industry against which new experimental fungicides are tested.

Zinc, A Safener

A drawback to the use of maneb on certain crops such as apples, is its phytotoxicity. This led to the addition of zinc which acts as a "safener"--reducing the phytotoxicity while retaining the antifungal activity. The latter formulation is known as zinc-maneb (Manzate D or Dithane M-22 Special).

Zinc Ion Maneb

Zinc-maneb led to the development of zinc ion maneb (Manzate 200 and Dithane M-45), formed as coordination products of zinc ion and maneb, which further reduces the phytotoxicity of maneb. A major feature of all maneb formulations is their good retention, or residual life, on treated foliage.

Benezene compounds: hexachlorobenzene (HCB), and pentachloronitrobenzene (PCNB or Terrachlor), have been used in the Pacific Northwest as seed treatments of wheat to control seed and soil-borne bunt. This is the only region in the United States where soil-borne bunt spores remain viable to infect germinating seeds and seedlings. These fungicides form vapor barriers in the soil, around the seed, and inhibit the germination of bunt spores.

PCNB

PCNB is also used as a soil fungicide and has long residual properties in the soil. Its activity appears to be mainly fungistatic, rather than fungicidal. It has a narrow spectrum of activity that includes *Rhizoctonia*, *Sclerotinia*, *Sclerotium*, and *Plasmodiophora*; but it is not effective for control of *Pythium*. When used as a soil treatment, it is usually applied in the furrow at planting time.

Chloroneb Is Systemic Seed Treatment

Chloroneb (1,4-dichloro-2,5-dimethoxybenzene), with the trade name of Demosan, is a seed treatment fungicide, that is used on beans, soybeans and cottonseed. Chloroneb is at least partially absorbed and translocated in seedlings, and acts as a systemic protectant against the ubiquitous *Rhizoctonia solani*.

Karathane

Karathane, is the trade name for the foliage fungicide 2-capryl-4,6-dinitrophenyl crotonate. This fungicide is highly specific for use against the powdery mildew fungi on many fruits, cucurbits, etc. It is not as effective as is sulfur, but is less phytotoxic, especially at high temperatures.

Botran

Botran is the trade name for 2,6-dichloro-4-nitroaniline. This compound is used as a foliar and soil fungicide on ornamentals and fruits; also, for postharvest treatment of some fruits.

Dexon

Dexon, the trade name for p-dimethylaminobenzenediazo sodium sulfonate, is a seed and soil treatment fungicide. It is highly effective against *Pythium* spp., that cause damping-off and root rots of vegetables, cotton, sugar beets, avocados, pineapples, etc.

Captan

Heterocyclic fungicides: the most prominent member of this group of fungicides is captan, N-trichloromethylthiotetrahydrophthalimide, which possesses a broad-spectrum of antifungal activity. Sold under trade names of Captan and Orthocide, it has proven to be a most versatile fungicide. Its range of usefulness extends from control of leafspots, blights, and fruit rots on: tree fruits, including citrus; small fruits, including grapes and cane fruits; vegetables, including potato and tomato; curcubits; ornamentals, including annuals and perennials; and turf. Captan is also an acknowledged seed treatment fungicide of vegetables,

Broad spectrum
Very safe - cancer
Bright pink colored seed
Good on a lot of crops

ornamentals, grasses, corn, cotton, cucurbits, potatoes, small grains, etc. It is also used as a root dip on asparagus, and as a preplant soil treatment for vegetables. Captan, however, does not preform well against rusts or powdery mildews and is somewhat weak in control of leaf-spots caused by *Alternaria* and *Septoria*, and against potato and tomato late blight (*Phytophthora infestans*).

One of the reasons for the success of captan is its low-order of phytotoxicity. Since its introduction, this fungicide has been a "standard" against which the performances of experimental fungicides are measured.

Folpet

Folpet (N-trichloromethylthiophthalimide), is structurally related to captan, and is sold under the trade name of Phaltan. Like captan, folpet also has a broad-spectrum of activity; unlike captan, folpet controls many powdery mildews. It is used as a foliage fungicide on numerous tree fruits, small fruits, melons, vegetables, and ornamentals.

Difolatan

Difolatan is the trade name for cis-N-(-(1,1,2,2-tetrachloroethyl)thiol)-4-cyclohexene-1,2-dicarboximide. It is structurally related to captan and folpet, and like these two compounds, Difolatan is also a broad spectrum fungicide. It is useful against numerous foliar diseases of tree fruit and nuts, pineapples, melons, potatoes and tomatoes. It is a liquid, which makes it easier to apply (no nozzle clogging), than most other protectant fungicides that are wettable powders. Difolatan possesses good retention for control of leaf blights.

SAT

RAM

Difolatan is usually applied at regular intervals, however, because of its low phytotoxicity, good retention and redistribution, it may be used on apples for scab control as a single application treatment (SAT) and on tomatoes for control of fungal foliar and fruit diseases in a reduced application method (RAM). For example, on apples, Difolatan may be applied at three times the normal dosage with the application made early in the growing season (from the late dormant stage of the buds until green tip). This treatment protects the foliage during the time span when most of the ascospores of *Venturia inaequalis* are released. At petal-fall (of the blossoms), additional sprays must be made for powdery mildew control or for scab control if the weather is favorable for infection. On tomatoes, using the RAM, Difolatan may be applied at a dosage of 2 or 3 times the single application rate, giving protection to the foliage (and fruit) from 14 to 20 days. Hence fewer applications need be made.

Dyrene

Dyrene (2,4-dichloro-6-o-chloroanilino-s-triazine) is a protectant fungicide that is registered for use against foliar diseases of small fruits, cucurbits, vegetables, ornamentals, and turf.

Carboxin Systemically Controls Loose Smut

Systemic fungicides: carboxin (trade name Vitavax), is 5,6-dihydro-2-methyl-1,4-oxathiin-3-carboxanilide. The development of this compound represented a significant breakthrough for control of loose smut of wheat and barley, when applied as a seed treatment. Carboxin is absorbed by the seed, translocated systemically, and inhibits or kills the smut mycelium within the seed. Prior to using carboxin, expensive and laborious hot water dips were employed to free the seed of loose smut. Carboxin was initially registered in the United States for use on foundation and registered barley and wheat seed in 1967. Its initial registration was only for its use on a non-food crop because no residue tolerances were established at that time. Foundation and registered seed is only harvested for seed purposes, not food purposes. In February, 1974, carboxin was officially registered for use on growers seed for wheat, barley, oats and corn.

Benomyl, First Systemic Fungicide For Control Of Wilt And Foliar Diseases

Benomyl (trade name Benlate), is methyl 1-(butylcarbamoyl)-2-benzimidazolecarbamate. The development of benomyl also represents a striking development, because it is the first systemic fungicide to be registered in the United States on a wide range of crop plants. On fruit trees, however, it does not appear to possess a practical level of systemic activity to provide disease control. Hence, its use on fruit is probably limited to that of a protectant fungicide. On vegetables and ornamentals the systemic nature of benomyl may be fully utilized. It may be applied to the foliage or soil (near the root zone), or as a seed or row treatment. The plant absorbs benomyl and translocates it upward as well as distally (away from the midrib) in the leaves. It is doubtful that any downward movement occurs. The concentration of benomyl in foliage decreases after absorption and translocation.

Benomyl is useful in control of soil fungi, such as *Rhizoctonia*, *Fusarium* and *Verticillium*; stem rotting fungi such as *Sclerotinia*; foliar leafspots caused by *Septoria*, *Cladosporium* and *Cercospora*; and fruit rotting fungi such as *Botrytis*. It is registered for use as both pre-and post-harvest treatments on apples and stone fruits. It is also used to control foliar diseases of bananas, pecans, pineapples, grapes, peaches, strawberries, pears, turf, snap beans, celery, melons, sugar beets and peanuts. Benomyl is also useful for control of Dutch Elm disease caused by the fungus *Ceratocystis ulmi*. It is either applied to elm foliage in the spring, or is injected into the tree trunk. In either case, it must be applied by a trained arborist in conjunction with sanitation and insect control programs. Some eradication properties are exhibited by benomyl. For example, if less than 5% of the crown foliage shows symptoms, recovery from the disease has been shown to be possible with proper application of benomyl.

Control Of Dutch Elm Disease

Thiabendazole

Thiabendazole (trade name Mertect) is 2-(4-thiazolyl) benzimidazole. This systemic fungicide is presently used for control of leafspots of sugar beets and ornamentals, and as a post harvest treatment for mold control on apples, bananas, citrus, pears and tobacco.

Dodine, A Standard On Apple And Cherry

Other Fungicides: dodine, (N-dodecylguanidine acetate), is sold under the trade name Cyprex. This a highly useful foliar fungicide that is a "standard" for control of apple scab (*Venturia inaequalis*) and cherry leafspot (*Coccomyces hiemalis*). It is also used on pears, peaches, strawberries, black walnuts and pecans. Dodine is unique in that it possesses a local systemic activity in leaves to which it is applied.

Chlorothalonil

Chlorothalonil (trade name Bravo), is tetrachloroisophthalonitril. It is a highly useful broad spectrum protectant fungicide that is registered for control of foliar diseases on lettuce, snapbeans, carrots, celery, crucifers, melons, onions, tomatoes, potatoes, sweet corn, peanuts, ornamentals and turf.

19-3.3 FUNGAL RESISTANCE TO FUNGICIDES

The development of resistance of human bacterial pathogens to antibiotics has long been recognized, so has the build-up of resistant insects to various insecticides. As might be expected, there has been a number of reports on the development of resistance by plant pathogens to several fungicides and bactericides. For example, resistance of the fire blight bacterium, *Erwinia amylovora*, to the antibiotic streptomycin is well known. Also well known is the resistance of the soil-borne fungus *Rhizoctonia solani* to the soil fungicide pentachloronitrobenzene. Kuiper, 1965, in Australia, reported on resistance of the common bunt of wheat (*Tilletia foetida*), to the seed treatment fungicide hexachlorobenzene. Since the use of these fungicides is rather limited, resistance to them has had little impact. Of more importance was the finding of Szkolnik and Gilpatrick, 1969, who reported on the resistance of the apple scab fungus, *Venturia inaequalis*, to the widely used fruit fungicide dodine.

Strain Of Apple Scab Fungus Is Resistant To Dodine

Although resistance has been reported against a few fungicides, no resistance has been reported aginst the broad spectrum (protectant) fungicides, several of which have been widely used for 10 to 20 years. Undoubtedly, this is because these fungicides have wide based modes of action whereby they block several metabolic pathways.

In order for a fungus to become resistant to such a broad spectrum fungicide it would have to undergo a multiple-gene shift (by mutation, hybridization, etc.)--an unlikely event. Examples of such broad spectrum fungicides are maneb, zineb, thiram, captan, Difolatan, etc.

Resistance To Protectant Fungicides Unlikely To Develop

The swift build-up of resistance to the narrow-spectrum systemic fungicides, however, is quite a different story. These systemic fungicides reportedly possess narrow modes of action that probably involve single metabolic pathways. Fungi only need a single gene shift in order to become resistant to most systemic fungicides. For example, in Greece, cercospora leafspot is a major disease of sugar beets. The fungicide benomyl was used extensively for control of this disease in 1970 and 1971, with excellent results. However, in 1972, resistant strains of the causal agent *Cercospora beticola* appeared, causing foliage destruction to about 25,000 acres of sugar beets, (Georgopoulos and Dovas, 1973). In this illustration, widespread fungal resistance occurred after only two years of use.

Fungal Resistance To Benomyl

It is obvious that in a disease control program where more than one application is required, a broad spectrum fungicide, (captan, maneb, Difolatan, zineb, etc.) should be alternated, or applied in conjunction with the systemic fungicide. Also, the area or areas treated with a systemic fungicide should be kept to a minimum size. In any case, constant and repeated use of systemic fungicides will allow the build-up of resistant fungal strains, and render further usage of these fungicides worthless. On the other hand, if the use of systemic fungicides is intermittent and minimal, resistant fungal strains will not become dominant in the fungal population, and the effectiveness of systemic fungicides can be maintained for many years.

19-3.4 ANTIBIOTICS

An antibiotic is a substance produced by one microorganism, that inhibits or kills other microorganisms. Most antibiotics are formed as metabolic by-products of various actinomycetes; their activity is mainly against bacteria, although a few possess antifungal activity. Although numerous antibiotics have been extremely successful in human and veterinary medicine, they have not proved as useful in plant disease control.

The antibiotics that are used in plant disease control are at least partially absorbed by the foliage and translocated within the host. However, they may also inhibit or kill pathogens upon contact.

Antibiotics Are Usually Systemic

Streptomycin is an antibiotic produced by *Streptomyces griseus*. It is formulated as the sulfate or nitrate salt and sold under trade names of Agrimycin, Phytomycin, and Agri-Strep. Streptomycin controls a number of bacterial

Streptomycin

diseases including fire blight of apple and pear, caused by *Erwinia amylovora*. However, it is registered for use only up to 50 days before harvest. It may also be used on tobacco, tomatoes, celery, and peppers while in plant beds; and on field tobacco.

Cycloheximide

Cycloheximide is also produced by *Streptomyces griseus* in the production of streptomycin. It is unusual in that it is useful to control a number of fungi including *Coccomyces hiemalis*, the cherry leafspot fungus; various powdery mildews; and some fungal pathogens of turf. It also has a higher level of toxicity than most antibiotics; therefore, it is applied to foliage at dilutions of only a few ppm active ingredients. However, it is also quite phytotoxic which restricts its application to a number of less sensitive plants. Cycloheximide is sold under trade names of Acti-dione and Actispray. It is also formulated in mixtures with ferrous sulfate, PCNB, and thiram. These mixtures extend the activity spectrum of cycloheximide alone.

19-3.5 SOIL FUMIGANTS

Soil Fumigants Are Generally Applied As Preplant Treatments

These chemicals are volatile, toxic liquids or gasses, or chemicals that decompose in the soil to form volatile toxic ingredients. The liquids and gasses are usually injected 6 to 10 inches deep into moist, well-worked soil prior to planting time by means of special chisels spaced 8-12 inches apart. The soil is smoothed and compacted immediately after application. Generally, these chemicals are applied in a broadcast pattern throughout the field, or are applied in the row to be planted. All soil fumigants are toxic to living plants; therefore, they must be applied prior to planting. Generally, a minimum of two weeks between application and planting must be allowed. The lethal vapors of fumigants spread through the upper layer of soil and become partially solubilized in the soil moisture where they inactivate soil-borne pests. Soil fumigants that destroy or inactivate nematodes are called "nematicides". Some fumigants inactivate soil-borne fungi, weed seeds, and insects in addition to nematodes. These latter chemicals are called broad-spectrum fumigants. Soil fumigants are dissipated by decomposition in the soil and through loss of vapor from the soil surface.

At recommended dosage rates, the soil fumigant nematicides are generally lower in cost than are the broad-spectrum soil fumigants. Low cash value crops such as small grains, soybeans, and corn cannot justify preplant soil fumigation; while moderate to high cash value crops, including vegetables and fruits, can be economically treated for nematode control. Generally, only the higher cash value crops, such as strawberries and plant beds, can withstand the cost of treatment with broad-spectrum soil fumigants.

The first successful soil fumigant was chloropicrin (trichloronitromethane). This chemical is a very potent lachrymator; because of this property, it was widely used as a tear gas in World War I. It is a broad-spectrum fumigant that is particularly obnoxious to handle. Although its vapor pressure is only similar to that of water, chloropicrin is often described as highly volatile. This is because concentrations in the air of 0.5 ppm can be detected, 3 ppm make the eyes water, and slightly higher concentrations are unbearingly uncomfortable. In fact, no one will willingly expose himself to anything approaching lethal concentrations. Chloropicrin has its own built-in warning agent. Therefore, despite its unpleasantness, it is a rather safe material. If it is used in confined quarters, the use of gas masks is a necessity. It should not be used in soils below 60° F., or applied to soils adjacent to living plants. It is used mainly for its fungicidal activity. Chloropicrin is sold under trade names of Larvicide, Picfume, etc.

Chloropicrin, A Broad-spectrum Soil Fumigant

Chloropicrin Is Its Own Warning Agent

Methyl bromide (bromomethane) is a broad-spectrum, high volatile fumigant. Because it is a gas at room temperature, it is marketed in pressurized containers or solubilized in a xylene carrier. Methyl bromide is extremely toxic and has only a slight odor so that a person could be inadvertantly subjected to a lethal exposure. Therefore, it is often formulated with 2 percent chloropicrin as a warning agent. Sold under trade names of Pestmaster, Dowfume MC2, Bed-Fume, etc., methyl bromide is used mainly for its fungicidal and herbicidal activity. Since it leaves bromide residues in treated soil, methyl bromide is restricted in its use to plant and seed beds for all crops other than strawberries or tomatoes. Mixtures of chloropicrin and methyl bromide are used in California to treat fields planted to strawberries.

Methyl Bromide, A Broad-spectrum Soil Fumigant

Methyl bromide must be applied to soils at least 60° F., or above, and covered with plastic tarpaulins while application is made. Pressurized methyl bromide may also be released beneath tarpaulins that are slightly raised above the soil surface and sealed at the edges.

In the 1940's, the first fumigant nematicide was developed; 1,3-dichloropropene and 1,2-dichloropropane. It was first sold under the trade name of D-D and later also as Vidden D. 1,3-Dichloropropene is the major active ingredient of the mixture; this chemical is sold alone as Telone. The commercial availability of D-D offered economical nematode control for many crops for the first time; a breakthrough of immense value. These fumigants are widely used for vegetables, tobacco, small fruits, sugar beets, etc. They are registered for use on a "non food" basis; which means there is no reasonable expectation of any residue on the food when used as directed on the labels.

D-D, The First Fumigant Nematicide

D-D, Vidden D, And Telone Are Registered On A Non Food Basis

EDB

Ethylene dibromide, or EDB, is sold under the trade name of Dowfume W-85. It is another preplant soil fumigant nematicide, which is also widely used.

Dibromo-chloropropane

1,2-Dibromo-3-chloropropane is sold under trade names of Nemagon and Fumazone 70E. This is a low volatile soil fumigant nematicide that is applied preplant; or postplant as a side dressing to certain tolerant species of perennial flowers, ornamental trees and shrubs.

Vapam

Sodium N-methyldithiocarbamate, with trade names of Vapam and VPM; and Mylone (3,5-dimethyl-1,3,5,2H-tetrahydrothiadiazine-2-thione), are broad-spectrum soil fumigants. They undergo decomposition in the soil to form the volatile, and active methyl isothiocyanate. Vapam is a water soluble liquid and Mylone is usually formulated as a wettable powder. These chemicals should be applied to soils at least 60° F., or warmer in order for them to decompose properly. These products are mainly used to control soil-borne fungi and weed seeds in plant beds and nursery soils.

Mylone

Vorlex

Vorlex is a synergistic mixture of methyl isothiocyanate and 1,3-dichloropropene (Kenaga and Seven, 1965). This is an unusual preplant soil fumigant in that at low dosage rates it is an economical nematicide. At higher concentrations, it controls soil-borne fungi, weed seeds, and soil insects in addition to nematodes. Vorlex requires well moistened soils, and performs well when applied to soils ranging in temperature from 33-70° F. Late fall application is common, this allows aeration to take place over winter and in the spring before planting. Vorlex, Vapam, and Mylone are registered on a "non-food" basis.

Vorlex, Vapam, And Mylone Are Registered On A Non Food Basis

19-3.6 OTHER PESTICIDES USED FOR DISEASE CONTROL

The non-fumigant nematicide Zinophos (0,0-diethyl 0-2-pyrazinyl phosphorothioate) is commonly used as a bare root dip for dormant nursery stock. This product is quite toxic and must be used with care.

Insecticides are usually applied to plants to protect them from the ravages of the insects themselves. However, they may also be used to control insect vectors of a number of pathogenic bacteria and fungi, and a great many viruses.

The use of herbicides for control of plant diseases is mainly for the destruction of unwanted plants that harbor various pathogens. These plants are generally in the fence rows or in adjoining fields. Another use of herbicides is to kill the foliage of diseased plants prior to harvest, thereby prohibiting the sporulation of the pathogen on live stems and foliage. For example, potato vines are often killed with herbicides prior to digging the tubers, in order to prevent infecting the tubers with spores of the late blight fungus.

19-4 BIOLOGICAL CONTROLS OTHER THAN RESISTANCE

Rule -- Predators And Parasites

There are few, if any, predators or parasites which under natural field conditions render a noticeable population control of foliar plant pathogens. This is in contrast with a number of insect pests of plants that have both parasites and predators which act to check their population naturally in the field. However, there are numerous reports of organisms that are parasitic on fungi, especially other fungi (mycoparasites). Although mycoparasites are frequently associated with diseased aerial plant issue, they are seldom present in sufficient numbers to effectively reduce disease incidence.

There is an altogether different story for soil-borne root pathogens. The microflora and fauna of soils is exceptionally rich and diverse in organisms that are both parasitic and antagonistic. Because of this, we shall see considerable efforts made to establish biological control procedures against root pathogens.

Promising areas of biological control include: mycorrhizae, parasites, compounds from decomposing crop residues and mulches, chemical manipulation of soil microflora, nematodes predaceous on plant nematodes, nematode attractants, etc. For sake of brevity, the following discussion will be limited to the first two categories mentioned above.

19-4.1 MYCORRHIZAE AS BIOLOGICAL DETERRENTS

Mycorrhiza Defined

Mycorrhiza is defined as the symbiotic association of mycelium of certain fungi with the roots of higher green plants and living in close association with root cells, especially those at the root surface. There are two major groups of mycorrhizae, ectomycorrhizae and endomycorrhizae A smaller group, the ectendomycorrhizae, an intermediate type is also recognized.

Ecto-mycorrhizae

The fungi involved in ectomycorrhizae penetrate cortical cells of feeder roots intercellularly, forming what is called the Hartig net. They also form a dense mycelial mat over the surface of invaded feeder roots, this is called the fungal mantle. Ectomycorrhizae are normally found on many trees such as pines, aspens and birches. Most ectomycorrhizal fungi are members of the Basidiomycetes. Ectomycorrhizae are less numerous than endomycorrhizae.

Endo-mycorrhizae

Endomycorrhizal fungi penetrate cortical cells of feeder roots intracellularly, or both intra- and intercellularly. They form a loose mycelial net over invaded roots. Endomycorrhizal hyphae coil within and between cortical cells producing swollen sac-like structures called vesicles, and haustorial-like structures called arbuscules within cortical cells. Hence, these fungi are often termed vesicular-

VA Mycorrhizae

arbuscular (VA) mycorrhizae. It is believed that arbuscules function in the exchange of nutrients between the host cell and the fungus.

Endomycorrhizae are divided into two groups, those caused by septate hyphae and those caused by non-septate hyphae. The latter are caused by members of the Phycomycetes (many in the genus *Endogone*), and compose the VA mycorrhizae. The VA mycorrhizae occur on more plant species than any other type. Several important crop plants which have VA mycorrhizae are: corn, redclover, soybean, cotton, tobacco, peas, apples, and citrus, (Gray, 1971).

Most Plant Roots Have Mycorrhiza

Gerdemann, 1971, states: "In nature, the mycorrhizal condition is the rule, the nonmycorrhizal condition, the exception. The roots of both cultivated and uncultivated plants are usually infected with mycorrhizal fungi." (Note: the underlining is the author's).

Mycorrhizae appear to be highly beneficial, if not required, for optimum growth and development of many plants.

The fungal mantle formed around roots by ectomycorrhizae appears to be a mechanical barrier to penetration by root pathogens. Also, the root cortical cells beneath the Hartig net also are resistant to infection. Furthermore, there is a different microbial population around mycorrhizal roots than nonmycorrhizal roots. This different microbial population probably competes with and may suppress the development of root pathogens. The biological control of root pathogens has been well reviewed by Marx, 1972.

Ecto-Mycorrhizae Protect Host Roots

Marx, 1973, reported that ectomycorrhizae were highly beneficial as biological deterrents in root infection of shortleaf pine seedlings by the fungal pathogen *Phytophthora cinnamomi*. This work supports the hypothesis that ectomycorrhizae are protective barriers to root infection by soil pathogens.

A great amount of research is currently being conducted with mycorrhizae, because it offers considerable potential for the biological control of root pathogens.

19-4.2 PARASITES AS BIOLOGICAL CONTROL AGENTS

The presence of numerous mycoparasites (fungi parasitic on other fungi) and bacterial parasites is well known. None-the-less, in the field, mycoparasites and bacterial parasites exert little economic population control of foliar pathogens. Therefore, if they are to be useful as biological control agents, they must be manipulated in some manner so as to maintain a high population. To illustrate,

Scherff, 1973, reported control of bacterial blight of soybean (*Pseudomonas glycinea*), by an isolate of *Bdellovibrio bacteriovorus*, a tiny, comma-shaped bacterium parasitic on larger gram-negative bacteria. Soybean foliage was inoculated with mixtures of *B. bacteriovorus* and *P. glycinea* at a ratio of 9:1, respectively. Nearly complete inhibition of *P. glycinea* was achieved. However, at ratios of 1:1, very little inhibition of *P. glycinea* was achieved.

A Bacterial Parasite

B. bacteriovorus is parasitic on a number of species of *Pseudomonas*, and may possibly be a future biological control agent.

The genus *Trichoderma* is perhaps the most well known of all mycoparasites. Species and strains of this genus have been reported to be antagonistic to a number of fungal pathogens, such as *Armillaria*, *Pythium*, *Rhizoctonia* and *Sclerotium*. Wells, et al., 1972, reported that an isolate of *T. harzianum* was pathogenic to *S. rolfsii* and *Botrytis cinerea*. Furthermore, these authors reported control of *S. rolfsii* on tomato, blue lupines and peanuts by applying inoculum of *T. harzianum* plus additional substrate over the soil surface and around the test plants, to assure a high population of the mycoparasite.

A fascinating study was conducted in France by Grosclaude, et al., 1973. They reported on using *T. viride* to prevent infection of young plum trees by the wood-inhabiting fungus *Stereum pupureum*, through pruning wounds. These authors applied *T. viride* as a spore suspension by spraying it on new wounds. Also, they adopted pruning shears so as to coat the cutting edge with a spore suspension which in turn coated the surface of the fresh wound.

These are only three examples of numerous studies concerning possible biological control procedures using parasites of plant pathogens.

Selected References

Erwin, D. C. 1973. Systemic fungicides: disease control, translocation, and mode of action. Annual Review of Phytopathology. 11:389-422.

First North American Conference on Mycorrhizae, Proceedings of. 1969. Mycorrhizae. Misc. publication 1189. U. S. Department of Agriculture - Forest Service.

Hollings, M. 1965. Disease control through virus-free stock. Annual Review of Phytopathology. 3:367-396.

Horsfall, J. G. 1956. Principles of fungicidal action. Cronica Botanica Co., Waltham, Mass.

Kincaid, R. R. 1960. Crop rotation and fallowing in relation to tobacco disease control. Botanical Review. 26:261-276.

Marsh, R. W., (ed). 1972. Systemic fungicides. John Wiley and Sons. New York.

Martin, H. 1964. The scientific principles of crop protection. Ed. 5. Edward Arnold and Co., London.

Marx, D. H. 1972. Ectomycorrhizae as biological deterrents to pathogenic root infections. Annual Review of Phytopathology. 10:429-454.

McClellan, W. D. 1973. Pest management requires national involvement. Nematology News Letter. Vol. 19:1-8.

National Academy of Sciences. 1972. Genetic vulnerability of major crops. National Academy of Sciences, Washington, D. C.

______________________________. 1972. Pest control strategies for the future. National Academy of Sciences, Washington, D. C.

Newhall, A. G. 1955. Disinfestation of soil by heat, flooding and fumigation. Botanical Review. 21:189-250.

Papavizas, G. C. 1966. Biological methods for the control of plant diseases and nematodes. pp. 82-94, In Pest control by chemical, biological, genetic, and physical means. A Symposium. U. S. Dept. of Agr. ARS 33-110.

Sharvelle, E. G. 1960. The nature and uses of modern fungicides. Burgess Publishing Co., Minneapolis.

_______________. 1969. Chemical control of plant disease. University Publishing, College Station, Texas.

Shertleff, M. C. 1966. How to control plant disease in home and garden. Ed. 2. Iowa State University Press, Ames, Iowa.

Stakman, E. C., and J. G. Harrar. 1957. Principles of plant pathology. The Ronald Press Co., New York pp. 412-418 and 431-458.

Stevens, R. B. 1960. Cultural practices in disease Control. pp. 357-429, In J. G. Horsfall and A. E. Dimond. Plant pathology. Vol. III. Academic Press, New York.

Torgeson, D., (ed.). 1969. Fungicides an advanced treatise. Vol. II. Chemistry and physiology. Academic Press, New York.

Walker, J. C. 1969. Plant pathology. Ed 3. McGraw-Hill Book Co., New York. pp 728-785.

Wittwer, S. H. 1974. Maximum production capacity of food crops. BioScience. 24:216-224.

CHAPTER 20

DISEASE CONTROL--QUARANTINES

20-1 INTRODUCTION

An important means of plant disease control is disease prevention through quarantines. In relation to plant pests, the term quarantine refers to the restraints and restrictions upon any and all commodities suspected of being carriers of plant pests. This may be on a local, state, federal, or international basis.

Prevent Entry Of New Pathogens

The purpose of a quarantine is to prevent the introduction of a disease agent into a disease-free area. Sometimes, it is the containment of a pathogen within the boundaries of its present area.

As has often been observed, plant quarantines do not lend themselves to experimental study. We cannot predict with any certainty, the extent of damage a given pathogen would cause if newly introduced into a country, or disease-free area. The question has often been raised; is it worth millions of dollars to try to prevent a new pathogen from entering this country just because scientists believe the pathogen is a dangerous one and potentially could cause severe losses? What proof is there that severe losses will occur? How can anyone be certain that the quarantine will prohibit the entry of the pathogen? These and similar questions have been asked by legislators and importers alike over the period of years during which our quarantine laws have been in existance. This has been and still is a matter of some controversy. It is true that quarantines are expensive.

Rule -- Host And Pathogen Coexistence

To obtain a better understanding of the necessity of plant quarantines, the first principle as given in the Introduction to Phytopathology is repeated: over a period of many years, naturally occurring infectious disease agents and their natural host plants establish a balanced relationship of coexistence. This balance of coexistence has often been upset by man's reliance upon plants for food. Travel by man and commodities between countries and across oceans, has resulted in the unwitting introduction of plant pathogens into new areas. When these exotic pathogens found a suitable host or hosts, the result was all too often a disastrous plant disease of epidemic proportions.

The new host or hosts had not been previously exposed to the pathogens, and therefore, had not evolved resistance. Past experiences with plant disease epidemics indicate that most of the devastating outbreaks of plant diseases among native plants result from the introduction of exotic plant pathogens.

Rule -- Devastation Through Introduction

There can be little doubt concerning the necessity of effective quarantines. The money paid for quarantines is but a fraction of the economic losses that would be suffered if plant pests were permitted free entry into this country and other countries.

Obviously, not all plant diseases lend themselves to restriction by quarantine. Then, what is the basis for establishing a quarantine? A plant quarantine must be founded upon sound biological knowledge of the pathogen and host. Furthermore, a quarantine is warranted only when, (1) there is a reasonable probability that it will prohibit the entry of the pathogen, (2) it is economically justifiable, and (3) its administration is feasible.

Justification Of Quarantine

A plant pathogen may be introduced into a new locality in a number of ways: (1) on the host, (2) as a contaminant in cargo or on the carrier, (3) in insects that are in or on cargo, (4) by travelers--on their person or in their luggage, (5) by flying insect vectors and birds, and (6) by air currents. Only pathogens with the first four modes of distribution lend themselves to control by quarantine measures, and the last two do not! For instance, it is unreasonable to think that it is possible to contain stem rust of wheat, because the spores are airborne for long distances. There is a high probability that pathogens that are airborne over long distances will eventually become distributed throughout a continent, and the world over, wherever a susceptible host is grown in an environment suitable for infection and disease development. Therefore, quarantines are aimed at diseases whose pathogens are carried with host parts or with nonhost carriers.

Mode Of Entry Of Foreign Pathogens

Quarantine Not Feasible For Some Pathogens

Congress passed the Plant Quarantine Act in 1912, authorizing the Secretary of Agriculture to condition, regulate, restrict and, if deemed necessary, to prohibit entry into the United States of nursery stock and other plants and plant products. The Plant Quarantine Act also applies to the interstate movements of plant commodities, with the regulation of such movement being left to individual states by mutual agreement. Inspection and certification of domestic products for export was authorized in 1926. Such inspection and certification was set up to meet plant quarantine laws of foreign countries. Additional legislation was enacted in the passage of the

Condition, Regulate, Restrict, Prohibit

Organic Act of 1944, regarding certification of domestic products for export purposes. The Federal Plant Pest Act was approved in 1957 and provided for additional controls over entry and movement of plant pests.

Inspection Service At 75 Ports Of Entry

The quarantine program in the United States is enforced by the Plant Quarantine Division of the Agricultural Research Service. Rainwater and Smith (1966), state: "Inspection service is maintained at 75 major ocean, Great Lakes, air and border ports of entry in the continental United States, Hawaii, Alaska, Puerto Rico, American Virgin Islands, Nassau, and Bermuda." A staff of over 500 inspectors examine cargoes of plants and plant products and if necessary, will post the required treatments or other safeguards to free them from pests. Incoming planes, ships, autos, trains, travelers, and their baggage are scrutinized along with mail and non-agricultural cargoes.

Postentry Quarantine

Imported plant propagative material is examined and treated by trained personnel at specially equipped inspection stations. A number of plant species are held in nurseries under a postentry quarantine for observation to see if any symptoms of disease develop. The state inspectors support the federal program by inspecting plants held under the postentry quarantine during the growing season.

20-2 INTRODUCED PLANT PESTS

An understanding of the potential dangers of introduced pathogens can be obtained from a backward glance at some foreign pathogens that have been introduced into the United States, and have caused epidemics. Some of the more "classical" of these diseases are: (1) white pine blister rust, (2) chestnut blight, (3) Dutch elm disease, (4) citrus canker, (5) potato black wart, and (6) golden nematode disease of potato.

20-2.1 WHITE PINE BLISTER RUST

Cronartium ribicola, the fungus that causes white pine blister rust was probably introduced into the United States in 1898 on nursery stock of eastern white pine brought into New England from Europe. As discussed in Chapter 19, this rust fungus produces uredial and telial stages on *Ribes* spp., and the aecial stage on five-needle pines. The aeciospores cannot reinfect pine, but can infect the *Ribes*. Therefore, the presence of both susceptible pines and *Ribes* is essential for the development and perpetuation of this disease.

Forests Of Primary Host

As fate would have it, there were several species of cultivated *Ribes* in addition to nine wild species growing everywhere the white pine grew. The alternate hosts were plentiful, and the white pine grew in dense forest stands along the northern states on the East Coast, in the Midwest, and on the West Coast. Last, but far from least, the environment was also favorable for infection and disease development.

Available Alternate Host

Favorable Environment

By 1910, the disease had spread from the New England states, west into Minnesota and south into North Carolina. Also, in 1910, the disease was introduced into British Columbia from infected white pine nursery stock imported from France. The disease then became widespread throughout the range of western white pine in British Columbia, Washington, Oregon, northern California, Idaho, and Montana. White pine blister rust has cost many millions of dollars in loss of stands of timber across the continent.

20-2.2 CHESTNUT BLIGHT

Chestnut blight, caused by the fungus *Endothia parasitica*, was first discovered in New York in 1904. Ascospores of the pathogen are spread locally by air currents, and the conidia are spread over long distances by birds. The fungus invades the tree and kills the bark and cambium. A canker is soon produced which girdles the branches and trunk, leading to death of the tree.

The disease quickly spread from New England through the Appalachians down into North Carolina and destroyed the entire chestnut forests of this country! Chestnut forests are now a thing of the past in North America. The dollar value represented by the loss of the entire chestnut forests cannot begin to be estimated.

Chestnut Forests Eliminated

Prior to its introduction into the United States, this fungal pathogen existed for many years in the Orient, where it was inconspicuous on native chestnuts and was not regarded as a destructive pathogen. Most likely, this was because oriental chestnut trees had probably become highly resistant through many years of natural selection. Once the pathogen was introduced into the United States, it became established in the highly susceptible chestnuts. The pathogen multiplied to epidemic stages very quickly because of the favorable climate and means of dissemination. There were no natural barriers against this pathogen, and no means of control. Resistant varieties may be available in the future, but this will not bring back the magnificent stands of chestnuts as they once were.

A valuable lesson is learned here. A pathogen, inconspicuous and of minor importance in its native habitat, may become an agent of destruction of the highest order when introduced into a new area.

Rule -- Introduced Minor Pathogen

20-2.3 DUTCH ELM DISEASE

The Dutch elm disease, caused by *Ceratocystis ulmi*, was found in Europe in 1920 and caused great damage, especially in Holland. The disease was first noticed in the United States in 1930, in Ohio. In the succeeding three years, outbreaks were found in New York, New Jersey, and Connecticut. By 1938, a number of cities reported serious outbreaks of the disease. The quarantine in the United States had been in effect for two decades, yet this disease was introduced from a foreign country! How had this pest slipped past the inspectors? What was the weak link in the quarantine that had permitted the introduction of this fungus? To understand these and other questions, it is important to understand the disease cycle.

Introduction Despite Quarantine!

The pathogen is an Ascomycete, which is transmitted from tree to tree principally by two species of bark beetles. One is the European bark beetle, itself an import from Europe prior to 1910, and the other is the native elm bark beetle. The bark beetles overwinter in dead and dying elm trees as larvae and immature adults. They emerge in the spring and migrate to healthy elm trees where they feed in the crotches of small twigs and branches. After mating, the females return to weakened elm trees, or trees of low vigor, and form galleries under the bark where they lay their eggs. Two such broods are usually laid each year, and the second brood is the overwintering stage. The fungus sporulates forming conidia freely in the galleries, and emerging beetles carry the sticky spores on their bodies. As the mature beetles feed upon healthy trees, the spores are deposited in the feeding wounds and the fungus begins to grow. Once inside the tree the fungus sporulates freely, producing conidia by budding. The spores are carried through the xylem vessels of the host and soon spread through most of the tree. Infected tissues become clogged, wilting of the crown foliage occurs, followed by chlorosis and defoliation of branches, and finally death of the entire tree.

It was discovered that elm logs had been imported from Europe several years prior to the outbreak of the disease. The ports of entry were through several Atlantic and Gulf of Mexico ports. Also, the logs were shipped by rail on open flatcars to veneer mills often located several hundred miles from the ports. It is estimated that over 12,000 miles of railway rights-of-way and 14 terminal areas were exposed. Undoubtedly, a number of the logs were still harboring the fungal pathogen and the adult borers were able to leave the logs to infect susceptible elms either at the terminals or the railway

Introduced On Inert Plant Parts

rights-of-way. Elm trees, although used for lumber, are highly prized as ornamental shade trees. The destruction of elm trees is almost complete in many eastern and midwestern communities of the United States. Control of this disease is based upon use of the systemic fungicide benomyl, and insecticides to control the bark beetles. However, this disease is still spreading, and at the present time threatens all of our native American elm trees.

This pathogen could have been stopped at the ports of entry if its disease cycle and insect vectors had been known. This is a prime example of a pest introduced through inert plant carriers.

20-2.4 CITRUS CANKER

Citrus canker, *Xanthomonas citri*, was introduced into Florida about 1912, probably from China or India, where it has been found and is probably native. At the time of its introduction, the bacterium was new to science, and so was the disease it caused. By 1914, the disease was recognized as a dangerous threat to the citrus industry. The pathogen seldom killed the trees, but cankers were formed on branches and scabby spots on the foliage and fruit.

Disease New To Science

The climatic factors most favorable for the spread of this disease are a relatively high temperature and precipitation during the period of most rapid growth of the citrus tree. These conditions do not occur in Arizona and California, which probably has been helpful in limiting the spread of this disease.

The appearance of citrus canker brought about immediate and thorough eradication measures (see Chapter 19), and now the pathogen has been eliminated from Florida. Since then, this pathogen has been intercepted numerous times at various ports of entry. For instance, in 1965 it was intercepted 245 times! Quarantine inspectors must always be on the lookout for this pathogen.

20-2.5 POTATO BLACK WART

This disease is caused by the fungus *Synchytrium endobioticum*, which attacks the subterranean stems and tubers of potato plants forming ugly, dark masses or warts on infected parts. The fungus persists for several years in infested soil. Dissemination of the pathogen is primarily by infected potato tubers. This disease was known in Europe before 1900 and became a major disease of potatoes there. Potato black wart first appeared in the United States about 1918. There is good circumstantial evidence that immigrant miners brought the disease from Europe in seed potatoes. The disease appeared in the miners' home gardens in West

Virginia and Maryland. After the disease was discovered, the infested areas were subject to strict quarantine that prevented the movement of potatoes. Even though a number of potato varieties were known to have a high level of resistance to this disease, they could still carry spores of the fungus to new localities. Therefore, all potatoes grown in the infested areas must be of the most resistant varieties and be consumed locally. This strict quarantine is still maintained. This is an excellent example of state and federal cooperation. No other infested areas in the United States have been found, although many of our potato varieties are highly susceptible.

Strict Local Quarantine

Local quarantine has served to restrict the spread of this potentially serious disease, and a strict quarantine on the national level has prevented further introduction.

20-2.6 GOLDEN NEMATODE DISEASE OF POTATO

The golden nematode, *Heterodera rostochiensis*, has been known in Europe since 1881, where it has since caused much damage throughout the northern potato-growing regions. In 1941, this pathogen was discovered in potato fields in Long Island. However, there is evidence to indicate that this pest was present in Long Island for 20 years prior to its discovery.

The golden nematode is so named because of its orange colored cysts. After potatoes are harvested from infected plants, thousands of cysts may remain behind in the soil from a single plant. Each cyst may contain up to 500 eggs from which larvae may hatch at any time up to 8 years or more. Dissemination is generally by means of the cysts either in the potato tubers, on dirty sacks and crates, or in infested soil clinging to boots and equipment.

By 1944, it was clear that the nematode was much more widespread on Long Island than was first thought. After World War II, a cooperative Federal-State Golden Nematode Control Project was set up. This group established that 17,151 acres were infested on Long Island (Spears, 1968).

A State quarantine was strictly enforced and the movement of potatoes and other crops within the quarantine area was closely supervised. The growing of potatoes and tomatoes on infested lands was prohibited. Undoubtedly, these measures, together with the fact that the fields were on an island, played a major role in prohibiting further spread of this pest. Growers whose income was in jeopardy were compensated for their losses, provided that they planted either small grains, field corn, or pasture crops.

Strict Local Quarantine

If the growers elected to plant high-value crops, such as vegetables, they were not elegible for compensation. In 1955, a field soil fumigation program was initiated, where two applications of fumigant were applied at 10 day intervals, and the soil was turned between applications. Today, the soil fumigants D-D and Vidden D may be applied in a split application of 45 gallons each for a total of 90 gallons per acre.

Golden Nematode Found In Canada

In Canada, the golden nematode was discovered in Newfoundland in 1962, and on Vancouver Island in 1965. The Canadian Government set up strict regulatory, quarantine, and soil fumigation practices. They also compensated the growers for their losses.

Golden Nematode Spread To Western New York And Delaware

In 1967, the golden nematode was found on a farm in western New York. This was the first infestation in the United States to be found away from Long Island. The pest had spread despite the local quarantine restrictions! It appears that some growers on Long Island, sold their farms to real estate developers and moved to western New York and brought the nematode with them. In January, 1969, golden nematode cysts were detected in routine soil samples taken in October, 1968, from two fields of a potato farm in New Castle County, Delaware.

In 1965, the golden nematode was intercepted by United States quarantine officials 101 times! One can only speculate on the damage that could occur if a constant vigil were not maintained to intercept these pests whenever and wherever they are brought into this country.

Selected References

Dean, H. S. 1953. Protection through quarantines. pp. 162-164, In Plant diseases. U. S. Dept. Agr. Yearbook.

Gram, E. 1960. Quarantines. pp. 313-356, In J. G. Horsfall and A. E. Dimond, (ed.), Plant pathology. Vol. III. Academic Press, New York.

Hunt, N. R. 1946. Destructive plant diseases not yet established in North America. The Botanical Review. 12:593-627.

Limber, D. P., and P. R. Frink. 1953. The inspection of imported plants. pp. 159-161, In Plant diseases. U. S. Dept. Agr. Yearbook.

McCubbin, W. A. 1944. Air-borne spores and plant quarantines. Scientific Monthly. 19:149-152.

______________. 1946. Preventing plant disease introduction. The Botanical Review. 12:101-139.

______________. 1950. Plant pathology in relation to federal domestic plant quarantines. U. S. Dept. Agr. Plant Disease Reporter, Supplement 191. pp. 67-91.

Rainwater, H. I., and C. A. Smith. 1966. Quarantines - first line of defense. pp. 216-224, In Protecting our food. U. S. Dept. Agr. Yearbook.

Spears, J. F. 1968. The golden nematode handbook survey, laboratory, control, and quarantine procedures. U. S. Dept. Agr. Agriculture Handbook No. 353.

Stakman, E. C., and C. M. Christensen. 1946. Aerobiology in relation to plant disease. The Botanical Review. 12:205-253.

______________, and J. G. Harrar. 1957. Principles of plant pathology. The Ronald Press Co., New York. pp. 339-411.

Walker, J. C. 1969. Plant pathology. Ed. 3. McGraw-Hill Book Co., New York. pp. 716-728.

LITERATURE CITED

Abeles, F. B., and H. E. Heggestad. 1973. Ethylene: an urban air pollutant. Journal of the Air Pollution Control Association. 23:517-521.

Applegate, H. G., and L. C. Durrant. 1969. Synergistic action of ozone-sulfur dioxide on peanuts. Environmental Science Technology. 3:759-760.

Beckman, C. H. 1966. Cell irritability and localization of vascular infections in plants. Phytopathology. 56:821-824.

______________, and S. Halmos. 1962. Relation of vascular occluding reactions in banana roots to pathogenicity of root-invading fungi. Phytopathology. 52:893-897.

______________, and G. E. Zaroogian. 1967. Origin and composition of vascular gel in infected banana roots. Phytopathology. 57:11-13.

Bennett, J. H., and A. C. Hill, 1973. Absorption of gaseous air pollutants by a standardized plant canopy. Journal of the Air Pollution Control Association. 23:203-206.

Bobrov, R. A. 1952. The anatomical effects of air-pollution on plants. pp. 129-134, In Proceedings of the National Air Pollution Symposium, 2nd. Pasadena, California.

Bracker, C. E. 1968. Ultrastructure of the haustorial apparatus of Erysiphe graminis and its relationship to the epidermal cell of barley. Phytopathology. 58:12-30.

Bradley, S. G. 1962. Parasexual phenomena in micororganisms. Annual Review of Microbiology. 16:35-52.

Breed, R. S., et al. 1957. Bergey's manual of determinative bacteriology. Ed. 7. The Williams and Wilkins Company, Baltimore.

Browning, J. A., and K. J. Frey. 1969. Multiline cultivars as a means of disease control. Annual Review of Phytopathology. 7:355-382.

Caldwell, R. M. 1966. Advances and challenges in the control of plant diseases through breeding. pp. 117-126, In Pest control by chemical, biological, genetic, and physical means. A Symposium ARS 33-110, U. S. Dept. Agr.

______________. 1968. Breeding for general and/or specific plant disease resistance. pp. 263-272, In K. W. Finlay and K. W. Shepherd, (ed.), Proceedings of the third international wheat genetics symposium, Australian Academy of Science. Butterworth and Co., Toronto.

Campbell, R. N. 1973. Personal communication.

Chen, T. A., and R. R. Granados. 1970. Isolation, maintenance, and infectivity studies of a plant-pathogenic mycoplasma, the causal agent of corn stunt diseases. Phytopathology (abstr.). 60:573.

Crosse, J. E., R. N. Goodman, and W. H. Shaffer, Jr. 1972. Leaf damage as a predisposing factor in the infection of apple shoots by Erwinia amylovora. Phytopathology. 62:176-182.

Cummins, G. B., G. A. Gries, S. N. Postlethwait, and F. W. Stearns. 1950. Textbook of intermediate plant science, Burgess Publishing Co., Minneapolis.

Cunningham, H. S. 1928. A study of the histologic changes induced in leaves by certain leaf-spotting fungi. Phytopathology. 18:717-751.

Daft, G. C., and C. Leben. 1972. Bacterial blight of soybeans: epidemiology of blight outbreaks. Phytopathology. 62:57-62.

Davis, R. E., and J. F. Worley. 1973. Spiroplasma: motile, helical micororganism associated with corn stunt disease. Phytopathology. 63:403-408.

___________, and R. F. Whitcomb, 1971. Mycoplasmas, Rickettsiae and Chlamydiae: possible relation to yellows diseases and other disorders of plants and insects. Annual Review of Phytopathology. 9:119-154.

___________, R. F. Whitcomb, and R. L. Steere. 1968a. Remission of aster yellows disease by antibiotics. Science. 161:793-794.

___________, ___________, and ___________. 1968b. Chemotherapy of aster yellows disease. Phytopathology (abstr.). 58:884.

Day, P. R. 1973. Genetic variability of crops. Annual Review of Phytopathology. 11:293-312.

Diener, T. O. 1963. Physiology of virus-infected plants. Annual Review of Phytopathology. 1:197-218.

___________. 1971. Potato spindle tuber "virus". IV. A replicating, low molecular weight RNA. Virology. 45:411-428.

___________, and R. H. Lawson. 1972. Chrysanthemum stunt, a viroid disease. Phytopathology (abstr.). 62:754.

Dienes, L. 1941. Isolation of L type growth from a strain of Bacteroides funduliformis. Proceedings of the Society for Experimental Biology and Medicine. 47:385-387.

Diamond, A. E., and J. G. Horsfall. 1960. Inoculum and the diseased population. pp. 1-22, In J. G. Horsfall and A. E. Diamond, (ed.), Plant Pathology. Vol. III, Academic Press, New York.

Dochinger, L. S., F. W. Bender, F. L. Fox, and W. W. Heck. 1970. Chlorotic dwarf of eastern white pine caused by an ozone and sulfur dioxide interaction. Nature (Lond.). 225:476.

Doi, Y., M. Teranaka, K. Yora, and H. Asuyama. 1967. Mycoplasma- or PLT group-like microorganisms found in the phloem elements of plants infected with mulberry dwarf, potato witches' broom, aster yellows, or paulownia witches' broom. Annals of the Phytopathological Society of Japan. 33:259-266.

Edward, D. G. 1960. p. 492, In J. B. Nelson (ed.), Biology of the pleuropneumonialike organisms. Annals of the New York Academy of Sciences. 79:305-758.

__________, and E. A. Freundt. 1956. The classification and nomenclature of organisms of the pleuropneumonia group. Journal of General Microbiology. 14:197-207.

__________, and __________. 1967. Proposal for Mollicutes as name of the class established for the order Mycoplasmatales. International Journal of Systemic Bacteriology. 17:267-268.

__________, and __________. 1969. Classification of the Mycoplasmatales. pp. 147-200, In L. Hayflick, (ed.), _The mycoplasmatales and the L-phase of bacteria_. Appleton-Century-Crofts, New York.

Faulkner, L. R., W. J. Bolander, and C. B. Skotland. 1970. Interaction of Verticillium dahliae and Pratylenchus minyus in verticillium wilt of peppermint: influence of the nematode as determined by a double root technique. Phytopathology. 60:100-103.

Fulton, R. W. 1964. Transmission of plant viruses by grafting, dodder, seed and mechanical transmission. pp. 39-67, In M. K. Corbett and H. D. Sisler, (ed.), _Plant virology_. University of Florida Press, Gainesville.

Georgopoulos, S. G., and C. Dovas. 1973. A serious outbreak of strains of Cercospora beticola resistant to benzimidazole fungicides in northern Greece. Plant Disease Reporter. 57:321-324.

Gerdemann, J. W. 1971. Fungi that form the vesicular-arbuscular type of endomycorrhiza. pp. 9-18, In _Mycorrhizae_. Proceedings of the first North American conference on mycorrhizae-April 1969. Misc. Publication 1189, U. S. Dept. Agr. Forest Service.

Ghemawat, M. S. 1968. Seedling and mature-plant resistance of wheat to powdery mildew. PhD. Thesis. Purdue University.

Giannotti, J., et al. 1971. Culture et transmission de mycoplasmes phytopathogenes. Comptes Rendus, Serie D. 272:1776-1178.

Granados, R. R., K. Maramorosch, and E. Shikata. 1968. Mycoplasma: suspected etiologic agent of corn stunt. National Academy of Sciences. 60:841-844.

Gray, L. E. 1971. Physiology of vesicular-arbuscular mycorrhizae. pp. 145-150, In Mycorrhizae. Proceedings of the first North American conference on mycorrhizae-April 1969. Misc. Publication 1189, U. S. Dept. Agr. Forest Service.

Grosclaude, C., J. Ricard, and B. Dubos. 1973. Inoculation of Trichoderma viride spores via pruning shears for biological control of Stereum purpureum on plum tree wounds. Plant Disease Reporter. 57:25-28.

Hampton, R. O., J. O. Stevens, and T. C. Allen. 1969. Mechanically transmissible mycoplasma from naturally infected peas. Plant Disease Reporter. 53:499-503.

Hanna, W. F. 1929. Studies in the physiology and cytology of Ustilago zeae and Sorosporium reilianum. Phytopathology. 19:415-442.

Hijmans, W., C. P. A. Van Boven, and H. A. L. Clasener. 1969. Fundamental biology of the L-phase of bacteria. pp. 67-143, In L. Hayflick, (ed.), The mycoplasmatales and the L-phase of bacteria. Appleton-Century-Crofts, New York.

Hildebrand, E. M. 1937. The blossom blight phase of fire blight, and methods of control. New York (Cornell) Agr. Exp. Sta. Mem. 207.

Hill, A. C. 1971. Vegetation: a sink for atmospheric pollutants. Journal of the Air Pollution Control Association. 21:341-346.

Hopkins, D. L., W. J. French, and H. H. Mollenhauer. 1973. Association of a rickettsia-like bacterium with phony peach disease. Phytopathology (abstr.). 63:443.

Huang, C. S., and A. R. Maggenti. 1969. Wall modifications in developing giant cells of Vicia faba and Cucumis sativus induced by root knot nematode, Meloidogyne javanica. Phytopathology. 59:931-937.

Hull, R. 1971. Mycoplasma-like organisms in plants. Review of Plant Pathology. 50:121-130.

Hussain, A., and A. Kelman. 1958. Relation of slime production to mechanism of wilting and pathogenicity of Pseudomonas solanacearum. Phytopathology. 48:155-165.

Ishiie, T., Y. Doi, K. Yora, and H. Asuyama. 1967. Supressive effects of antibiotics of tetracycline group on symptom development of mulberry dwarf disease. Annals of the Phytopathological Society of Japan. 33:267-275.

Jones, R. A. C., and B. D. Harrison. 1969. The behaviour of potato mop-top virus in soil, and evidence for its transmission by Spongospora subterranea (Wallr.) Lagerh. The Annals of Applied Biology. 63:1-17.

Kehr, A. E. 1966. Current status and opportunities for the control of nematodes by plant breeding. pp. 126-138, In Pest control by chemical, biological, genetic, and physical means. A Symposium. ARS 33-110, U. S. Dept. Agr.

Keil, H. L., and T. van der Zwet. 1972a. Recovery of Erwinia amylovora from symptomless stems and shoots of Jonathan apple and Bartlett pear trees. Phytopathology. 62:39-42.

__________, and ______________. 1972b. Aerial strands of Erwinia amylovora: structure and enhanced production by pesticide oil. Phytopathology. 62:355-361.

Kenaga, C. B., and R. P. Seven. 1965. Synergistic nematocide. United States patent No. 3,205,129.

Klieneberger-Nobel, E. 1960. L-forms of bacteria. pp. 361-386, In I. C. Gunsalus, and R. Y. Stainer, (ed.), The bacteria, a treatise on structure and function. Vol. I. Academic Press, New York.

__________________. 1967. Mycoplasma, a brief historical review. Annals of the New York Academy of Sciences. 143:713-718.

Kontaxis, D. G. 1962. Leaf trichomes as avenues for infection by Corynebacterium michiganense. Phytopathology. 52:1306-1307.

Kuiper, J. 1965. Failure of hexachlorobenzene to control common bunt of wheat. Nature (Lond.). 206:1219-1220.

Layne, R. E. C. 1967. Foliar trichomes and their importance as infection sites for Corynebacterium michiganense on tomato. Phytopathology. 57:981-985.

Leben, C. 1973. Survival of plant pathogenic bacteria. 2nd International Congress of Plant Pathology. University of Minnesota, Minneapolis, Minnesota (abstr.). 0326.

LeClerg, E. L. 1964. Crop losses due to plant diseases in the United States. Phytopathology. 54:1309-1313.

Lewis, S., and R. N. Goodman. 1965. Mode of penetration and movement of fire blight bacteria in apple leaf and stem tissue. Phytopathology. 55:719-723.

Lin, K. H. 1939. The number of spores in a pycnidium of Septoria apii. Phytopathology. 29:646-647.

Lister, R. M. 1969. Tobacco rattle, NETU, viruses in relation to functional heterogeneity in plant viruses. Federation Proceedings. 28:1875-1889.

___________, and A. F. Murant. 1967. Seed-transmission of nematode-borne viruses. The Annals of Applied Biology. 59:49-62.

Martin, G. W. 1961. Key to the families of fungi. pp. 497-517, In G. C. Ainsworth, Ainsworth and Bisby's dictionary of the fungi. Ed. 5. Commonwealth Mycological Institute, Kew Surrey.

Martin, J. T. 1964. Role of cuticle in the defense against plant disease. Annual Review of Phytopathology. 2:81-100.

Marx, D. H. 1972. Ectomycorrhizae as biological deterrents to pathogenic root infections. Annual Review of Phytopathology. 10:429-454.

__________. 1973. Growth of ectomycorrhizal and nonmycorrhizal shortleaf pine seedlings in soil with Phytophthora cinnamomi. Phytopathology. 63:18-23.

Mayol, P. S., and G. B. Bergeson. 1970. The role of secondary invaders in Meloidogyne incognita infection. Journal of Nematology. 2:80-83.

McCallan, S.E.A., and R. H. Wellman. 1943. A greenhouse method of evaluating fungicides by means of tomato foliage diseases. Contributions from Boyce Thompson Institute. 13:93-134.

McGee, Z. A., M. Rogul, and R. G. Wittler. 1967. Molecular genetic studies of relationships among mycoplasma, L-forms and bacteria. Annals of the New York Academy of Sciences. 143:21-30.

McWhorter, F. P. 1965. Plant virus inclusions. Annual Review of Phytopathology. 3:287-312.

Menser, H. A., and H. E. Heggestad, 1966. Ozone and sulfur dioxide synergism: injury to tobacco plants. Science. 153:424-425.

Miles, P. W. 1968. Insect secretions in plants. Annual Review of Phytopathology. 6:137-164.

Mills, W. D., and A. A. Laplante. 1951. Diseases and insects in the orchard. Cornell Extension Bulletin 711 (revised), Cornell University, Ithaca, New York.

Murant, A. F., and R. M. Lister. 1967. Seed-transmission in the ecology of nematode-borne viruses. The Annals of Applied Biology. 59:63-76.

National Academy of Sciences. 1972. Genetic vulnerability of major crops. National Academy of Sciences. Washington, D. C.

__________________________. 1968. Principles of plant and animal pest control. Volume I. Plant-disease development and control. Publication 1596. National Academy of Sciences. Washington, D. C.

Nayar, R. 1971. Etiological agent of yellow leaf disease of Areca cathecu. Plant Disease Reporter. 55:170-171.

Nelson, P. E., and R. S. Dickey. 1966. Pseudomonas caryophylli in carnation. II. Histological studies of infected plants. Phytopathology. 56:154-163.

O'Brien, T. P., N. Feder, and M. E. McCully. 1965. Polychromatic staining of plant cell walls by toluidine blue O. Protoplasma. 59:368-373.

Panopoulos, N. J., and M. N. Schroth. 1973. The role of motility in leaf penetration by bacterial pathogens. 2nd International Congress of Plant Pathology. University of Minnesota, Minneapolis, Minnesota. (abstr.). 0184.

Pitcher, R. S. 1963. Role of plant-parasitic nematodes in bacterial diseases. Phytopathology. 53:35-39.

Porter, D. M., and N. T. Powell. 1967. Influence of certain Meloidogyne species on fusarium wilt development in flue-cured tobacco. Phytopathology. 57:282-285.

Powell, N. T. 1971. Interactions between nematodes and fungi in disease complexes. Annual Review of Phytopathology. 9:253-274.

_____________, P. L. Melendez, and C. K. Batten. 1971. Disease complexes in tobacco involving Meloidogyne incognita and certain soil-borne fungi. Phytopathology. 61:1332-1337.

Presley, J. T., H. R. Carns, E. E. Taylor, and W. C. Schnathorst. 1966. Movement of conidia of Verticillium albo-atrum in cotton plants. Phytopathology. 56:375.

Rainwater, H. I., and C. A. Smith. 1966. Quarantines-first line of defense. pp. 216-224, In Protecting our food. U. S. Dept. Agr. Yearbook.

Rao, A. S., and M. K. Brakke. 1969. Relation of soil-borne wheat mosaic virus and its fungal vector, Polymyxa graminis. Phytopathology. 59:581-587.

Rasmussen, R. A. 1972. What do the hydrocarbons from trees contribute to air pollution. Journal of the Air Pollution Control Association. 22:537-543.

Robinson, R. A. 1971. Vertical resistance. Review of Plant Pathology. 50:233-239.

Rohde, R. A. 1965. The nature of resistance in plants to nematodes. Phytopathology. 55:1159-1162.

Saaltink, G. J., and A. E. Dimond. 1964. Nature of plugging material in xylem and its relation to rate of water flow in fusarium-infected tomato stems. Phytopathology. 54:1137-1140.

Sadasivan, T. S. 1961. Physiology of wilt disease. Annual Review of Plant Physiology. 12:449-468.

Samuel, G. 1927. On the shot-hole disease caused by Clasterosporium carpophilum and on the "shot-hole" effect. Annals of Botany. 41:375-404.

Schafer, J. F. 1971. Tolerance to plant disease. Annual Review of Phytopathology. 9:235-252.

Scherff, R. H. 1973. Control of bacterial blight of soybean by Bdellovibrio bacteriovorus. Phytopathology. 63:400-402.

Schlegel, D. E., S. H. Smith, and G. A. de Zoeten. 1967. Sites of virus synthesis within cells. Annual Review of Phytopathology. 5:223-246.

Semancik, J. S., and L. G. Weathers. 1972. Exocortis virus: an infectious free-nucleic acid plant virus with unusual properties. Virology. 47:456-466.

Smith, P. F. 1971. The biology of Mycoplasmas. Academic Press, New York.

Somerson, N. L., P. R. Reich, R. M. Chanock, and S. M. Weissman. 1967. Genetic differentiation by nucleic acid homology. III. Relationships among mycoplasma, L-forms and bacteria. Annals of the New York Academy of Sciences. 143:9-20.

Spears, J. F. 1968. The golden nematode handbook survey, laboratory, control, and quarantine procedures. U. S. Dept. Agr. Agriculture Handbook No. 353.

Suit, R. F., and H. W. Ford. 1950. Present status of spreading decline. Proceedings of the Florida State Horticultural Society. 63:36-42.

Szkolnik, M., and J. D. Gilpatrick. 1969. Apparent resistance of Venturia inaequalis to dodine in New York apple orchards. Plant Disease Reporter. 53:861-864.

Taylor, O. C., and F. M. Eaton. 1966. Suppression of plant growth by nitrogen dioxide. Plant Physiology. 41:132-136.

Thatcher, F. S. 1942. Further studies of osmotic permeability relations in parasitism. Canadian Journal of Research, Section C. XX, 283-311.

Tingey, D. T., R. A. Reinert, J. A. Dunning, and W. W. Heck. 1971. Vegetation injury from the interaction of nitrogen dioxide and sulfur dioxide. Phytopathology. 61:1506-1511.

Tinline, R. D., and B. H. MacNeill. 1969. Parasexuality in plant pathogenic fungi. Annual Review of Phytopathology. 7:147-170.

Tweedy, B. G. 1969. Elemental sulfur. pp. 119-145, In D. C. Torgeson, (ed.), Fungicides, an advanced treatise. Vol. II. Academic Press, New York.

Ushijama, R., and R. E. F. Matthews. 1970. The significance of chloroplast abnormalities associated with infection by turnip yellow mosaic virus. Virology. 42:293-303.

Vakili, N. G. 1967. Importance of wounds in bacterial spot (Xanthomonas vesicatoria) of tomatoes in the field. Phytopathology. 57:1099-1103.

Van der Plant, J. E. 1959. Some epidemiological consequences of systemic infection. pp. 566-573, In C. S. Holton, et al., (ed.), Plant Pathology problems and progress 1908-1958. The University of Wisconsin Press, Madison.

__________. 1963. Plant diseases: epidemics and control. Academic Press, New York.

Van Gundy, S. D. 1965. Factors in survival of nematodes. Annual Review of Phytopathology. 3:43-68.

van Kammen, A. 1972. Plant viruses with a divided genome. Annual Review of Phytopathology. 10:125-150.

Waggoner, P. E., and A. E. Diamond. 1954. Reduction in water flow by mycelium in vessels. American Journal of Botany. 41:637-640.

Walker, H. L., and T. P. Pirone. 1972. Particle numbers associated with mechanical and aphid transmission of some plant viruses. Phytopathology. 62:1283-1288.

Wallace, H. R. 1971. A reply to Father Timm. Nematology News Letter. 17:4-6.

Wheeler, H., and H. H. Luke. 1963. Microbial toxins in plant disease. Annual Review of Microbiology. 17:223-242.

Wood, F. A. 1971. Plant pathogenic air pollutants. Mimeograph. Presented at 1971 Short Course of College Teachers of Introductory Plant Pathology. Madison, Wisconsin.

__________. 1968. Sources of plant-pathogenic air pollutants. Phytopathology. 58:1075-1084.

Woolley, D. W., G. Schaffner, and A. Braun. 1955. Studies on the structure of the phytopathogenic toxin of Pseudomonas tabaci. Journal of Biological Chemistry. 215:485-493.

GLOSSARY

Abcission layer.
A layer of parenchyma cells in dicotyledons and gymnosperms that become separated from one another through dissolution of middle lamellae before leaf-fall. Such a layer is also formed by some dicotyledonous plants around the edges of local lesions causing the inner necrotic tissue to fall out, thus forming a shot-hole.

Acervulus (pl. acervuli).
An erumpent, or bed-like mat of hyphae bearing short conidiophores and conidia, and sometimes setae.

Actinomycetes.
A group of microorganisms apparently intermediate between bacteria and fungi.

Active ingredient.
The actual toxic agent present in a pesticide.

Acute disease or injury.
Attended with symptoms of some severity and coming speedily to a crisis; opposed to chronic.

Adventitious roots.
Roots that appear in an unusual place or position.

Aeciospore.
A dikaryotic spore produced in an aecium.

Aecium (pl. aecia).
A more or less cup-like sorus of the rust fungi, in which alternating aeciospores and disjunctor cells are borne in chains.

Agar.
A gelatin-like material extracted from seaweed. It is an ingredient used in making culture media on which microorganisms are grown.

Air pollutant.
Any factor mediated by the atmosphere that causes an unwanted effect. Often restricted to particles, gases, dusts, odors, radiation and noise.

Allele.
One of a pair of corresponding but contrasting genes on homologous chromosomes; also allelomorph.

Allelic pair.
A pair of genes, or series of genes, that occur at similar loci of homologous chromosomes.

Alternate host.
One or other of the two unlike hosts of a heteroecious rust.

Amphitrichous.
Denotes a bacterial cell with one or more flagella at each pole.

Antheridium (pl. antheridia).
A male gametangium.

Anthracnose.
Diseases or symptoms with limited necrotic lesions, caused by fungi producing asexual spores in acervuli.

Antibiosis.
Antagonistic association between two organisms or between one organism and a metabolic product of another organism, to the detriment of one of the organisms.

Antibiotic.
Damaging to life; especially a substance produced by one microorganism that inhibits or kills other microorganisms.

Antibody.
A specific protein which is produced in the tissues or fluids of animals that opposes the action of another substance, an antigen; the antigen is normally foreign to the animal tissue, and the antibody combines chemically to it. Antibodies are important defense mechanisms in vertebrates and some invertebrates against invasion by bacterial and viral pathogens.

Antigen.
A substance capable of stimulating formation of an antibody; a foreign protein.

Antiserum.
The blood serum of a warm-blooded animal that contains antibodies.

Apical growth.
Growth at the point or tip.

Apothecium (pl. apothecia).
An open ascocarp.

Appressorium (pl. appressoria).
A swelling on a fungus germ tube for attachment to host in early stage of infection, and from which a minute infection peg usually grows to penetrate the cuticle and epidermal cell of the host; found especially in anthracnose and rust fungi.

Ascocarp.
A fruiting body containing asci.

Ascogonium (pl. ascogonia).
A female gametangium of the Ascomycetes.

Ascomycetes.
One of the main classes of fungi; characterized by bearing sexual spores in asci.

Ascospore.
A sexually produced spore formed in an ascus.

Ascus (pl. asci).
A minute sac-like structure typically containing eight ascospores, which are formed in a sexual manner and are characteristic of the class Ascomycetes.

Aseptate.
Without cross-wall.

Asexual reproduction.
Reproduction without the union of two nuclei.

Assimilative phase.
See vegetative phase.

Attachment disc.
An organ formed upon the germination of most mistletoe seeds; it is disc-shaped and adheres to the branch of a susceptible host, a primary haustorium forms on the underside of the disc and penetrates through a lenticel or axillary bud into the vascular system.

Autoecious.
Condition whereby a parasitic fungus is able to complete its life-cycle on one host; especially of the rusts.

Avirulent.
Non-pathogenic.

Axil.
The angle formed by the petiole of a leaf and the stem above its attachment.

Bactericide.
A chemical that kills or inhibits bacteria.

Bacteriology.
The science of bacteria.

Bacteriophage.
A virus with the ability to infect bacteria, and eventually cause lysis. Also known as phage.

Bacterium (pl. bacteria).
A unicellular microscopic plant that lacks chlorophyll and multiplies by fission.

Basidiocarp.
A fruiting body which bears basidia.

Basidiomycetes.
One of the main classes of fungi; characterized by bearing sexual spores on basidia.

Basidiospore.
A sexually produced spore, formed on the outside of a basidium; also called a sporidium in the smut fungi.

Basidium (pl. basidia).
A minute structure which bears on its surface typically four basidiospores, which are formed in a sexual manner and are characteristic of the class Basidiomycetes. Also see promycelium.

Biennial plant.
One that develps seed at the end of its second year of growth and then dies.

Binomial.
See Latin binomial.

Binucleate cell.
A cell having two nuclei. Also see dikaryotic.

Biocide.
A compound that is toxic to all forms of life.

Biological control.
Disease control by means of predators, parasites, competitive microorganisms, and decomposing plant material, which restrict or reduce the population of the pathogen.

Biology.
The science of living organisms.

Biotype.
A population of life forms that are identical in all inheritable traits; a subdivision of a race of fungi.

Blast, blasting.
A failure to produce fruit or seeds, or the sudden death of buds, flowers or young fruit.

Blight.
A disease with sudden, severe leaf damage and often with general killing of flowers and stems.

Blotch
Foliage symptom involving necrosis of large, irregular areas of leaf tissue.

Botany.
The science of plants.

Bract.
A reduced leaf associated with a flower.

Breaking.
Disease symptom caused by a virus; loss of flower or fruit color in a variegated pattern.

Broad-spectrum fungicide.
A fungicide that controls a wide range of diseases when applied correctly.

Budding.
Arising by extrusion from a parent cell; an asexual method of reproduction.

Bulb.
An underground storage organ of plants, the major part of which is composed of fleshy leaf bases.

Callus.
Tissue overgrowth around a wound or canker. Develops from cambium or other exposed meristem.

Calyx.
The sepals, considered collectively.

Cambium.
A meristematic layer of cells lying between the xylem and phloem; forms additional xylem and phloem elements in the process known as secondary thickening. The cambium of a vascular bundle is "facicular" cambium.

Canker.
A lesion on a stem, often swollen or sunken, surrounded by living tissues.

Capsid.
The protective sheath of protein that surrounds the inner core of nucleic acid of a virus.

Cardinal temperatures.
The minimum, optimum, and maximum termperatures for fungus spore germination.

Carrier.
Infected plant showing no marked symptoms, but is a source of infection for other plants.

Causal agent.
An agent (bacterium, fungus, virus, nematode, etc.) that produces a given disease.

Certified seed.
Seed produced and sold under inspection control to maintain varietal purity and freedom from pathogens, insect pests, and weed seed.

Chlamydospore.
A thick-walled asexual resting spore formed by the rounding-off of an ordinary hyphal cell; also see teliospore.

Chlorosis.
Yellowing of normally green tissue due to the destruction of the chlorophyll or the partial failure of the chlorophyll to develop.

Chromosome.
Microscopic, chromatin body detectable in the nucleus of a cell at the time of nuclear division; contain the units of inheritance called genes.

Chronic disease.
Lasting a long time; a disease of long duration; opposed to acute.

Circulative.
Characterizes those plant viruses that are acquired through mouthparts of insect vectors, accumulated internally, passed through the insect tissues, and then trasmitted into plants through the vectors' mouthparts.

Cleistothecium (pl. cleistothecia).
A completely closed ascocarp.

Clone.
The descendants produced vegetatively from a single parent plant.

Coalesce.
The coming together of two or more lesions to form a large spot or blotch.

Coccus (pl. cocci).
A spherical bacterium.

Coenocytic.
Refers to the condition whereby nuclei are embedded in the cytoplasm without being separated by cross-walls.

Complementary gene action.
Two or more genes acting together that cause the expression of a character.

Conidiophore.
A specialized hypha bearing conidia.

Conidium (pl. conidia).
A spore formed asexually, usually at the tip or side of a specialized hypha (conidiophore). Also conidiospore.

Conjugation.
In fungi--the sexual reproduction following fusion of similar gametes. In bacteria--a process whereby two compatible bacteria come in contact and a fraction of genetic material is passed from one bacterium to the other.

Conk.
A forestry term for fruiting bodies of wood-rotting fungi formed on tree stumps, branches, or trunks.

Cork.
Protective tissue of dead, impermeable cells formed by phellogen (cork cambium), which, with increase in diameter of young stems and roots, replaces the epidermis.

Cork cambium.
See phellogen.

Corm.
Organ of vegetative reproduction. A short, erect underground stem, bulb-like but solid, e.g. crocus and gladiolus.

Cotyledon.
A seed leaf, or the first leaf of a seedling.

Crossing-over
The exchange of corresponding segments between each member of paired (homologous) chromosomes.

Cross-protection.
The condition whereby a normally susceptible host is infected with an avirulent pathogen (usually a virus), and thereby becomes resistant to the infection by a virulent pathogen.

Crowd disease.
A disease that is likely to reach epidemic proportions only when the host plants are crowded together into fields.

Culturing.
The propagation of organisms on nutrient media or living plants.

Cuticle.
In higher plants, forms a continuous superficial, non-cellular layer over aerial parts, broken only by stomata and lenticels; protects against mechanical injury, but chief function is preventing excessive water loss; secreted by epidermis.

Cutin.
A waxy substance that waterproofs the walls of epidermal cells of many plants.

Cyst.
A protective capsule.

Cytoplasm.
All of the protoplasm of a cell excluding the nucleus.

Damping-off.
Seed decay in the soil or seedling blight, usually caused by soil-borne fungi.

Decay.
The disintegration of plant tissue.

Defoliate.
To strip or become stripped of leaves.

Desiccation.
Drying out.

Dieback.
Progressive death of branches or shoots beginning at the tips.

Dikaryotic.
Pertaining to a cell that contains a pair of closely associated nuclei, each usually derived from a different parent cell, usually written as (n + n). During growth and cell division of dikaryotic cells, the nuclei undergo simultaneous division perpetuating the dikaryotic condition. A specialized binucleate condition.

Dioecious.
Male and female organs on different individuals of the same species.

Diploid.
The condition whereby each nucleus has the chromosomes in pairs, the members of which are homologous, so that twice the haploid number is present; often referred to as the double (2n) number of chromosomes.

Disease.
See plant disease.

Disease complex.
A plant disease that is caused by the interaction of two or more species or types of pathogens.

Disease-cycle.
The sequence of events that occur between the time of infection and the final expression of disease.

Disease endurance.
See disease tolerance.

Disease escape.
Capacity of a susceptible plant to avoid infectious disease through some character of the plant or other factor that prevents infection by the pathogen.

Disease range.
The geographic distribution of a disease.

Disease tolerance.
The ability of a host plant to survive and give satisfactory yields at a level of infection that causes economic loss to other varieties of the same host species.

Dissemination.
The transport or spread of inoculum.

Dominant gene.
One member of an allelic pair of genes that has the quality of manifesting itself wholly or largely to the exclusion of the other member.

Dormant.
In a resting condition, alive but with relatively inactive metabolism. It may be independent of external conditions, or imposed by environmental conditions that are unfavorable for growth.

Dry rot.
See rotting.

Ectomycorrhiza (pl. ectomycorrhizae).
A type of mycorrhiza whose fungi penetrate cortical cells of feeder roots intercellularly, forming what is known as the Hartig net. They also form a dense mycelial mat over the surface of invaded roots, called the fungal mantle.

Ectoparasite.
Term used to characterize plant parasitic nematodes that feed on the root surface and normally do not enter the root tissue of the host. It is also descriptive of many parasitic seed plants that attach themselves to their hosts with a specialized absorbing root, and form an aerial system of stems and leaves.

Egg.
A female gamete. In nematodes an egg may contain a zygote or a larva.

Elementary bodies.
See Mycoplasma.

Enation.
A disease symptom whereby there is a mass of hypertrophied tissue on the surface of a leaf or stem.

Encyst.
To form a cyst.

Endemic.
Natural to a country, geographical region or area.

Endodermis.
A thick-walled layer of cells that surrounds the vascular tissues of the roots.

Endomycorrhiza (pl. endomycorrhizae).
A type of mycorrhiza whose fungi penetrate cortical cells of feeder roots intracellularly, or both intra- and intercellularly. They also form a loose mycelial net around invader roots.

Endoparasite.
Term used to characterize plant parasitic nematodes that invade, feed, and become established in the host root tissue or permanently attach themselves to it.

Endospore.
An asexual spore, developed within a cell, highly resistant, thick-walled, as in some bacteria. Capable of surviving exposure to high temperatures or for long periods of dryness. The only plant parasitic bacteria to form endospores are in the genus *Streptomyces*.

Entomology.
The science of insects.

Epidemic.
The extensive development of a disease in a geographical area or community, also called an epiphytotic.

Epidemiology.
The branch of plant pathology that treats of disease in plant populations.

Epidermis.
The superficial layer of cells.

Epinasty.
Downward curling or bending of a leaf-blade.

Epiphyte.
A plant attached to another plant, not growing parasitically upon it, but merely using it for support.

Eradicant fungicide.
Fungicide that destroys a fungus in an established lesion.

Eradication.
Control of disease by eliminating the pathogen after it is already established.

Ergot.
(1) Disease of inflorescence of cereals and wild grasses caused by the fungus *Claviceps purpurea*. (2) Dark sclerotium developing in place of a healthy grain in a diseased inflorescence. Ergots contain substances poisonous to man and animals. Eating of bread made from ergot infected cereals, and of ergot infected fodder by animals, gives rise to a disease called ergotism. The alkaloid poisons (ergotamine, etc.), are used in medicine as drugs.

Erumpent.
Breaking through the host tissue.

Escape.
See disease escape.

Etiolation.
A yellowish condition of normally green tissue and elongation of stems caused by insufficient light.

Evolution.
Irreversible change in the characteristics of organisms, occurring in successive generations related by descent.

Extracellular.
Outside the cells.

Exudate.
A liquid discharge (often in droplet form) from diseased or healthy tissue.

Factor.
See gene.

Facultative parasite.
An organism that is normally saprophytic but may live parasitically under certain conditions.

Facultative saprophyte.
An organism that is normally parasitic but may live saprophytically under certain conditions.

Fallow.
Maintenance of land with no plant growth. Weeds are controlled by cultivation or the application of herbicides and no crop is planted.

Fasciation.
Plant disease symptom expressed as flattened and usually curved or curled stems.

Filamentous.
Threadlike, filiform, or composed of threads.

Fission.
Asexual reproduction, expecially in microorganisms, in which the parent divides into two equal daughter organisms.

Flagellum (pl. flagella).
A thread-, hair-, or whip-like structure that serves to propel a motile cell.

Free-living.
Of an organism that lives freely, not as a parasite.

Frost resistance.
Ability of plants to withstand a lower temperature than they originally could, by being conditioned through repeated exposures to temperatures just above the freezing point; not all plants exhibit frost resistance. Same as hardiness.

Fruiting body.
Fungus structure containing or bearing spores.

Fumigant.
A volatile pesticide, dispersing and often killing pests by means of vapor.

Fungicide.
A chemical agent that kills or inhibits fungi.

Fungi Imperfecti.
A class of fungi for which no sexual stage has been found.

Fungistatic.
A chemical agent that inhibits fungal growth without killing the fungus.

Fungus (pl. fungi).
An organism with no chlorophyll, reproducing by sexual or asexual spores, usually with mycelium with well-marked nuclei.

Gall.
Outgrowth or swelling, often more or less spherical, of unorganized plant cells, usually a result of attack by insects, bacteria, fungi or nematodes.

Gametangium (pl. gametangia).
A structure that contains gametes. Gametangia are the sex organs of fungi.

Gamete.
A sex cell or a sex nucleus that fuses with another in sexual reproduction.

Gene.
The smallest unit of inheritance; genes are located in a linear fashion on chromosomes, and may act independently or by interaction with one or more other genes together with the environment in the development of a character.

General resistance.
A uniform resistance against all races of a pathogen. The level of resistance is usually only moderate and often influenced by the temperature. It is usually inherited polygenically. Same as horizontal resistance.

Genome.
One haploid set of chromosomes with the genes they contain.

Genus.
See Latin binomial.

Germination.
The process of starting growth, as in a seed or spore.

Germ tube.
Hypha produced by a germinated fungus spore.

Giant cells.
See nurse cells.

Gram stain.
A stain used in classifying bacteria. The term gram-negative denotes bacteria not being stained purple, while the term gram-positive denotes bacteria being stained purple.

Guard cells.
Cells surrounding the slit-like opening (stoma) of a leaf or stem.

Gummosis.
Exuding of sap, gum, or latex from inside of a plant. Often caused by a pathogen within the plant, or by unfavorable environmental conditions.

Guttation.
Loss of water by exudation through hydathodes.

Haploid.
Having a single set of unpaired chromosomes in each nucleus, often referred to as the reduced (n) number of chromosomes.

Hardiness.
See frost resistance.

Haustorium (pl. haustoria).
(1) A special filament (hypha) of a fungus that penetrates a cell of a host plant and absorbs food from it. (2) A root-like absorbing organ connecting a parasitic seed plant to the food conducting system of its host.

Herbicide
A chemical that is toxic to weeds or weed seeds.

Heteroecious.
Condition whereby a parasitic fungus needs two host species for the completion of its life-cycle; especially of the rusts.

Heterokaryosis.
Refers to the multinucleate condition of hyphae or individual cells of a hypha, wherein the nuclei possess different genetic factors; this condition can result when hyphal fusion occurs between similar, but genetically different fungi.

Heterothallic.
Refers to fungi that require the union of two compatible thalli for sexual reproduction.

Heterozygous.
Having a pair of unlike genes.

Holocarpic.
Refers to an organism whose thallus is entirely converted into one or more reproductive structures.

Homologous chromosomes.
Pairs of chromosomes, or parts thereof, that contain identical sets of loci and have a strong attraction for each other during early stages of meiosis and undergo pairing.

Homothallic.
Refers to fungi in which sexual reproduction takes place in a single thallus, meaning that the fungus is self-compatible.

Homozygous.
Having paired genes alike; of pure breed.

Horizontal resistance.
See general resistance.

Hormone.
A growth regulator. Often termed an auxin.

Host
Any plant attacked by a parasite or pathogen.

Host indexing.
See indexing.

Host range.
The various kinds of plants that may be attacked by a pathogen.

Hyaline.
Colorless, transparent.

Hybridization.
The production of offspring from two different parent races, varieties, or species of genera.

Hybrid vigor.
The increased vigor of progeny from the cross of two different parental lines, which results in better growth than either parent.

Hydathode.
Water-excreting gland or pore occurring on the edges or tips of the leaves of many plants.

Hymenium (pl. hymenia).
A fertile layer consisting of asci or basidia.

Hyperparasitism.
The parasitism of a parasite by another organism. Also see mycoparasitism.

Hyperplasia.
The production of an abnormally large number of cells.

Hypersensitivity.
The sudden localized host response to the presence of a pathogen that involves the quick death of host cells around the pathogen and the restriction and/or death of the pathogen.

Hypertrophy.
An abnormal increase in the size of cells.

Hypha (pl. hyphae).
A single thread or filament of a fungus thallus, composed of one or more tubular or cylindrical cells with thin transparent walls; increases in length by apical growth and gives rise to new hyphae by lateral branching.

Hypobiotic.
The condition of reduced metabolism.

Hypoplasia.
The production of a subnormal number of cells.

Hypotrophy.
A subnormal decrease in the size of cells.

Immunity.
Condition of the host whereby it is completely free from attack and injury by an infectious pathogen.

Imperfect stage.
The period of life of a fungus during which spores are produced asexually.

Incubation period.
Interval of time between infection and appearance of symptoms.

Indexing.
A procedure to determine whether a given plant is a carrier of a pathogenic virus. Material is taken from one plant and transferred to another plant that will develop characteristic symptoms if affected by the virus in question. Also host indexing.

Indigenous.
Of organisms that are native to a particular area, not introduced.

Infection.
The establishment of the pathogen in the host after entering.

Infection cushion.
An aggregation of hyphae, usually closely knit, that takes on a dome-shaped appearance, and functions essentially as a multiple appressorium. A structure often formed by soil-borne fungi on the surface of a host root.

Infection peg.
See penetration peg.

Infective.
Capacity of a pathogen to attack a living organism.

Infestation.
The introduction of a pathogen into the environment of the host; external contamination by a pathogen, as opposed to infection.

Injury (of plants).
A harmful disorder that occurs swiftly, over a short time period and is a one-time phenomenon. It is often caused by agents such as fire, wind, hail, lightning, animals, machinery and chemicals.

Inoculate.
Placement of a pathogen in contact with, or inside, a host plant or plant part.

Inoculum.
Portions of a pathogen capable of being disseminated and causing infection. For fungal pathogens, this includes those portions which after dissemination form structures (zoospores, germ tubes, appressoria, penetration pegs, etc.) that lead to infection.

Inoculum potential.
The number of independent infections and amount of tissue invaded per infection that may occur in a population of susceptible hosts at any time or place.

Intercalary.
Formation within hyphal segments; as opposed to hyphal tips.

Intercellular.
Between cells.

Interveinal chlorosis.
The condition whereby the tissue between veins becomes chlorotic.

Intracellular.
Within cells.

In vitro.
Term expressing under artificial conditions; outside the host.

In vivo.
Term expressing under natural conditions; inside the host.

Isolation.
Process of placing an organism in pure culture.

Larva (pl. larvae); (of a nematode).
An immature stage between the embryo and adult.

Latent infection.
Condition of a pathogen that has invaded, but not substantially colonized the host, and the pathogen may be inactive or only slightly active. The pathogen may resume growth and subsequently colonize the host forming necrotic lesions; also when an infectious disease agent infects a host that it does not incite disease in.

Latin binomial.
The scientific name of an organism. It is composed of two names, the first designating the genus, and the second, the species.

Lenticel.
A small raised pore, usually elliptical in shape, formed in woody stems when the epidermis is replaced by cork; packed with loosely arranged cells and allowing exchange of gasses between the interior of the stem and the atmosphere.

Lesion.
A localized area of diseased tissue of a host plant.

L-form.
See L-phase.

Life-cycle.
The progressive series of changes undergone by an organism or lineal succession of organisms, including sexual and asexual stages. In many fungi, this is the stage or series of stages between one spore form and the development of the same spore again.

Lignification.
The formation of lignin in cell walls through the transformation or impregnation of elements of the cell wall.

Linkage.
Tendency of genes of the same chromosome to be passed on together to the offspring.

Linked genes.
Genes that determine characters, but are located in the same chromosome. The group of genes on the same chromosome are called a linkage group.

Local infection.
Infection involving only a limited part of the host.

Local lesion.
An area of diseased tissue that is small and restricted in size.

Lophotrichous.
Denotes a bacterial cell with more than one flagellum at one pole.

L-phase.
Independent growth variants of bacteria that lack a rigid cell wall and have a potential reversibility to the parent strain. The L stands for Lister Institute in London.

Lumen.
The cavity or space within the walls of a cell.

Lysis.
The cellular breakdown or destruction of cell walls.

Masked.
The condition whereby a host is infected but does not exhibit disease symptoms because the environment is not favorable for disease expression and/or development.

Medium (pl. media).
Substrate of organic or inorganic composition used for growing micro-organisms in the laboratory; may be liquid or solid.

Meiosis.
Two nuclear divisions occurring in quick succession, one of which is reductional; the final result being four haploid (n) nuclei.

Meristem.
Undifferentiated tissue whose cells undergo division and differentiate into specialized tissues.

Mesophyll.
The parenchyma cells of a leaf between the epidermal layers.

Microorganism.
A minute organism.

Miscroscopic.
Too small to be seen except with the aid of a microscope.

Mildew.
Plant disease caused by a fungus whose mycelium grows over the surface of the host.

Mitosis.
The usual process of nuclear division. Each chromosome duplicates and pulls apart resulting in each daughter nucleus containing an identical complement of chromosomes.

Modifying gene.
A gene that affects the expression of another nonallelic gene; may cancel or promote the effect.

Molds.
Generally saprophytic fungi with conspicuous mycelium or spore masses.

Mollicutes.
A class of primative plants containing the order Mycoplasmatales.

Molt.
The shedding off of the cuticle.

Monoculture.
The growing of a single crop.

Monogenic.
A character governed by a single gene.

Monokaryotic.
Pertaining to a cell or cells that contain one nucleus.

Monotrichous.
Denotes a bacterial cell with a single polar flagellum.

Mosaic.
A term applied to variegated patterns of shades of greens and yellows in normally green leaves; caused by a disarrangement of the chlorophyll content.

Mottle.
Foliar symptom of irregular light and dark areas.

Multinucleate cell.
A cell having more than two nuclei.

Multiple-cycle disease.
A disease whose pathogen possesses a secondary infection cycle.

Mummy.
Dried shriveled fruit resulting from infection by a pathogen.

Mutation.
A sudden departure from the parent type, as when an individual differs from its parents in one or more inheritable characteristics; also, an individual species, or the like, resulting from such a departure.

Mycelium (pl. mycelia).
A mass of fungus hyphae; the vegetative body of a fungus.

Mycology.
The science of fungi.

Mycoparasitism.
The parasitism of one fungus by another fungus. Also see hyperparasitism.

Mycoplasma.
A genus assigned to the class Mollicutes, order Mycoplasmatales. Mycoplasma species are soft, pleomorphic, unicellular organisms that lack cell walls. The word mycoplasma means "fungus form", which is descriptive of the branching filaments they often assume. These filaments break up into tiny round cells called "elementary bodies" ranging in size from 125 to 250 millimicrons. They also form "large bodies" which approach the dimensions of rigid bacteria.

Mycorrhiza (pl. mycorrhizae).
The symbiotic association of mycelium of certain fungi with the roots of higher green plants and living in close association with root cells, especially those at the root surface.

Natural selection.
The natural process by which the best-fitted types of organisms survive and produce offspring, the unfit being eliminated; the theory of Darwin.

Necrosis.
Death of plant cells, usually resulting in the affected tissue turning dark.

Necrotic.
An adjective, dead.

Nectary cells.
Specialized glands in many insect pollinated flowers, which secrete drops of nectar.

Nematicide.
A chemical that kills or inhibits nematodes.

Nematode.
Unsegmented roundworms, most are tiny to microscopic, usually thread-like, free-living or parasites of plants or animals.

Nonpathogenic.
Incapable of inciting disease.

Nonpersistent
A term formerly used to characterize those plant viruses that insect vectors were capable of transmitting quickly after being acquired, but this capability was soon lost unless the virus was re-acquired.

Nonrestricted virus.
A virus that is capable of inducing infection when introduced into any type of living cell of its host.

Non-septate.
Without cross-walls.

Nucleus (pl. nuclei).
Body containing the chromosomes, present in living cells of plants and animals, bounded by a nuclear membrane (lacking in bacteria).

Nurse cells.
Abnormally large, multinucleate host cells formed by repeated mitosis and enlargement without cell division of several parenchyma cells near the heads of certain endoparasitic nematodes, and in response to the nematodes which require these cells for further development. Also giant cells.

Obligate parasite.
A parasite that can develop only in living tissues.

Obligate saprophyte.
An organism that can develop only on dead organic matter and inorganic materials.

Oogonium (pl. oogonia).
A female gametangium.

Oosphere.
A large, naked, non-motile female gamete.

Oospore.
A thick-walled spore that develops from an oosphere through fertilization or parthenogenesis.

Ostiole.
A pore-like opening of a perithecium or pycnidium through which their spores escape.

Ovary.
A female reproductive organ that contains or produces the egg.

Ozone (O_3).
A highly reactive form of oxygen. In relatively high concentrations, ozone may injure plants. A common air pollutant.

Palisade layer.
Layer of elongated parenchyma cells that is found just beneath the upper epidermis of leaves, and contains chloroplasts.

PAN.
The abbreviation for peroxyacyl nitrates - toxic air pollutants formed by photochemical reactions in the air during daylight.

Pandemic.
An epidemic that occurs over an extended geographical area or areas.

Paraphysis (pl. paraphyses).
A sterile hyphal element in a hymenium.

Parasexualism.
The process of genetic recombination occurring within vegetative heterokaryotic hyphae.

Parasite.
An organism that is partially or wholly dependent upon living tissue for its existence.

Parasitism.
The situation whereby one organism (the parasite) grows at the expense of another (the host).

Parenchyma.
Tissue composed of thin-walled cells that often have intercellular spaces between them.

Pathogen.
A specific agent that incites infectious disease.

Pathogenesis.
The period in disease from the time of infection to the final reaction in the host.

Pathogenic.
Causing, or capable of causing disease.

Pathogenicity.
The capacity of a pathogen to incite disease.

Penetration.
The entrance of the parasite into the host.

Penetration peg.
A thin strand of hypha developing on the underside of an appressorium and penetrating between closed guard cells into the substomatal cavity, through the cuticle, or through the cuticle and epidermal cell wall into an epidermal cell. Also infection peg.

Perfect stage.
The period of life of a fungus during which spores are produced sexually.

Perithecium (pl. perithecia).
A closed ascocarp with a pore, an ostiole, at the top and a wall of its own.

Peritrichous.
Denotes a bacterial cell with many flagella around its surface.

Persistent.
A term formerly used to characterize those plant viruses that after being acquired by insect vectors could not be transmitted until after a latent period, but after this period maintained the ability to transmit the viruses over a period of many days.

Pesticide.
A chemical that destroys pests (fungi, insects, nematodes, weeds, etc.).

Phellogen.
Cambium that gives rise to external cork and sometimes also to internal phelloderm. Also called cork cambium.

Phloem.
Vascular tissue that conducts synthesized foods through the plant; characterized by the presence of sieve-tubes.

Phloem dependent virus.
A virus that is capable of inducing infection only when introduced into a phloem cell of its host.

Physiologic race.
One or a number of biotypes of the same species and variety that possess similar characteristics of pathogenicity.

Physiologic specialization.
The occurrence of physiologic races within a species or variety of a pathogen.

Phytopathology.
The science of plant disease. Also plant pathology.

Phytotoxic.
Toxic to green plants.

Plant disease.
(1) A harmful alteration of the normal physiological and biochemical development of a plant. (2) An alteration in one or more of the ordered, sequential series of physiological processes, culminating in a loss of coordination of energy utilization in a plant as a result of the continuous irritations from the presence or absence of some factor or agent.

Plant pathology.
The science of plant disease. Also phytopathology.

Plasmodesma (pl. plasmodesmata).
Cytoplasmic strands through cell walls of adjoining cells, interconnecting the living protoplasts.

Plasmolysis.
Contraction or shrinking of the cytoplasm in a living cell due to loss of water by exosmosis.

Polygenic.
A character governed by two or more genes.

PPLO.
An abbreviation of pleuropneumonia-like organisms. No longer in current use.

Predisposition.
The tendency of nongenetic conditions, acting before infection, to increase the susceptibility of a plant or plants to disease.

Primary air pollutant.
An air pollutant that originates at its source in a form that is toxic to plants.

Primary infection.
First infection by a pathogen after going through a resting or dormant period.

Primary inoculum.
Pathogen or its parts, as spores, fragments of mycelium, etc., which can incite the primary infection.

Proliferation.
The development of an abnormal number of flowers or fruits in the position normally occupied by a single organ.

Promycelium (pl. promycelia).
A basidium, issuing from a teliospore, on which basidiospores (sporidia) are formed.

Propagative.
A term used for those plant disease viruses that are capable of propagating inside their insect vectors.

Protoplasm.
Substance within and including plasma-membrane of a cell (or protoplast), but usually understood to exclude large vacuoles.

Protoplast.
The actively metabolizing part of a cell bounded by a membrane.

Pustule.
A blister-like, frequently erumpent, spot or spore mass.

Pycnidiospore.
A conidium borne in a pycnidium.

Pycnidium (pl. pycnidia).
An asexual, hollow, globose or flask-like fruiting body, lined inside with conidiophores.

Pycniospore.
See spermatium.

Pycnium (pl. pycnia).
See spermagonium.

Quarantine.
The restraints and restrictions upon any and all commodities suspected of being carriers of plant pests.

Race
See physiologic race.

Range.
Geographic distribution. Not to be confused with "host range".

Receptive hypha.
A specialized hypha protruding out of a spermagonium or pycnium which functions as a female gamete or gametangium.

Recessive.
Applied to a hereditary character that does not manifest itself in the offspring when its contrasting allele is present; opposite of dominant.

Recessive gene.
One member of an allelic pair of genes that lacks the ability to manifest itself wholly or in part when the other or dominant member is present.

Reduction division.
The cell division in the maturation of germ cells in which the number of chromosomes is reduced to the haploid number.

Replacement.
A plant disease symptom in which a pathogen ramifies and replaces host tissue within an organ of the host, usually a floral organ. Also called transformation.

Resident bacteria.
Plant pathogenic bacteria that multiply in association with all parts of the healthy plant.

Resident phase.
The stage of growth in the life cycle of pathogenic bacteria that takes place in association with all parts of the healthy plant, such as in the rhizosphere, within the plant, on the shoot surface and between bud scales.

Resistance.
The inherent ability of a host plant to suppress, retard, or prevent entry or subsequent activity of a pathogen or other injurious factor.

Resting spore.
A spore, often thick-walled, that can remain alive in a dormant condition for some length of time, later germinating and growing when environmental conditions permit it.

Rhizome.
Underground stem, bearing buds in axils of reduced scale-like leaves; serving as a means of vegetative reproduction.

Rhizomorph.
A thick interwoven strand of vegetative hyphae in which the hyphae have lost their individuality; the dense string-like or cord-like mass acting as a single unit.

Rhizosphere.
The soil zone or ecologic habitat around plant roots and subject to the specific influence of the plant roots.

Ringspot.
Disease symptom caused by many viruses; a circular chlorotic ring around a green center.

Rogue.
To remove undesired individual plants from a planting.

Rosette.
Symptom characterized by stems shortened to produce a bunchy growth habit.

Rotting.
The disintegration and decomposition of plant tissue accompanied by discoloration; usually a firm dry decay often termed a dry rot; or if soft, usually watery and odoriferous, and called a wet rot.

Russet.
Brownish roughened areas on the skin of fruit resulting from diseases, insects, or spray injuries that cause abnormal production of cork.

Rust.
One of many fungi belonging to the order Uredinales, class Basidiomycetes; also, the common name of diseases caused by these fungi, since they often appear "rusty".

Sanitation.
Term used for a number of cultural methods that reduce or may reduce inoculum production; clean cultivation, burial of crop refuse by plowing, removal and burning of infected plants or plant parts, sterilizing tools and machinery, etc.

Saprophyte.
An organism which subsists upon dead organic matter and inorganic materials.

Scab.
Crust-like disease lesion; or a disease in which scabs are prominent symptoms.

Sclerenchyma.
Strengthening tissue, whose cell walls are usually thick and often lignified; may or may not be devoid of a protoplast at maturity.

Sclerotium (pl. sclerotia).
Dense, compacted aggregates of hyphae, resistant to unfavorable conditions and capable of remaining dormant for long periods of time, and able to germinate upon the return of favorable conditions.

Scorch.
Symptom from infection or weather conditions, expressed as burning of leaf tissue, starting as a marginal necrosis.

Secondary air pollutant.
One that is formed in the atmosphere, usually photochemically, by reaction among pollutants.

Secondary infection.
An infection resulting from inoculum arising from a primary infection or from another secondary infection, without an intervening inactive period.

Secondary inoculum.
Pathogen or its parts, as spores, fragments of mycelium, etc., arising from a primary or secondary infection without an intervening inactive period, and can incite a secondary infection.

Seed-certification.
Seed production and marketing under control to maintain varietal purity and freedom from seed-borne pests.

Segregation.
Separation into different gametes, and then into different offspring, of the two members of any pair of allelomorphs possessed by an individual.

Septate.
Refers to the existance of cross-walls.

Septum (pl. septa).
Wall, cross-wall, or partition.

Serology.
A method for the detection and identification of antigenic (protein) substances and the organism that carries them (usually a bacterium or virus), that involves the highly specific antigen-antibody reaction.

Serum.
The liquid portion of the blood remaining after coagulation.

Sexual reproduction.
Reproduction involving the union of two compatible nuclei.

Shot-hole.
Advance leafspot symptom in which the necrotic lesions disintegrate and fall out.

Sieve plate.
Perforated wall area between phloem cells through which their protoplasts are connected.

Sign.
Characteristic structures of the causal agent appearing on or in the diseased plant. Not disease symptom.

Single-cycle disease.
A disease whose pathogen lacks a secondary infection cycle.

Sinker.
An anchorage and absorption organ of mistletoes; actually a modified root which is haustoria-like, and penetrates into the vascular system of the host; a number of sinkers are formed radiating from the primary haustorium, which develops from the attachment disc.

Smut.
One of many fungi belonging to the order Ustilaginales, class Basidiomycetes; also the common name of diseases caused by these fungi.

Soft rot.
Disease symptom involving primarily the parenchyma tissue of various plant organs, resulting in a slimy softening and decay.

Soil inhabitant.
A microorganism that is usually strongly competitive with other normal microflora of the soil and which often survives many years in the complete absence of suitable hosts.

Soil invader.
A microorganism that is poorly competitive with normal soil microflora and seldom survives over one or two years in the soil in the complete absence of suitable hosts.

Sorus (pl. sori).
A fruiting structure in certain fungi, especially the spore mass in the rusts and smut fungi.

Source of inoculum.
The place or object on or in which inoculum is produced.

Specific resistance.
Resistance against only certain races of a pathogen. It is usually inherited monogenecially. The level of resistance it imparts is usually high, frequently prohibiting colonization and reproduction of the pathogen. It is often hypersensitive in character. Same as vertical resistance.

Spermagonium (pl. spermagonia).
A pycnidium-like haploid fruit body of rust fungi which produces gametes or gametangia; also called pycnium.

Spermatium (pl. spermatia).
A non-motile, uninucleate, spore-like male gamete or gametangium that empties its contents into a receptive female structure prior to the fusion of protoplasts; also pycniospore.

Spirillum (pl. spirilla).
A corkscrew or spiral bacterium.

Sporangiophore.
A specialized hypha bearing one or more sporangia.

Sporangiospore.
An asexual spore borne within a sporangium.

Sporangium (pl. sporangia).
An asexual fungus cell containing one or more asexual spores.

Spore.
A minute propagative unit that functions as a seed, but differs from it in that a spore does not contain a preformed embryo.

Sporidium (pl. sporidia).
See basidiospore.

Sporodochium (pl. sporodochia).
A cushion-like mass of fungal tissue that breaks through the host tissue and bears conidiophores over its surface.

Sporulate.
To produce spores.

Stele.
The central cylinder inside the cortex of roots and stems of vascular plants.

Sterigma.
A tiny protuberance on a basidium upon which a basidiospore is formed.

Stoma (pl. stomata).
(1) Of plants--a pore and two guard-cells that surround it, located in the epidermis of plants, usually present in large numbers, particularly in leaves and through which gaseous exchange takes place. The crescent-shaped guard-cells govern the opening and closing of the pore by movements due to changes in turgidity. (2) Of nematodes--the mouth or buccal cavity.

Stroma (pl. stromata).
A mass of vegetative hyphae with or without tissue of the host or substratum, sometimes sclerotium-like in form, in or on which spores are produced.

Stunted.
An unthrifty plant reduced in size and vigor due to infectious or non-infectious diseases.

Stylet.
Of nematodes--a relatively long, slender, axially located, hollow feeding structure of sclerotized cuticle located in the buccal cavity.

Stylet-borne.
Characterizes those plant viruses that are carried on the stylet of insect vectors, and not taken internally by the vector.

Suberin.
A waxy substance found in the walls of cork cells.

Suberization.
Conversion of exposed plant surfaces into tough corky tissue.

Suboxidation.
Insufficient oxygen during storage.

Substrate.
The substance or object on which an organism lives and from which it gets nourishment.

Susceptibility.
The condition of a host plant in which it is normally subject to attack by one or more pathogens, or to be affected by a disease.

Susceptible.
The inherent capacity of a host plant to become infected by a pathogen or damaged by an injurious factor.

Symbiosis.
The living together in close association of two dissimilar organisms; usually applied to cases where the relationship is useful to one or both.

Symbiotic.
Term applied to the relationship of two dissimilar organisms living together in close association; usually applied to cases where the relationship is useful to one or both.

Symptom.
Any visible reaction of a host plant to disease.

Symptomless carrier.
An infected host plant that does not, or will not, show disease symptoms.

Syndrome.
The pattern of symptoms and signs in a disease.

Synergism.
The effect produced by the interaction of two objects, chemicals, or forces that is greater than the sum of the actions of the two objects, chemicals, or forces.

Synnema (pl. synnemata).
A dense cemented fascicle of erect conidiophores, with a head of conidia-bearing branches.

Systemic.
Term applied to disease in which a single infection can lead to general spread of the pathogen throughout the plant body. Also applied to chemicals that are transported throughout the plant in the vascular system.

Teliospore.
Thick-walled dikaryotic resting spore in rust and smut fungi in which nuclear fusion occurs just prior to meiosis; part of the basidial apparatus. Also called chlamydospore in the smut fungi.

Telium (pl. telia).
The fruiting body of a rust fungus in which teliospores are formed.

Temperature lapse.
The decrease of air temperature with height.

Thallus (pl. thalli).
Denotes a simple plant body that does not have stems, roots, or leaves.

Thermal inactivation point.
The upper temperature limit at which death of an organism occurs after a selected interval of time.

Tolerance.
See disease tolerance.

Top necrosis.
A disease symptom whereby there is a dead bud, branch, or entire top of a plant.

Toxin.
A poison formed by an organism.

Transformation.
(1) Acquisition by a bacterial clone of heretible characteristics of another clone by growing the former in an extract or filtrate of the latter; the agent of transformation appears to be deoxyribonucleic acid (DNA). (2) A plant disease symptom in which a pathogen ramifies and replaces host tissue within an organ of the host, usually a floral organ.

Translocation.
The segmental exchange between non-paired (nonhomologous) chromosomes to form two new chromosomes. Also transfer of nutrients or viruses through a plant.

Transmission.
The transfer of an infectious agent from one plant to another.

Transpiration.
Loss of water vapor from the surface of plant leaves.

Trichome.
Single-celled or many-celled outgrowth from an epidermal cell, may or may not be glandular (secretory). A hair.

Tylosis (pl. tyloses).
A cell outgrowth of a parenchyma cell into the cavity of an adjacent water-conducting (xylem) vessel that partially or completely plugs the vessel, (adj. tylose).

Ultramicroscopic.
Denotes objects so small they cannot be detected with a compound microscope.

Uninucleate cell.
A cell having one nucleus.

Urediospore (also uredospore).
A dikaryotic, repeating spore of the rust fungi.

Uredium (pl. uredia).
A fruiting body of a rust fungus in which urediospores are formed. It comprises a group of spore-bearing filaments crowded together on which masses of spores are formed.

Vacuole.
The cavity in the cytoplasm of a plant cell that is filled with a watery solution.

Vascular.
Refers to the conducting system of plants.

Vector.
An organism (usually an animal) that can transmit a pathogen.

Vegetative phase.
Refers to the growth stage of an organism in contrast to the reproductive stage, and involves the absorption and building up of food-stuffs into complex constituents of the organism. The same as assimilative phase.

Veinbanding.
A disease symptom whereby there is a band of chlorotic tissue along the veins.

Veinclearing.
Leaf symptom in which the veins are lighter green than normal.

Vertical resistance.
See specific resistance.

Viable.
Able to live.

Virulent.
A strong capacity to produce disease. Highly pathogenic.

Virus.
Ultramicroscopic infectious entities that contain only one of the two forms of nucleic acid, are synthesized only within suitable host cells by utilizing the synthetic mechanisms of the cells to produce the viral substances, and are unable to increase in size.

Water-parasite.
See hemi-parasite.

Wet rot.
See rotting.

Wilt.
Loss of freshness or drooping of plants due to inadequate water supply or excessive transpiration; a vascular disease interfering with utilization of water.

Witches'-broom.
Disease symptom with abnormal brushlike development of many weak shoots.

Wound parasite.
A pathogen that can enter a host only through wounds.

Yellows.
A term applied to diseases in which yellowing or chlorosis is a principal symptom.

Xylem.
The part of a vascular bundle that consists of woody-walled water conducting cells.

Zoosporangium (pl. zoosporangia).
A sporangium that contains zoospores.

Zoospore.
A motile, asexually produced spore.

Zygospore.
The sexual (often resting) spore of Zygomycetes produced by the fusion of similar gametes.

Zygote.
The cell in which nuclear fusion occurs, the fertilized ovum; before it undergoes further differentiation.

SUMMARY OF GENERAL PRINCIPLES

	Page
Nematodes often alter the epidemiology of other plant diseases.	120
Lack of moisture, high temperatures, dry winds, and bright sunlight are interrelating factors of drought.	134
Symptoms of nutrient deficiency tend to be similar in different plants.	137
Vegetation plays an enormous role in acting as a natural "sink" for the removal of pollutants from the atmosphere.	162
Most bacterial and fungal foliar pathogens, require free surface moisture in order to enter their hosts.	176
Temperature and moisture are the main factors influencing the development of most infectious diseases.	178
Either moisture or temperature can be limiting in the infection and establishment of disease. If one is favorable, the other is generally limiting.	178
Infectious diseases cannot occur without inoculum.	185
The largest inoculum resevoir resides in living plants.	186
The sources of the secondary inoculum for plant diseases that have a secondary infection cycle are diseased, living or dying hosts.	187
A passive release is characteristic of all bacteria.	191
The major limiting factor of spread of virus inoculum is the population of efficient insect vectors.	201
Many plant pathogenic fungi survive unfavorable growing periods in a dormant stage, which is more resistant than the vegetative stage.	201
Even though host plants are susceptible, most, if not all, still possess a limited and variable measure of resistance.	204
Many diseases require numerous points of infection on a susceptible host plant in order to cause disease in that plant.	206
In respect to the mode of entry into host plants, parasitic seed plants, nematodes, and most fungi are active, forceful, and dynamic; many bacteria enter actively, but not forcefully; viruses, viroids and mycoplasmas are completely passive.	206
Watersoaking of host leaves greatly facilitates host entry and subsequent infection by motile pathogenic bacteria.	211

	Page
In respect to the host, the lack of susceptible varieties and the presence of resistant varieties, increase race development in pathogens.	217
Conditions of the pathogen that accelerate race development are: obligate parasitism; sexual reproduction; inability to complete its life-cycle, or multiply in the field as a saprophyte; and a narrow host range.	217
For each gene for resistance in the host there is a corresponding gene in the pathogen capable of overcoming that resistance.	219 and 288
Most plants are immune to most pathogens.	227
A number of fungi enter many immune plants even though they cannot incite disease.	227
Hypersensitivity is a common defense reaction of green plants to invasion by many pathogens.	237
Most, if not all, plant pathogenic bacteria are able to multiply in the fluid that is contained in the intercellular spaces and in the conducting vessels of host plants.	247
Most, if not all, infectious diseases increase respiration.	261
Plant diseases cause a reduced photosynthesis.	263
The transpiration rate of infected tissues is increased in most plant diseases.	263
Plant cells adjacent to cells infected with either obligate or facultative parasites, have increased permeability to water and solutes.	264
Normal translocation is usually decreased by infectious diseases that often increase the movement of metabolites towards affected tissues.	264
An accumulation of phenolic compounds around centers of infection is a characteristic host response to infection.	267
Vascular discoloration is characteristic of vascular infection.	268
Crop devastation is incidental to the pathogen, but paramount to man.	274
Pathogens with a short survival time generally multiply more rapidly than those with a long survival time.	274

Page

	Page
Minor diseases may become major diseases if susceptible hosts are grown on large acreages.	293
Specific resistance is likely to be of greater value against a single-cycle disease than against a multiple-cycle disease.	293
Specific resistance is likely to be of greater value in crops that are easy to breed than in perennial crops or annual crops that are hard to breed.	294
Specific resistance is likely to be of greater value when combined with useful levels of general resistance than when used alone.	294
Any factor that reduces hybridization of a pathogen will also reduce its race development.	303
The effectiveness of pesticides increases when there is a decrease of the inoculum potential.	311
There are few, if any, predators or parasites which under natural field conditions render a noticeable population control of foliar plant pathogens.	327
Most of the devastating outbreaks of plant diseases among native plants result from the introduction of exotic plant pathogens.	333
A pathogen, inconspicuous and of minor importance in its native habitat, may become an agent of destruction of the highest order when introduced into a new area.	335

For an additional listing of principles, the reader is referred to:

Robinson, R. A. 1971. Vertical resistance. Review of Plant Pathology. 50:233-239.

Yarwood, C. E. 1962. Some principles of plant pathology. Phytopathology. 52:166-167.

_____________. 1973. Some principles of plant pathology. II. Phytopathology. 63:1324-1325.

LIST OF PLANT SPECIES SENSITIVE AND TOLERANT TO SPECIFIC AIR POLUTANTS

(Courtesy of Donald H. Scott)

Sensitivity of selected plants to ozone.

Sensitive

Alder (Alnus sp.)
Alfalfa (Medicago sativa)
Apple, crab (Malus baccata)
Ash, green (Fraxinus pennsylvanica)
Ash, white (F. americana)
Aspen, quaking (Populus tremuloides)
Barley (Hordeum vulgare)
Basswood (Tilia americana)
Bean (Phaseolus vulgaris)
Boxelder (Acer negundo)
Bridal wreath (Spirea prunifolia)
Carnation (Dianthus caryophyllus)
Carrot (Daucus carota v. sativa)
Catalpa (Catalpa speciosa)
Chickory (Cichorium intybus)
Chickweed (species not given)
Chrysanthemum (Chrysanthemum sp.)
Clover, red (Trifolium pratense)
Corn, sweet (Zea mays)
Cotoneaster, rock (Cotoneaster sp.)
Cucumber (Cucumis sativus)
Endive (Cichorium endiva)
Forsythia (Forsythia sp.)
Grape (Vitis vinifera)
Grass, bent (Agrostis sp.)
Grass, brome (Bromus inermis)
Grass, crab (Digitaria sanquinalis)
Grass, Kentucky Blue (Poa pratensis)
Grass, orchard (Dactylis glomerata)
Hemlock, eastern (Tsuga canadensis)
Honeylocust (Gleditsia triacanthos)
Larch, European (Larix decidua)
Larch, Japanese (L. leptolepis)
Lettuce (Lactuca sp.)
Lilac (Syringa vulgaris)
Maple, silver (Acer saccharinum)
Milkweed (Asclepias sp.)
Mountain ash (species not given)
Muskmelon (Cucumis melo)
Oak, gambel (Quercus gambelii)
Oak, pin (Q. palustris)
Oak, scarlet (Q. coccinea)
Oak, white (Q. alba)
Oat (Avena sativa)
Onion (Allium cepa)
Parsley (Petroselium crispum)
Peanut (Arachis hypogaea)
Petunia (Petunia hybrida)
Pine, Austrian (Pinus nigra)
Pine, jack, (P. banksiana)
Pine, pitch (P. rigida)
Pine, ponderosa (P. ponderosa)
Pine, scotch (P. sylvestris)
Pine, Virginia (P. virginiana)
Pine, white (P. stobus)
Potato (Solanum tuberosum)
Privet (Ligustrum vulgare)
Radish (Raphanus sativus)
Rhododendron (species not given)
Rye (Secale cereale)
Snowberry (Symphoricarpos albus)
Spinach (Spinacea oleracea)
Squash (Cucurbita maxima)
Sunflower (Helianthus spp.)
Sweetgum (Liquidambar styraciflua)
Sweet pea (Lathyrus odoratus)
Swiss chard (Beta vulvaris var. cicla)
Sycamore (Platanus occidentalis)
Tobacco (Nicotina tabacum)
Tomato (Lycopersicum esculentum)
Tulip tree (Liriodendron tulipifera)
Turnip (Brassica rapa)
Watermelon (Citrullus vulgaris)
Wheat (Triticum aesativum)
Willow, weeping (Salix babylonica)

Tolerant

Arborvitae (Thuja occidentalis)
Birch, European white (Betula pendula)
Dogwood, white (Cornus florida)
Euonymus (Euonymus sp.)
European Mountain Ash (Sorbus aucuparia)
Fir, balsam (Abies balsamea)
Fir, Douglas (Pseudotsuga menziesii)
Fir, white (Abies concolor)
Grass, bermuda (Cynodon dactylon)
Grass, Zoysia (Zoysia japonica)
Juniper, Pfitzer (Juniperus chinensis)
Linden, little leaf (Tilia cordata)
Maple, Norway (Acer platanoides)
Maple, sugar (Acer saccharum)
Oak, English (Quercus robur)
Oak, Red (Q. rubra)
Oak, Shingle (Q. imbricaria)
Pine, red (Pinus resinosa)
Spruce, Black Hills (Picea glauca var. densata)
Spruce, Colorado blue (P. pungene)
Spruce, Norway (P. abies)
Spruce, white (P. glauca)
Walnut, black (Juglans nigra)

Sensitivity of selected plants to sulfur dioxide.

Sensitive

Alder (Alnus sp.)
Alfalfa (Medicago sativa)
Apple (Malus sp.)
Apple, crab (species not given)
Apricot (Prunus armeniaca)*
Aspen, quaking (Populus tremuloides)
Aster (Aster bigelovii)
Bachelor's button (Centarea cyanus)
Barley (Hordeum vulgare)
Bindweed (Convolvulus arvensis)
Bean (Phaseolus vulgaris)
Beet (Beta vulgaris)
Birch (Betula spp.)
Birch, white (B. pendula)
Broccoli (Brassica oleracea)
Brussel sprouts (B. oleracea)
Buckwheat (Fagopyrum sagittatum)
Careless weed (Amaranthus palmeri)
Carrot (Daucus carota)
Catalpa (Catalpa speciosa)
Cherry, black (Prunus serotina)
Chickweed (species not given)
Clover (Trifolium and Melilotus spp.)
Cosmos (Cosmos bipinnatus)
Cotton (Gossypium sp.)
Curly dock (Rumex crispus)
Dahlia (Dahlia sp.)
Endive (Cichorium endivia)
Elm, American (Ulmus americana)
European Mountain Ash (Sorbus aucuparia)
Fern, bracken (Pteridium sp.)
Fir (Abies spp.) (species not given)
Fleabane (Erigeron canadensis)
Four-O'clock (Mirabilis jalapa)
Grass, annual blue (Poa annua)
Grass, Kentucky bluegrass (Poa pratensis)
Grass, bent (Agrostis palustris)
Grass, fescue (Festuca rubra)
Grass, rye (Lolium sp.)
Hemlock, Mountain (Tsuga mertensiana)
Hickory (species not given)
Larch (Larix sp.)

*Reported by some workers to be intermediate in sensitivity.

(Sensitivity of selected plants to sulfur dioxide, *sensitive*-continued)

Lettuce (Lactuca sativa)
Lettuce, prickly (L. scariola)
Maple, red (Acer rubrum)*
Mallow (Malva parviflora)
Morningglory (Ipomoea purpurea)
Mountain laurel (Kalmia latifolia)
Mulberry (Morus microphylla)
Mustard, wild (species not given)
Ninebark (Physocarpus capitatus)
Oat (Avena sativa)
Okra (Hibiscus esculentus)
Pear (Pyrus communis)
Pepper (Capsicum frutescens)
Petunia (Petunia sp.)
Pigweed (Amaranthus restroflexus)
Pine, jack (Pinus banksiana)
Pine, scotch (Pinus sylvestris)*
Pine, Western white (P. monticola)
Pine, white (P. strobus)
Plantain (Plantago major)
Poplar, Lombardy (Populus nigra 'italica')
Pumpkin (Cucurbita pepo)
Radish (Raphanus sativus)
Ragweed (Ambrosia artemisiifolia)
Rhubarb (Rheum rhaponticum)
Rye (Secale cereale)
Safflower (Carthamus tinctorious)
Salt bush (Atriplex sp.)
Smartweed (Polygonum sp.)
Shadblow, Serviceberry (Amelanchier sp.)
Soybean (Glycine max)
Spinach (Spinacea oleracea)
Spruce, Engelmann (Picea engelmannii)
Squash (Curcurbita maxima)
Sumac (Rhus sp.)
Sunflower (Helianthus sp.)
Sweet pea (Lathyrus sp.)
Sweet potato (Ipomoea batatas)
Swiss chard (Beta vulgaris)
Tomato (Lycopersicum sp.)
Tulip (Tulipa gesneriana)
Turnip (Brassica rapa)
Velvet weed (Gaura parviflora)
Verbena (Verbena canadensis)
Violet (Viola sp.)
Walnut, English (Juglans regia)
Wheat (Triticum aesativum)
Willow (Salix sp.)
Witch hazel (Hamamelis sp.)
Zinnia (Zinnia elegans)

*Reported by some workers to be intermediate in sensitivity

Intermediate

Aspen, bigtooth (Populus grandidentata)
Begonia (Begonia sp.)
Cabbage (Brassica sp.)*
Cauliflower (Brassica sp.)
Cocklebur (Zanthium sp.)
Dandelion (Taraxacum officinale)
Douglas fir (Pseudotsuga menzieii)
Fir, balsam (Abies balsamea)
Fir, silver (A. pectinata)
Fir, white (A. concolor)
Gooseberry (Ribes sp.)
Grape (Vitis sp.)
Grass, orchard (Dactylis glomerata)
Hollyhock (Althea sp.)
Hydrangea (Hydrangea sp.)
Iris (Iris sp.)
Leek (Allium porrum)
Linden (Tilia sp.)
Lambsquarters (Chenopodium sp.)
Nightshade (Solanum sp.)
Pine, lodgepole (Pinus contorta latifolia)
Poplar, balsam (Populus balsamifera)

*Reported by some workers to be tolerant in sensitivity.

Tolerant

Arborvitae (Thuja occidentalis)
Ash, green (Fraxinus pennsylvanica)
Beech, European (Fagus sylvatica)
Cedar, red (Thuja plicata)
Cedar, white (T. occidentalis)
Cottonwood, Eastern (Populus deltoides)
Cypress, Lawson (Cupressus lawsoniana)
Dogwood, white (Cornus florida)
Ginko (Ginko biloba)
Grass, bermuda (Cynodon dactylon)
Grass, zoysia (Zoysia japonica)
Gum, black (Nyssa sylvatica)
Holly, English (Ilex aquifolium)
Hornbeam, European (Carpinus betulus)
Juniper (Juniperus sp.)
Lilac (Syringa sp.)
Locust, black (Robinia pseudoacacia)
Maple, hedge (Acer campestre)
Maple, mountain (A. spicatum)
Maple, sugar (A. saccharum)
Oak, English (Quercus robur)
Oak, Live (Q. sp.)
Oak, red (Q. rubra)
Pine, dwarf mugho (Pinus mugo mughus)
Pine, Austrian (P. nigra)
Planetree, Oriental (Platanus orientalis)
Planetree or Sycamore (P. occidentalis)
Spruce (Picea glauca)
Sourwood (Oxydendron arboreum)
Tulip tree (Liriodendron tulipifera)

Sensitivity of selected plants to fluorides.

Sensitive

Apricot, chinese & royal (Prunus armeniaca)
Boxelder (Acer negundo)
Blueberry (Vaccinium sp.)
Buckwheat (Fagopyrum sagittatum)
Corn, sweet (Zea mays)
Douglas fir (Pseudotsuga monziesii)
Gladiolus, Snow Princess (Gladiolus sp.)
Grape, European (Vitis vinifera)
Grape, Oregon (Mahonia repens)
Iris (Iris sp.)
Larch, Western (Larix occidentalis)
Lily-of-the-Valley (species not given)
Milo (Sorghum sp.)
Peach, fruit (Prunus persica)
Pigweed (Amaranthus sp.)
Pine, lodgepole (Pinus contorta latifolia)
Pine, mugho (P. mughus)
Pine, ponderosa (P. ponderosa)
Pine, Scotch (P. sylvestris)
Pine, white (P. strobus)
Plum, Bradshaw (Prunus domestica)
Prune, Italian (Prunus domestica)
Sorghum (Sorghum sp.)
Smartweed (Polygonum sp.)
Spruce, Colorado blue (Picea pungens)
Tulip (Tulipa gesneriana)

(Sensitivity of selected plants to fluorides-continued)

Intermediate

Apricot, Moorpark & Tilton (Prunus armenica)
Arborvitae (Thuja sp.)
Ash, green (Fraxinus pennsylvania var. lanceolata)
Ash, European (F. excelsior)
Aspen, quaking (Populus tremuloides)
Aster (Aster sp.)
Beech, European (Fagus sylvatica)
Birch, European white (Betula pendula)
Catalpa (Catalpa bignoniodes)
Cherry, Bing & Royal Ann (Prunus avium)
Cherry, choke (P. virginiana)
Chickweed (Cerastium sp.)
Citrus, lemon & tangerine (Citrus sp.)
Fir, Douglas (Pseudotsuga menziesii)
Geranium (Geranium sp.)
Goldenrod (Solidago sp.)
Grape, Concord (Vitis labrusca)
Grass, crab (Digitaria ischanemum)
Holly, English (Ilex aquifolium)
Hornbeam, European (Carpinus betulus)
Lambsquarters (Chenopodium album)
Lilac (Syringa vulgaris)
Linden, European (Tilia europea)
Locust, black (Robinia pseudoacacia)
Maple, hedge (Acer campestre)
Maple, silver (A. saccharinum)
Mulberry (Morus rubra)
Narcissus (Narcissus sp.)
Oak, English (Quercus robur)
Paeonia (Paeonia sp.)
Peach (Prunus persica)
Pine, Austrian (Pinus nigra)
Pine, lodgepole (P. contorta)
Pine, scotch (P. sylvestris)
Pine, white (P. strobus)
Planetree, Oriental (Platanus orientalis)
Poplar, lombardy (Populus nigra)
Poplar, Caroline (P. deltoides)
Ragweed (Ambrosia sp.)
Rhododendron (Rhodendron sp.)
Rose (Rosa sp.)
Rasberry (Rubus idaeus)
Shadblow, Serviceberry (Amelanchier canadensis)
Sumac, smooth (Rhus glabra)
Sunflower (Helianthus sp.)
Spruce, white (Picea glauca)*
Veratrum (Veratrum californicum)
Vicia (Vicia sp.)
Violet (Viola sp.)
Walnut, black (Juglans nigra)
Walnut, English (J. regia)
Willow (Salix sp.)
Yew, Japanese (Taxus cuspidata)

*Reported by some workers to be tolerant in sensitivity.

Tolerant

Alfalfa (Medicago sativa)
Apple (Malus sylvestris)
Aquilegia (Aquilegia sp.)
Ash, American Mountain (Sorbus comestica)
Ash, European Mt. (S. aucuparia)
Ash, Modesto (Fraxinus velutina)
Asparagus (Asparagus sp.)
Bean (Phaseolus vulgaris)
Bean, soya (Glycine max)
Birch, cutleaf (Betula pendula var. gracilis)
Bridal wreath (Spiraea prunifolia)
Burdock (Anctium sp.)
Cherry, flowering (Prunus serrulata)
Cornelian cherry (Cornus mas)
Cotton (Gossypium hirsutum)
Current (Ribes sp.)
Elderberry (Sambucus sp.)
Elm, American (Ulmus americana)
Juniper (Juniperus sp.)
Linden, American (Tilia americana)
Linden, European (T. europaea)
Linden little leaf (T. cordata)
Onion (Allium sp.)
Pear (Pyrus communis)
Pigweed (Amaranthus retroflexus)
Planetree (Platanus sp.)
Plum flowering (Prunus cerasifera)
Pyracantha, firethorn (Pyracantha sp.)
Russian Olive (Elaeagnus angustifolia)
Serviceberry (Amelanchier alnifolia)
Squash, summer (Cucurbita pepo)
Strawberry (Fragaria sp.)
Tomato (Lycopersicum esculentum)
Tree-of-Heaven (Ailanthus altissima)
Virginia creeper (Parthenocissus quinqefolia)
Yew (Taxus cuspidata)
Wheat (Triticum aesativum)

Sensitivity of selected plants to peroxyacetyl nitrate (PAN)

Sensitive

Bean, pinto (Phaseolus vulgaris)
Chard, Swiss (Beta chilensin)
Chickweed (Stellaria media)
Dahlia (Dahlia sp.)
Grass, annual blue (Poa annua)
Lettuce (Lactuca sativa)
Linden, little leaf (Tilia cordata)
Mustard (Brassica juncea)
Nettle, little-leaf (Urtica ureans)
Oat (Avena sativa)
Oak, pin (Quercus palustris)
Petunia (Petunia hybrida)
Tomato (Lycopersicum esculentum)
Tulip tree (Liriodendron tulipifera)

Tolerant

Azalea (Rhododendron sp.)
Bean, lima (Phaseolus limensis)
Begonia (Begonia sp.)
Broccoli (Brassica oleracea)
Chrysanthemum (Chrysanthemum sp.)
Corn (Zea mays)
Cotton (Gossypium hirsutum)
Cucumber (Cucumis sativus)
Douglas fir (Pseudostuga menziesii)
Hemlock, Eastern (Tsuga canadensis)
Maple, Sugar (Acer saccharum)
Onion (Allium cepa)
Periwinkle (Vinca sp.)
Pine, Austrian (Pinus nigra)
Pine, Eastern white (P. strobus)
Pine, Jack (P. banksiana)
Pine, Pitch (P. rigida)
Radish (Raphanus sativus)
Sorghum (Sorghum vulgare)
Spruce, Colorado blue (Picea pungens)
Spruce, white (P. glauca)
Touch-me-not (Impatiens sp.)

Sensitivity of selected plants to nitrogen oxides.

Sensitive

Apple (Malus sp.)
Azalea (Rhododendron sp.)
Bean, pinto (Phaseolus vulgaris)
Beech, European (Fagus sylvatica)
Brittlewood (Malalenca leucadendra)
Ginko (Ginko biloba)
Hibiscus (Hibiscus rosasinensis)
Hornbeam, European (Carpinus betulus)
Lettuce, head (Lactuca sativa)
Linden, large leaf (Tilia grandiflora)
Linden, little leaf (T. cordata)
Locust, black (Robinia pseudoacacia)
Maple, Japanese (Acer palmatum)
Maple, Norway (A. platanoides)
Mustard (Brassica sp.)
Pear (Pyrus communis)
Petunia (Petunia hybrida)
Pine, Austrian (Pinus nigra)
Pine, dwarf mugho (P. mugo mughus)
Pine, Eastern white (P. strobus)
Sunflower (Helianthus annuus)
Spruce, Colorado (Picea pungens)
Spruce, white (P. glauca)
Tobacco (Nicotiana glutinosa)

Intermediate

Larch, European (Larix decidua)

Tolerant

Asparagus (Asparagus officinalis)
Bean, bush (Phaseolus vulgaris)
Carissa (Carissa carandas)
Croton (Codiaeum sp.)
Grass, Kentucky blue (Poa pratensis)
Heath (Erica sp.)
Ixora (Ixora sp.)
Lambsquarters (Chenopodium album)
Nettleleaf goosefoot (Chenopodium sp.)
Pigweed (Chenopodium sp.)

Sensitivity of selected plants to chlorine.

Sensitive

Alfalfa (Medicago sativa)
Apple, crab (Malus baccata)
Blackberry (Rubus sp.)
Boxelder (Acer negundo)
Buckwheat (Fagopyrum esculentum)
Chestnut, horse (Aesculus hippocastanum)
Chickweed (Stellaria media)
Coleus (Coleus sp.)
Corn (Zea mays)
Cosmos (Cosmos sp.)
Gomphrena (Gomphrena sp.)
Grass, Johnson (Holcus halepensis)
Johnny-jump-up (Viola palmata)
Maple, sugar (Acer saccharum)
Mustard (Brassica sp.)
Oak, pin (Quercus palustris)
Onion (Allium cepa)
Orchid (species not given)
Pine, white (Pinus strobus)
Primrose (Primula vulgaris)
Privet (Ligustrum sp.)
Radish (Raphanus sativus)
Rose, tea (Rosa odorata)
Sassafras (Sassafras albidum)
Sunflower (Helianthus annuus)
Sweetgum (Liquidambar styraciflua)
Tobacco (Nicotiana tabacum)
Tree-of Heaven (Ailanthus altissima)
Tulip (Tulipa sp.)
Venus-looking-glass (Specularia perfoliata)
Virginia creeper (Parthenocissus quinquefolia)
Witch Hazel (Hamamelis virginiana)
Zinnia (Zinnia sp.)

Intermediate

Azalea (Rhododendron sp.)
Bean, pinto (Phaseolus vulgaris)
Cheeseweed (Malva rotundifolia)
Cherry, black (Prunus serotina)
Cowpea (Vigna sinensis)
Cucumber (Cucumis sativus)
Dahlia (Dahlia sp.)
Dandelion (Taraxacum officinale)
Fern, braken (Pteridium aquilinium)
Geranium (Geranium sp.)
Grape (Vitis sp.)
Grass, annual blue (Poa annua)
Gum, black (Nyssa sylvatica)
Halesia (Halesia sp.)
Nasturtium (Tropaeolum sp.)
Nettleleaf goosefoot (Chenopodium murale)
Orange, mock (Philadelphus sp.)
Peach (Prunus persica)
Petunia (Petunia hybrida)
Pine, Jack (Pinus banksiana)
Pine, loblolly (Pinus taeda)
Pine, shortleaf (Pinus echinata)
Pine, slash (Pinus caribaea)
Rhodotypos (Rhodotypos sp.)
Sassafras (Sassafras albidum)
Squash (Cucurbita moschata)
Tobacco (Nicotiana tabacum)
Tomato (Lycopersicum esculentum)
Wandering Jew (Zebrina sp.)

Tolerant

Begonia (Begonis rex)
Corn, field (Zea mays)
Eggplant (Solanum melongena)
Grass, Kentucky blue (Poa pratensis)
Hemlock (Tsuga sp.)
Holly, Chinese (Ilex cornuta)
Lambsquarters (Chenopodium album)
Oak, red (Quercus rubra)
Olive, Russian (Elaeagnus angustifolia)
Oxalis (Oxalis sp.)
Pepper (Capsicum sp.)
Pigweed (Amaranthus retroflexus)
Polygonum (Polygonum sp.)
Soybean (Glycine max)
Tobacco (Nicotiana tabacum)
Yew (Taxus sp.)

Sensitivity of selected plants to hydrogen chloride.

Sensitive		Intermediate
Beet, sugar (Beta vulgaris)	Maple (Acer sp.)	Begonia (Begonia rex)
Cherry (Prunus sp.)	Tomato (Lycopersicum esculentum)	Rose (Rosa sp.)
Larch (Larix sp.)	Viburnum (Viburnum)	Rosebud (Rosa sp.)
		Spruce (Picea sp.)

Tolerant		
Beech (Fagus sp.)	Maple, Norway (A. platanoides)	Pear (Pyrus communis)
Birch (Betula sp.)	Maple, sugar (A. saccharum)	Pine, Austrian (Pinus nigra)
Cherry, black (Prunus serotina)	Oak (Quercus sp.)	Pine, Eastern white (P. strobus)
Fir, balsam (Abies balsamea)	Oak, red (Q. rubra)	Spruce, Norway (Picea abies)
Maple (Acer sp.)		

Sensitivity of selected plants to ethylene.

Sensitive		Tolerant
		Beet (Beta vulgaris)
		Cabbage (Brassica oleracea)
Arborvitae (Thuja sp.)	Marigold, African (Tagetes erecta)	Clover (Trifolium sp.)
Azalea (Rhododendron sp.)	Pea (Pisum sativum)	Endive (Chichorium endivia)
Bean, black valentine (Phaseolus vulgaris)	Peach (Prunus persica)	Oat (Avena sativa)
Carnation (Dianthus caryophyllus)	Philodendron (Philodendron cordatum)	Onion (Allium cepa)
Cotton (Gossypium hirsutum)	Privet (Liqustrum sp.)	Radish (Rhaphanus sativus)
Cowpea (Vigna sinensis)	Rose (Rosa sp.)	Rye grass (Lolium multiflorum)
Cucumber (Cucumis sativus)	Sweet potato (Ipomoea batatas)	Sorghum (Sorghum vulgare)
Holly, Japanese (Ilex crenata)	Tomato (Lycopersicum esculantum)	

I N D E X

Most terms defined in the Glossary are not included in the Index.

B

C

D

E

F

I

J

K

L

M

N

O

P

T

W

X

Z